AF839291

ELEMENTS OF ECOLOGY

Edited by

Lindsay Patterson

Toronto Academic Press

ELEMENTS OF ECOLOGY

Editor: **Lindsay Patterson**

Toronto Academic Press
4164 Lakeshore Road
Burlington ON L7L 1A4
Canada
www.tap-books.com

Email: orders@arclereducation.com

ISBN: 978-1-77956-298-2

Toronto Academic Press publishes wide variety of books and eBooks. For more information about Toronto Academic Press and its products, visit our website at www.tap-books.com.

ABOUT THE EDITOR

Lindsay Patterson is an ecologist, wildlife researcher, and copy editor passionate about animals, conservation, and the power of the written, and the spoken word. She has an undergraduate degree in English and Philosophy studies, an honors degree in Environmental Sciences, and a doctorate in Ecology, which has deepened her understanding of working hard for what you love. Her passion for wildlife research began with the Chacma baboons of South Africa's Table Mountain National Park and led her to the Cederberg Mountain's elusive Cape leopards, Madagascar's unique diversity of lemurs, frogs, and snakes, and KwaZulu-Natal's precocious vervet monkeys. In her 10-year career as a natural history researcher, she worked on over 200 nature documentaries for award-winning production houses, including Love Nature, National Geographic, Discovery, and Smithsonian. In her downtime, she loves camping in remote spaces and is an avid reader of everything.

Table of Contents

List of Figures

List of Tables

PREFACE

Ecology encompasses the study of how individual organisms come together to form populations, how these populations interact within communities, and how these communities, in turn, fit into the larger framework of ecosystems. A specialized branch of ecology, known as autecology, focuses specifically on the ways in which single species adapt to their surroundings. This adaptation is achieved through a variety of mechanisms, including biochemical, morphological, and physiological changes, tailored to meet the demands of their environment.

In the plant kingdom, the evidence of adaptation is abundant and diverse. For instance, certain desert plants have evolved thorns in place of leaves, a modification aimed at reducing water loss in arid conditions. The development of a wide spectrum of complex chemical defenses is another testament to plant adaptation, with examples including the painful sting of nettles, the toxic principles of hemlock, and the irritant properties of poison ivy. Additionally, plants such as succulents have adapted to survive in water-scarce environments by developing specialized structures for water storage, showcasing the range of adaptative strategies from the molecular level to the whole-plant level.

The animal kingdom is equally rich in examples of adaptation to environmental pressures. Antarctic fishes, for example, have evolved antifreeze proteins to survive in icy waters, while seals have developed specialized cardiovascular systems that enable them to dive for extended periods without oxygen. The natural world is also replete with chemical strategies for survival, including the potent tetrodotoxin used for defense and the venomous attacks of snakes. Furthermore, the discovery of giant variants of marine species near deep-sea hydrothermal vents highlights the adaptability of life in extreme environments.

The distribution of organisms across the globe is largely influenced by environmental factors. While it might seem that a single factor can determine the success or distribution of a species, the reality is often more complex, involving a multitude of interacting elements. The study of ecological races in plants, for instance, has revealed that adaptations to environmental variables such as altitude involve a complex interplay of multiple genes, affecting traits like frost tolerance and flowering time. Advances in our understanding of these adaptations are being propelled forward by integrating whole-organism studies with research at the cellular and subcellular levels, aided by the latest developments in molecular biology and subcellular physiology.

The impact of these scientific advancements extends beyond academic interest. They hold the potential to revolutionize crop productivity by applying a deeper understanding of physiological and genetic adaptation mechanisms. Moreover, these insights contribute significantly to our comprehension of evolutionary processes and the broader principles of ecology.

The scope of ecology includes not only the relationship between organisms and their physical surroundings but also their interactions with members of their own and other species. These interactions are crucial in defining the roles organisms play within the natural world. Weaving together the study of these dual aspects, ecology provides a comprehensive picture of how plant and animal populations are regulated and structured within their environments, offering a holistic view of the natural world's complexity and interconnectivity.

The aim of "Elements of Ecology" is to offer a comprehensive introduction to the field of ecology, presenting the essential concepts, principles, and processes that govern the interactions within and among organisms and their environments. This book is designed primarily for undergraduate students majoring in biology, environmental science, and related disciplines who are embarking on their exploration of ecological systems and the complex relationships that sustain life on Earth.

Integrating the latest topics with foundational ecological theories, the book seeks to enhance students' understanding of how ecosystems function, the role of biodiversity, the impact of human activities on natural habitats, and the strategies for conservation and sustainable management of ecosystems. "Elements of Ecology" covers a wide range of topics, including population dynamics, community interactions, ecosystem energetics, biogeochemical cycles, and global environmental changes, aiming to provide a solid foundation in ecology that is both scientifically rigorous and accessible.

The readers will benefit from clear explanations, illustrative diagrams, and real-world examples that demonstrate the relevance of ecological principles in addressing environmental challenges. Through this book, students are encouraged to develop a deeper appreciation for the complexity of ecological systems and the importance of preserving the natural world for future generations.

CHAPTER

1 Introduction to Ecology

LEARNING OBJECTIVES

After studying this chapter, you will be able to:

- Explain ecology: scope and objectives
- Define ecology's historical background
- Describe the development of ecology
- List the level of study in ecology
- Explain the relevance and application of ecology

INTRODUCTORY EXAMPLE

Ecology is the study of organisms and how they interact with the environment around them. An ecologist studies the relationship between living things and their habitats. To learn about the natural world, ecologists must study multiple aspects of life ranging from the moss that grows on rocks to the wolf population in Yellowstone National Park, United States. In order to research the environment, scientists ask questions, such as: How do organisms interact with the living and nonliving factors around them? What do organisms need to survive and thrive in their current environments? To find the answers to these questions, ecologists must study and observe all forms of life and their ecosystems throughout our world.

In addition to examining how ecosystems function, ecologists study what happens when ecosystems do not function normally. Changes in ecosystems can result from many different factors, including diseases among the organisms living in the area, increases in temperature, and increased human activities. Understanding these changes can help ecologists anticipate future ecological challenges and inform other scientists and policymakers about the challenges facing their local ecosystems.

Ecology became popular in the 1960s when environmental issues were increasing. Scientists had been studying nature for centuries, but modern ecology began in the 19th century. European and American scientists observed how plants function and affect habitats and assessed how animals interact with plants and each other within their various ecosystems. Today, ecologists use historical data and share worldwide information on ecosystems. building on the information that shapes our planet's future.

Human actions affect ecosystems on a global scale. Pollution from fuels and factories can upset food chains, the translocation of invasive species to new habitats can harm local wildlife, and adding or removing one species can be detrimental for the health of others, including the spread of disease and overhunting.

INTRODUCTION

Ecology is the scientific study of how organisms interact with their surroundings, including how living things relate to each other and interact with nonliving elements. Ecology is the study of the distribution and abundance of organisms, the variables that affect these patterns, and the ecological processes that control the exchange of nutrients and energy within ecosystems.

1.1. DEFINITION OF ECOLOGY

Learning Objectives

- Describe the three pervasive definitions of ecology

The Greek word "oekologie" (oikos, which means "house," and logos, which means "learning") refers to the study of the habitation of living things. Different authors have given different definitions of the word "ecology". Some would rather refer to it as "the study of biotic communities," "the science of community population," or "scientific natural history."

"A study of animals and plants in their relation to each other and to their environment" is the most thorough definition of ecology. Ernst Haeckel coined the term "ecology" in 1869, but many other contributions to the field were made much earlier. However, much later, in the 1900s, ecology was acknowledged as a separate branch of science.

It was first roughly split into two categories: plant and animal ecology. However, as time went on, concepts such as the food chain, biotic communities, and material cycling were understood, aiding in developing the fundamental theory for the field of general ecology as a whole.

Until recently, academic circles regarded ecology as a subfield of biology, along with genetics, molecular biology, developmental biology, evolution, etc., and it was not always regarded as a topic exclusive to the biological sciences.

Currently, though, the focus is on studying the environmental systems of the entire "household," which is related to the fundamental meaning of the term. As a result, ecology has developed from a branch of the biological sciences to a significant field of study that bridges the social, physical, and biological sciences.

The terms "plant ecology" and "animal ecology" refer to the specific references to plants or animals that are studied in ecology. Because of the close relationship between plants and animals, studying just plant or animal ecology will always be faulty and insufficient.

For this reason, it is preferable to study both plant and animal ecology under the umbrella of bioecology and to give them equal weight. Ecological studies conducted at the species level are referred to as autecology, whereas ecological studies conducted at the community level are referred to as synecology.

The scientific study of the processes influencing the distribution and abundance of organisms, the interactions among organisms, and the interactions between organisms and the transformation and flux of energy and matter.

Our definition is unique in that it emphasizes several factors :

- A starting focus on organisms, aggregations of organisms, or systems incorporating organisms or their by-products
- The bounding of ecology by both the biological and physical sciences
- The breadth of subject matters within ecology
- The joint consideration of both biotic and abiotic aspects of nature
- Depending on the ecological specialty, the focus can be on different proportions of biotic or abiotic aspects of nature
- The relationships between organisms and the physical world can be bidirectional, although different specialties may emphasize the effect of the organisms (and systems containing them) on the physical world or the effect of the physical world on the organisms.
- The boundary between the abiotic and the biotic aspects of ecology blurs.
- The disciplinary focus is on "processes," "interactions," and "relations" rather than on the physical entities.

When ecology was first defined in the middle of the 19th century, biology was a very different field from what it is today.

Ernst Haeckel provided the initial definition, characterizing ecology as the study of how organisms interact with their surroundings. Other definitions of ecology have been put forth for the last 150 years in an effort to define new fields of study, reflect the discipline's expansion, or delineate disciplinary boundaries.

There are three pervasive definitions of ecology. The Haeckelian form, which studies the interaction between organisms and their surroundings, is where the first definition originates.

According to the second definition—possibly the one that is most frequently used—ecology is the study of the abundance and distribution of organisms (Andrewartha and Birch, 1954).

According to Odum (1971), the third definition of ecology centers on the study of ecosystems.

Every one of the three types of definitions has advantages and limitations. Ecology is distinguished by an all-encompassing, synthetic perspective of nature rather than a piecemeal one.

We combine the second and third definitions of ecology to create our own definition. The aim of this new comprehensive definition is to facilitate synthesis and integration by attempting to bridge the gap between various ecological perspectives.

1.1.1. 1st Definition

In the classical Haeckelian definition, both the living and nonliving constituents of the natural world are emphasized. Nevertheless, it underscores that organisms are the pertinent manifestation of the biotic world, reflecting its historical context.

1.1.2. 2nd Definition

In 1954, Andrewartha and Birch reaffirmed that the organism is the central unit of ecology. Their research demonstrates that variables affecting distribution and abundance include both the biotic and abiotic environments.

Their acknowledgment of the significance of climate variations, for instance, demonstrates this. Nevertheless, the definition proposed by Andrewartha and Birch has frequently been linked to a primarily biological emphasis in its application.

This definition is now used as a kind of catchphrase for ecology that is population- and community-centered. It is evident that this definition has not encouraged research on the intersection of ecology and the physical environment sciences.

1.1.3. 3rd Definition

Odum (1971) started with the Haeckelian definition, but he deviated from it more than most because he wanted to create an entirely new field of ecology called ecosystem ecology.

He gave various definitions of ecology, one of which was the ambiguous claim that ecology was just environmental biology. His definition of ecology as the study of the structure and function of nature was the epitome of his brand of ecosystem thinking.

Odum's loosening of the reins holding Haeckel's focus on the organism is helpful, even though many ecologists find his excessive reliance on emergent properties and revival of "superorganismic" thinking problematic.

1.1.3.1. The Limits and Advantages of the Three Definitions of Ecology

The first definition has the advantage of being straightforward and emphasizing both biotic and abiotic elements of nature.

However, its excessive emphasis on the organism as the focal point is a drawback. It is always best to interpret Haeckelian statements as the study of relationships as opposed to the study of organisms in their environments. Though seemingly insignificant, the shift in emphasis highlights the shortcomings in Haeckel's definition.

The second definition is deficient in that it leaves out several important ecological topics, despite its emphasis on measurable and clear parameters.

To its credit, the third definition acknowledges that ecology is about processes and is not limited to patterns or organisms.

Organisms serve as the foundation for every definition. However, in some instances, it is not made clear that ecology can consider any kind of system (in the widest sense) that involves living things and the products they produce.

Did you get it?

1. **What is the first definition of ecology?**
2. **What is the advantage of the first definition of ecology?**

1.2. HISTORICAL BACKGROUND

Learning Objectives

- Learn about 18th and 19th-century ecological murmurs
- Understand the early 20th-century expansion of ecological thought
- Discuss the influence of ecology on the social sciences and humanities

Since ecology did not gain widespread recognition until the latter part of the 20th century, it is typically considered a relatively new science. More specifically, it is generally acknowledged that ecology became a separate field of study at the turn of the 20th century and came to the attention of the general public in the 1960s as a result of growing environmental concerns. Nevertheless, ecological thought has been around for a while in some capacity, and the field's guiding principles have progressively evolved alongside other biological fields. Because of their shared interest in a wide variety of animal species, Aristotle and his pupil Theophrastus may have been among the first ecologists. As early as the 4th century BC, Theophrastus wrote about the interactions between animals and their surroundings.

1.2.1. 18th and 19th Century ~ Ecological Murmurs

The major naval powers, including Britain, Spain, and Portugal, conducted numerous global exploration expeditions in the 18th and early 19th centuries in an effort to expand maritime trade with other nations and find and catalog new natural resources. About 20,000 plant species were known at the start of the 18th century, compared to 40,000 at the start of the 19th century and nearly 400,000 today.

Numerous scientists, including botanists like the German explorer Alexander von Humboldt, participated in these expeditions. Many people refer to Humboldt as the father of ecology. He was a pioneer in tackling the study of the interaction between living things and their surroundings. Through the use of latitude and altitude, he revealed the relationships that currently exist between observed plant species and climate, giving rise to the field of geobotany.

For example, in 1804, Humboldt reported an astounding number of species, mostly plants, for which he attempted to use geological data to explain their geographic distribution. One of Humboldt's famous works was "Idea for a Plant Geography" (1805). Other important botanists of the time included Aimé Bonpland.

1.2.1.1. The Notion of Biocenosis : Wallace and Möbius

Darwin's rival and contemporary, Alfred Russel Wallace, was the first to suggest a "geography" of animal species. At the time, many writers acknowledged that species were interdependent and classified them into plant and animal species, and subsequently into biocenosis, or communities of living things. The term "société" was first used in 1877, according to Karl Möbius, but it was in fact first used in 1825 by French naturalist Adolphe Dureau de la Malle to describe a grouping of individuals of different species of plants.

1.2.1.2. Warming and the Foundation of Ecology as a Discipline

Darwin exclusively considered competition to be a selective force, but Eugen Warming created a new field of study that included abiotic elements like fire, salt, cold, and drought, in the formation of biotic communities, as seriously as biotic factors. Before Warming, most biogeography was descriptive in nature, either floristic or faunistic. Warming sought to achieve this goal by examining the morphology and anatomy of organisms, i.e., adaptation, which explains a species' existence in a particular set of environmental circumstances. Furthermore, the new field sought to explain why, despite frequently having very different phylogenetic histories, species that shared similar environments and dangers would approach problems in comparable ways. Drawing from his personal observations in the Brazilian Cerrado, Denmark, Norway's Finnmark, and Greenland, Warming taught the first ecological plant geography course at a university. His book "*Plantesamfund*," which was based on his lectures, was quickly translated into German, Polish, and Russian before being published in English as "Oecology of Plants." The book had a significant impact on British and American scientists like Arthur Tansley and Henry Chandler Cowles through its German translation.

Keyword

Geobotany is the study of the relationship between plants and the geological substrate.

1.2.1.3. Darwinism and the Science of Ecology

A common belief is that Darwin is the father of scientific ecology. This claim might seem plausible at first because "On the Origin of Species" contains numerous observations and theories that neatly fall inside the parameters of contemporary ecology (e.g., the cat-to-clover chain, an ecological cascade), and because Ernst Haeckel, a fervent Darwinist, coined the term "ecology" in 1866. Even in his most "ecological" writings, like the preface to Hermann Müller's "The Fertilization of Flowers" (1883) or his treatise on earthworms and mull formation in forest soils ("The Formation of Vegetable Mold through the Action of Worms," 1881), Darwin never used the word again after this year. Moreover, the pioneers that founded ecology as a scientific discipline, such as Eugen Warming, A. F. W. Schimper, Gaston Bonnier, F.A. Forel, S.A. Forbes and Karl Möbius, made almost no reference to Darwin's ideas in their works. This was not due to ignorance or a lack of awareness of Darwin's writings; rather, ecology has always been preoccupied with the connection between organism morphology.

1.2.2. Early 20th Century ~ Expansion of Ecological Thought

1.2.2.1. The Biosphere – Eduard Suess, Henry Chandler Cowles, and Vladimir Vernadsky

Due to new chemical discoveries made by Lavoisier and de Saussure, particularly the nitrogen cycle, ecology flourished by the 19th century. The term "biosphere" was first proposed in 1875 by Austrian geologist Eduard Suess, who noted that life could only have developed within the strict boundaries of each compartment that comprises the atmosphere, hydrosphere, and lithosphere. The conditions that support life, such as those found on Earth, including flora, fauna, minerals, matter cycles, et cetera, were given the term "biosphere" by Suess.

Keyword

Deforestation is the intentional or accidental removal of trees from an area of land.

In his work "The Biosphere" (1926), Russian geologist Vladimir Vernadsky—who had defected to France—explained the basic concepts of the biogeochemical cycles and expanded on the concept of the biosphere. As a result, he redefined the biosphere to include all ecosystems combined.

The first records of ecological damage date back to the 18th century, when colonial expansion led to deforestation. Concerns about how human activity affects the environment have become more and more urgent since the Industrial Revolution in the 19th century. Ecologist is a word that has been around since the late 1800s.

1.2.2.2. The Ecosystem: Arthur Tansley

Zoogeography and botanical geography came together during the 19th century to create the foundation of biogeography. This branch of science that studies habitats of species aims to explain why a particular species exists in a particular area.

The term ecosystem was first used in 1935 by British ecologist Arthur Tansley to describe the dynamic relationship that exists between a biocenosis (a group of living things) and its biotope (the environment in which they live). Therefore, ecology was born as the science of ecosystems.

Eugene P. Odum, a dynamic and well-known biology teacher, took up Tansley's idea of the ecosystem. In addition to Howard Odum, his brother, Eugene authored a textbook published in 1953, which taught biologists and ecologists across multiple generations in North America.

1.2.2.3. Ecological Succession – Henry Chandler Cowles

Henry Chandler Cowles was a pioneer of the then-emerging field of "dynamic ecology" at the turn of the 20th century. He did this by studying ecological succession at the Indiana Dunes, which are sand dunes at the southernmost point of Lake Michigan. Here, Cowles discovered signs of age-related ecological succession in the soil and vegetation. Cowles was well aware of the concept's historical origins and the work of his (primordial) forebears. He thus credits the Finnish botanist Ragnar Hult (1885) with conducting the first thorough investigation of successional processes, and the French naturalist Adolphe Dureau de la Malle (1855) with describing the vegetation development following forest clear-felling as the origin of the word 'succession'.

1.2.2.4. Timeline of Ecologists

A list of founders, innovators, and their significant contributions to ecology, from romanticism onward.

Notable figure	Lifespan	Major contribution and citation
Antoni van Leeuwenhoek	1632–1723	First to develop the concept of food chains.
Carl Linnaeus	1707–1778	Influential naturalist, and inventor of science on the economy of nature.
Alexander Humboldt	1769–1859	First to describe the ecological gradient of latitudinal biodiversity increase toward the tropics in 1807.
Charles Darwin	1809–1882	Founder of evolution utilizing natural selection, and founder of ecological studies of soils.
Herbert Spencer	1820–1903	Early founder of social ecology, who coined the phrase 'survival of the fittest'.
Karl Möbius	1825–1908	First to develop the concept of an ecological community, biocenosis, and living community.
Ernst Haeckel	1834–1919	Invented the term ecology, and popularized research links between ecology and evolution.
Victor Hensen	1835–1924	Invented the term plankton, and developed quantitative and statistical measures of productivity in the seas.
Eugenius Warming	1841–1924	Early founder of Ecological Plant Geography.
Ellen Swallow Richards	1842–1911	Pioneer and educator who linked urban ecology to human health.

Stephen Forbes	1844–1930	Early founder of entomology and ecological concepts in 1887.
Vito Volterra	1860–1940	Independently pioneered mathematical population models around the same time as Alfred J. Lotka.
Vladimir Vernadsky	1869–1939	Founded the biosphere concept.
Henry C. Cowles	1869–1939	Pioneered studies and conceptual development in studies of ecological succession.
Jan Christian Smuts	1870–1950	Coined the term holism in the 1926 book *Holism and Evolution*.
Arthur G. Tansley	1871–1955	First to coin the term ecosystem in 1936 and a notable researcher.
Charles Christopher Adams	1873–1955	Animal ecologist, biogeographer, and author of the first American book on animal ecology in 1913, who founded ecological energetics.
Friedrich Ratzel	1844–1904	German geographer who first coined the term biogeography in 1891.
Frederic Clements	1874–1945	Author of the first influential American ecology book in 1905.
Victor Ernest Shelford	1877–1968	Founder of physiological ecology, who pioneered food-web and biome concepts, and founded The Nature Conservancy.
Alfred J. Lotka	1880–1949	First to pioneer mathematical population models explaining trophic (predator-prey) interactions using a logistic equation.
Henry Gleason	1882–1975	Early ecology pioneer, quantitative theorist, author, and founder of the individualistic concept of ecology.
Charles S. Elton	1900–1991	'Father' of animal ecology, who pioneered food-web and niche concepts and authored influential *Animal Ecology* texts.
G. Evelyn Hutchinson	1903–1991	Limnologist who conceptually advanced the niche concept.
Eugene P. Odum	1913–2002	Co-founder of ecosystem ecology and ecological thermodynamic concepts.
Howard T. Odum	1924–2002	Co-founder of ecosystem ecology and ecological thermodynamic concepts.
Robert MacArthur	1930–1972	Co-founder of the Theory of Island Biogeography and innovator of ecological statistical methods.

1.2.3. Influence of Ecology on the Social Sciences and Humanities

1.2.3.1. Human Ecology

In the 1970s, this field was recognized as a separate field of study. This was the first time that the fact that humans had colonized every continent on Earth was acknowledged as a significant ecological factor. Through habitat development (especially urban planning), intensive exploitation activities like fishing and logging, and byproducts of mining, industry, and agriculture, humans significantly alter the environment. In addition to ecology and biology, this field encompassed a wide range of other natural and social sciences, including economics, demography, anthropology, architecture, urban planning, psychology, and medicine. The growth of human ecology has resulted in the growing significance of ecological science in the planning and administration of urban areas.

Human ecology has piqued the interest of organizational researchers in recent years. According to Hannan and Freeman (Population Ecology of Organizations *(1977)*, American Journal of Sociology) argue that organizations do not only adapt to an environment. Rather, populations within organizations are also chosen or rejected based on their surroundings. In any given environment (in equilibrium) there will only be one form of organization (isomorphism). One well-known theory for explaining the diversity of organizations and how their makeup changes over time is organizational ecology.

1.2.3.2. James Lovelock and the Gaia Hypothesis

The idea that the Earth should be viewed as a single living macro-organism was advanced by James Lovelock's Gaia theory, which was put forth in his book "Gaia: A New Look at Life on Earth". Specifically, it contended that the community of living things has collectively evolved the capacity to regulate the world's environment — by affecting significant physical parameters like the chemistry of soils and oceans, the composition of the atmosphere, and the rate of evaporation — in order to preserve conditions that are conducive to life.

This vision was essentially a reflection of the times, especially the growing belief following World War II that human endeavors like nuclear energy, industrialization, pollution, and overuse of natural resources, driven by exponential population growth, posed a threat to global catastrophes. As a result, even though Lovelock's Gaia hypothesis divided scientists, many environmental movements accepted it as a motivating theory, believing that Gaia, their mother Earth, was "becoming sick from humans and their activities."

1.2.3.3. Conservation and Environmental Movements

Ecology and other scientific fields have been utilized by environmentalists and conservationists (e.g., climatology) to bolster their stances on advocacy. For political or financial reasons, environmentalist opinions are frequently divisive. Because of this, some ecological science has a direct impact on political discourse and policy, which in turn frequently guides ecological research.

It is important to distinguish between the history of ecology and the history of environmental thought. The only sources for ecology as a modern science are Darwin's "Origin of Species" and Haeckel's subsequent nomenclature of the sciences required to investigate Darwin's theory. Gilbert White, a resident of Selborne, England in the 18th century, is credited with raising awareness of how humans affect the environment. Even further back in time is the awareness of nature and its interactions. However, before Darwin, ecology was comparable to medicine before Pasteur recognized that disease was an infectious process. Although the history exists, it is not very important.

Neither Darwin nor Haeckel indeed declared themselves to be ecologists. The same can be said for scientists from various disciplines who made contributions to ecological theory long after they stopped identifying as ecologists. Population studies conducted by Raymond Pearl serve as an example. Studies conducted by botanists and plant geographers in the late 19th and early 20th centuries, which ironically lacked Darwinian evolutionary perspectives, gave rise to ecology as a field of study and methodology. Darwinism lost credibility until Mendel's pea experiments were rediscovered and incorporated into the Modern Synthesis. Darwin occasionally shared the Lamarckian perspective on inheritance held by many early plant ecologists. However, ecological research on plants and animals—ideally in the field and in real life—kept up speed.

The Ecological Society of America (ESA) was founded with a conservationist viewpoint in 1915. One of the objectives of Victor M. Shelford, a pioneer in the establishment of the society, was to protect the natural areas that ecologists were studying at the time but were at risk of being damaged by human encroachment. Publications like "The Control of Pneumonia and Influenza by the Weather," "An Overlook of the Relations of Dust to Humanity," "The Ecological Relations of the Polar Eskimo," and "City Street Dust and Infectious Diseases" in the early pages of Ecology and Ecological Monographs show that human ecology was also a visible component of the ESA at its inception. Ellsworth Huntington, the second president of the ESA, specialized in human ecology. Another early president, Stephen Forbes, advocated in 1921 for "humanizing" ecology because it was obvious that humans were the dominant species on the planet.

In reality, the auspicious inception of the new science marked the beginning of a series of unfortunate advancements and regressions in the field of conservation. Human ecology was compelled to focus on environments impacted by humans and the real-world issues they present. Nevertheless, the collective aim of ecologists was to establish ecology as a fundamental science, one esteemed enough to be admitted to Ivy League universities. It was believed that disrupted environments would not reveal the secrets of nature.

The American Dust Bowl ignited widespread interest in the environment, prompting a surge of requests in 1935 for ecology to address practical problems. The pioneering ecologist, C. C. Adams, sought to reintroduce human ecology into scientific investigation. At that time, Clements, the foremost authority on plant ecology, examined the land use issues related to the Dust Bowl in the context of his theories on plant succession and climax. Paul Sears's book, "*Deserts on the March*," gained significant readership. Perhaps World War II contributed to the neglect of the issue.

After World War II, there was a growing conflict between applied ecology, which aimed to describe pure ecology. Adams made another attempt to expand the ESA's scope by asking it to establish an endowment to support ecology. He said that because of ecology's propensity for integration, "a great expansion of ecology" was about to occur. However, ecologists were cognizant of the idea that ecology was still not regarded as a rigorous, quantitative science. Again, those who advocated for applied research and hands-on participation in conservation were politely turned down. Sociology eventually absorbed human ecology. Lewis Mumford, a sociologist, was the one who first introduced George Perkins Marsh's theories to the contemporary world at the 1955 conference "Man's Role in Changing the Face of the Earth." Social scientists predominated at that esteemed gathering. It was then claimed that ecology "lacked experimental methods" and that "man as an ecological agent" was ignored. An individual disregarded ecology, calling it "sterile and archaic." Shelford's Committee on Preservation of Natural Conditions was unable to continue its work within the ESA.

However, two things caused ecology's trajectory to diverge from applied issues. The Manhattan Project was one reason. After the war, it changed its name to the Nuclear Energy Commission. The Department of Energy (DOE) is what it is now. Studies on the effects of producing and using nuclear weapons were part of its substantial budget. That made ecology a part of the problem and turned it into a "Big Science." Theoretical ecology (then known as evolutionary ecology and also mathematical ecology) started to challenge ecosystem science, both basic and applied. Eugene Odum, the author of a widely read ecology textbook published in 1953, rose to prominence as an ecosystem advocate. Odum advocated for ecology to have an applied and ecosystem-focused focus in his publications.

The release of "Silent Spring" was the second event. The public was introduced to ecology as a term and idea through Rachel Carson's book. Her impact was felt right away. Encouraged by the book's release, a study committee informed the ESA that their science was not yet prepared to handle the assigned responsibilities.

Eugene Odum greatly influenced Carson's understanding of ecology, leading to the dominance of ecosystem science during the International Biological Program of the 1960s and 1970s. This brought significant financial support and elevated the reputation of ecology. Additionally, Carson's "Silent Spring" played a crucial role in inspiring environmental protection initiatives in the Kennedy and Johnson administrations, which were enacted into law just before the inaugural Earth Day. Ecologists were actively sought after for their expertise, as exemplified by Stanley Cain, a former President of the ESA, who was appointed as an Assistant Secretary in the Department of the Interior.

According to one environmental lawyer, the 1970 National Environmental Protection Act's (NEPA) requirement for environmental assessments "legitimized ecology.". It was dubbed "an ecological Magna Carta" by the president of the ESA. An eminent ecologist from Canada called it a "boondoggle.". If nothing else, state statutes similar to NEPA gave ecologists a lot of work. That was the problem. Ecology and ecologists were not prepared for this kind of work. Because there weren't enough ecologists working on impact assessments outside of DOE laboratories, "instant ecologists" with shaky credentials and abilities became more prevalent. Proposals for the professionalization of ecology started to surface. Particularly, innovative scientist Frank Egler committed his incisive writing to the endeavor. In the ESA, there was once more division between basic and applied scientists, which was made worse by the issue of environmental advocacy. The controversy, whose past has not received enough attention, raged through the 1970s and 1980s before the ESA's voluntary certification process and lobbying arm in Washington brought it to an end. After Earth Day, in addition to advocacy and professionalism issues, ecology had to address issues pertaining to its fundamental ideas. Numerous theoretical tenets and techniques from the fields of evolutionary ecology and ecosystem science began proving ineffective when it came to environmental analysis and evaluation. Under the weight of their newfound prominence, ecologists generally began to doubt the procedures and reasoning behind their work. In the meantime, employees of environmental advocacy organizations and government agencies were charged with applying dubious principles in their conservation work on a religious basis. The dispute erupted over the management of endangered Spotted Owl populations.

The conservation movement posed challenges similar to those faced by former scientists of the Manhattan Project who worked on nuclear power. Both groups had to navigate the difficult task of reconciling science with their personal political views, religious beliefs, and broader worldviews. While some ecologists successfully maintained a separation between their scientific work and their activism, others unabashedly embraced their role as environmental advocates.

1.2.3.4. Ecology and Global Policy

As early as 1971, when the United Nations Educational, Scientific and Cultural Organization (UNESCO) started a research program named Man and Biosphere to learn more about how humans and nature interact, ecology played a significant role in global politics. It clarified the idea of a Biosphere Reserve a few years later.

Rene Dubos and other specialists prepared for the first international conference on the human environment, which was hosted by the United Nations (UN) in Stockholm in 1972. At this conference, the expression "Think Globally, Act Locally" first appeared. The next significant developments in ecology were the emergence of the term "biosphere" and the concept of "biological diversity," which is now more widely used to refer to biodiversity. These terms were developed on during the 1992 Rio de Janeiro Earth Summit, where the major international organizations officially recognized the concept of the biosphere and acknowledged the risks associated with declining biodiversity.

Then, in 1997, at the conference that resulted in the Kyoto Protocol, the threats facing the biosphere were acknowledged from a global perspective. This conference focused in particular on the growing risks associated with the greenhouse effect, which is caused by an increase in greenhouse gas concentrations in the atmosphere and consequent global climate change. In Kyoto, most of the world's nations recognized the importance of looking at ecology from a global point of view, on a worldwide scale, and to consider the impact of humans on the Earth's environment.

Did you get it?

1. **Who is known as the father of ecology?**
2. **Why was the Park Grass Experiment founded?**

1.3. ECOLOGY: SCOPE AND OBJECTIVES

Learning Objectives

- Describe the objectives of ecology
- State the classifications of ecology

Ecology is a scientific field that examines the interactions between biotic and abiotic elements of the environment, as well as the relationships between biotic component individuals. Many scientists and ecologists have given different definitions of ecology.

Ecology is a science that studies organisms in relation to their environment, according to Woodbury (1954). Ecology is "the study of structure and function of nature," according to E. P. Odum (1969). Ecology was defined by R. Margalef (1968) as the "study of ecosystems." As a result, contemporary ecologists have expanded the definition of ecology to include all essential facets of the field. In any case, ecology is essentially a biological science since it examines how organisms interact with their surroundings. Our daily lives are significantly impacted by ecology. It is focused on horticulture, agriculture, soil, wildlife, forests, water resources, and other conservation efforts. Its domain is too large to be contained in a brief discussion. Nonetheless, learning about ecological principles gives you a foundational understanding of issues with soil, surface water, forests, and other natural resources.

1.3.1. Objectives of Ecology

The key ideas covered below shed light on a number of different ecological topics.

(i) The abundance and distribution of organisms on a local and regional scale (habitat niche, community, biogeography).

(ii) Seasonal, annual, successional, and geological changes in the occurrence, abundance, and activities of organisms over time.

(iii) Population ecology, or the relationship between organisms in populations and communities.

(iv) The way organisms modify their structure and function in response to their physical surroundings.

(v) An organism's behavior in its natural environment (ethology).

(vi) The way in which these relationships have evolved over time (evolutionary ecology).

(vii) The relationship between nature's biological productivity and humans.

(viii) Systems analysis, which is the creation of mathematical models to relate parameter interactions and forecast outcomes.

(ix) Applied ecology: the preservation and control of pollution and natural resources.

Keyword

Natural resources are resources that are drawn from nature and used with few modifications.

1.3.2. Classifications of Ecology

In general, plant and animal ecology are the two major subfields of ecology. Autecology and synecology are other classifications for it. Autecology is the study of one species of organism in terms of ecology. Studying the ecology of individual communities or entire ecosystems is the focus of synecology.

However, ecology can be divided into the following branches:

(i) Habitat Ecology: It examines the ecological aspects of the planet Earth's various habitats and how those habitats affect the creatures that inhabit them.

(ii) Community Ecology: It focuses on the investigation of animal distribution locally in diverse environments. It also has to do with succession planning, community unit recognition, and composition.

(iii) Population Ecology: Another name for this is demography. Its focus is on the structure, growth pattern, and regulation of population organisms. Interactions between populations of various species within a community are another area of study for population ecology.

(iv) Evolutionary Ecology: It is focused on the issues of speciation and niche segregation.

(v) Taxonomic Ecology: It has to do with the ecology of various taxonomic classes of living things.

(vi) Human Ecology: It mostly covers population ecology. It examines how humans and the environment interact with one another. Human ecology studies the effects that humans have on the biosphere and how these effects affect humanity.

(vii) Applied Ecology: It is a significant area of ecology. It addresses the application of ecology. The field of this study is primarily the application of ecological concepts to human needs. Among these are land use, population ecology, aquaculture, agriculture, horticulture, animal husbandry, forestry, conservation, and insect control, as the practical applications of ecology.

Keyword

Animal husbandry is the branch of agriculture concerned with animals that are raised for meat, fiber, milk, or other products.

(viii) Ecosystem Dynamics: It covers the ecological investigation of soil formation, energy transfer, productivity, and other related processes.

(ix) Production Ecology: It examines the total and net output of various ecologies, including freshwater, marine, agricultural, and horticultural. Production ecology aims to maximize production by efficiently managing ecosystems.

(x) Ecological Energetics: It addresses the conservation of energy and its impact on the creatures in the ecosystem.

(xi) Physiological Ecology (Eco-physiology): The functional aspects of an organism are directly linked to environmental factors. The survival of the population due to the functional adaptation of organisms to different ecological conditions is the focus of eco-physiology.

(xii) Chemical Ecology: It involves how animals adapt to the preferences of specific organisms, such as insects, to particular chemicals.

(xiii) Ecological Genetics: Ecologists identify the type of genetic variability present in all living things. Only organisms that receive environmental support can thrive in any given environment. Therefore, the study of species variations based on their genetic potential falls under the umbrella of genecology.

(xiv) Paleoecology: It is the study of historical living and environmental circumstances.

(xv) Geographic Ecology (Eco-geography): It sheds light on the study of plant and animal geographic distribution.

(xvi) Space Ecology: It focuses on creating ecosystems that can regenerate entirely or partially to support human life during protracted space travel.

(xvii) Pedology: It is concerned with the study of soil, with a focus on factors such as acidity, mineral content, human content, and soil types, and how they affect organisms.

(xviii) Radiation Ecology: It investigates the wide-ranging effects that radiation and radioactive materials have on the environment and living things.

(xix) Ethology: It investigates how animals behave in their natural environments.

(xx) Sociology: It is the study of human ethology and ecology.

(xxi) System Ecology: It deals with the application of advanced statistical methods, mathematical models, applied mathematics, and other tools to analyze and comprehend the structure and function of ecosystems.

Did you get it?

1. **What are the classifications of ecology?**
2. **What is habitat ecology?**

1.4. DEVELOPMENT OF ECOLOGY

Learning Objectives

- Describe the relationship between ecology and sustainable development
- Learn about the consequences of development on ecology and the environment

The relationship between ecology, environment, and development is introduced here through a discussion of the concepts and their evolution. It talks about how development affects the environment and ecology and also touches on the topics of social pollution, environmental degradation, and forest destruction. Here, too, the main issues raised by ecological movements are covered. The final portion of this unit covers a few development projects that alarmed environmentalists worldwide.

1.4.1. Ecology and Sustainable Development

The growing ecological awareness at local, national, and international levels among communities and , as well as among nation-states and governments has given the term "sustainable development" widespread international currency in recent years. Retaining the productive potential of natural ecosystems for future generations and attaining sustainable development patterns are necessary to ensure environmental sustainability, which is the seventh Millennium Development Goal.

Ecology was not taken seriously as a major issue until the early 1980s in many countries around the world because it was not integrated as an essential element of development planning. Concerning effects on the environment around the world have resulted from the last 150 years of economic growth. The rate at which the ozone layer is being destroyed, air pollution, forest and biodiversity loss, animal and plant extinction, loss of marine life, and pollution of the water and soil is alarming. The concept of ecology gained popularity in the 1980s as a result of growing awareness of the significance of environmental variations, the problems they cause, and their effects on human settlement, quality of life, developmental issues, and variations in fertility, mortality, and morbidity. It led to the realization that maintaining the ecosystem was necessary to improve life as a whole.

The remarkably high rates of resource utilization in the recent era of human history set it apart from the previous one. There are growing demands on the global resource flow and stock as a result of the continuous expansion and intensification of industrial and agricultural production. The commercialization of natural resources through development interventions necessitates a significant change in how resource rights are understood and used. Through either direct resource transfer away from their basic needs or the destruction of the vital ecological process that ensures the renewability of life-supporting natural resources, the resource base available to the economically disadvantaged and powerless has been reduced as a result of development's resource demands. In order for development to be sustainable, it must consider the long- and short-term benefits and drawbacks of the living and nonliving resource base, as well as social, cultural, ecological, and economic factors.

Remember

The word "Ecology" may be used interchangeably with the geographic environment, and consequently, ecological studies are often limited to the study of the direct effect of the environment on the material culture of people with simple technologies.

Did you know?

Aristotle was an early influence on the philosophical development of ecology. He and his student Theophrastus made extensive observations on plant and animal migrations, biogeography, physiology, and their behavior, giving an early analog to the modern concept of an ecological niche.

1.4.1.1. Sustainable Development

Sustainable development, as defined by the Brundtland Commission, meets present needs without compromising future generations. It prioritizes the basic needs of the world's poor and considers environmental constraints. Sustainability should guide economic and social development globally, with diverse interpretations aligning with common fundamentals. Development involves societal and economic transformation, aiming for physical sustainability despite social and political challenges. Policies must address resource access and cost-benefit distribution for sustainability.

1.4.1.2. Colonial Domination over National Resources

Village communities have governed and utilized essential natural resources like land, water, and forests for centuries, guaranteeing the sustainable use of these renewable resources. Colonial domination of this region was linked to the first significant shift in resource control and the emergence of significant conflicts over natural resources caused by non-local factors. The common essential resources were methodically turned into commodities by colonial dominance in order to increase profits and revenue. This conversion of resources into commodities allowed European industries access to South African resources, which in turn fueled the first Industrial Revolution. The conversion of commons into commodities has two consequences: first, it takes away the politically marginalized groups' access to commons and thus their right to survival; second, it takes away nature's capacity for self-renewal and sustainability by removing the social constraints on resource use that form the cornerstone of common property management. It was expected that the international conflict over natural resources would decrease and be replaced by resource policies driven by comprehensive national interests with the decline of the international colonial structure and the rise of sovereign countries in the region. However, resource use policies continued in the same colonial fashion, and recently, in response to demands from Third World elites and international obligations, a second, drastic shift in resource use began, leading to another sharp conflict between various interests. The politically weak and socially disorganized group, whose resource needs are minimal and whose survival depends primarily on natural products outside the market system, seems to be the most seriously threatened interest in this conflict. The survival requirements of these groups have been largely overlooked by recent changes in resource utilization. The demands of the elites in the North and the South are the main drivers of these changes.

Remember

Disturbance and resilience are two important ecological concepts necessary to understand changing environments. Different communities and species react differently when disturbed; however, the system crashes when it reaches a certain threshold.

1.4.1.3. Expansion of Global Market Fares

Ideological development makes indirect access to the dominance of the world market possible. It generates the need for foreign debt and international assistance, which supply the funding for these resource-commercialization or resource-privatizing development projects. As a result, local resources are increasingly in the hands of international financial institutions, rather than local governments or even national governments. International investment and aid are linked to the resources of even the most remote village through forestry, dam, and fisheries projects. Multilateral development organizations, like the World Bank, provide loans for ecologically delicate sectors like forestry, agriculture, and irrigation. By doing so, they elevate the market economy and establish the necessity of the natural and survival economies. The terms of the loan dictate how natural resources are used; the rates of return on investments in irrigation projects make it necessary to cultivate cash crops and waste water, even though doing so leaves the land parched or in an arid desert. Development projects financed by international organizations drive conflicts between tribal and peasant communities over natural resources.

1.4.2. Consequences of Development on Ecology and Environment

The enormous effects of human ecology are clearly visible in the air, water, land, and energy life-support systems. The increasing human population leads to a disruption in food production and a space crunch, resulting in environmental pollution and adverse changes to our surroundings. Under such conditions, human survival necessitates both the expansion of oneself or relocation, and the destruction of the natural forest for human habitation. Over time, these human activities cause a variety of unfavorable environmental changes, or pollution, which is currently a global concern. Here, we talk about these human endeavors.

a) Water Pollution

The major sources of water pollution are:

i) Industrial effluents (wastes) or toxic by-products.

ii) Sewage wastes: This contains decomposable organic matter and pathogenic agents directly discharged into rivers, streams, lakes, etc.

iii) Agricultural pollutants: there are excessive agricultural nutrients such as fertilizers, disease-controlling chemicals (pesticides, herbicides, insecticides, and fungicides).

These pollutants harm not just people but also animals and plants. Industrial effluents, however, pollute far more than the other two do.

b) Air Pollution and Noise

The sources are:

i) Industrial manufacturing processes: steel, chemical plants, oil refineries, fertilizer factories, etc.

ii) Combustion: Industrial and domestic combustion of coal, oil, forest fires, etc., emit smoke, dust, carbon dioxide, sulfur dioxide, etc.

iii) Automobiles emit carbon monoxide, nitrogen oxide, and suspended particulate matter.

iv) Miscellaneous sources include agricultural activities such as crop spraying for pest control, nuclear energy programs, etc.

v) Radiation pollution, for example, incidents like the one in Chernobyl or the Gas tragedy in Bhopal.

The primary source of pollution in this case is the release of gases from manufacturing, industry, and radiation.

Keyword

Radiation is energy that comes from a source and travels through space at the speed of light.

c) Soil Pollution

i) Solid fallout from the explosion of nuclear devices.

ii) Agricultural activities — over application of inorganic manure and different chemical-based pesticides.

d) Destruction of Forests

Deforestation on a large scale causes flooding, soil erosion, river silting, agricultural area contraction, and desertification. Compared to the local villagers who use the wood for fuel, forest contractors are the ones responsible for more deforestation.

All of these pollution types have an impact on productivity, the environment, other flora and fauna, recreational facilities, and human health either directly or indirectly.

Remember

Ecological areas of study include topics ranging from the interactions and adaptations of organisms within an ecosystem to the abiotic processes that drive the development of those ecosystems.

Did you get it?

1. **What are the major sources of water pollution?**
2. **What happens due to large-scale deforestation?**

1.5. LEVEL OF STUDY IN ECOLOGY

Learning Objectives

- Describe the differences between organismal ecology and population ecology
- Explain what community ecology and ecosystem ecology entail

Ecology is the study of relationships and interactions between organisms, populations, and communities within the ecosystems they inhabit.

Ecology is the study of how living things interact with their surroundings. Ecology is a discipline where researchers operate at four distinct levels, which can overlap or operate discretely. The organism, population, community, and ecosystem are at these levels (Figure 1.1). According to ecology, ecosystems are made up of constituents that interact dynamically: living things, the communities they belong to, and the inanimate (abiotic) elements of their surroundings. The flow of energy and matter through an environment is regulated by ecosystem processes like primary production, pedogenesis (the creation of soil), nutrient cycling, and different niche construction activities. Only certain types of organisms with particular life histories can support these processes. Certain ecosystem services are improved by the diversity of organisms, or biodiversity, which refers to the various species, genes, and ecosystems.

In essence, ecologists seek to explain:

- life processes
- interactions, interrelationships, behaviors, and adaptations of organisms
- the movement of materials and energy through living communities
- the successional development of ecosystems
- the abundance and distribution of organisms and biodiversity in the context of the environment

Ecology has many real-world applications in fields such as conservation biology, wetland management, agroecology, agriculture, forestry, fisheries, urban ecology, community health, economics, basic and applied science, and human ecology (the study of human social interaction). Ecosystems are made up of resources and living things. Ecosystems, in turn, support biophysical feedback mechanisms that control processes affecting both living (biotic) and nonliving (abiotic) elements of the earth. In addition to producing natural capital, ecosystems maintain life-supporting processes like biomass production (food, fuel, fiber, and medicine), climate regulation, global biogeochemical cycles, water filtration, soil formation, erosion control, flood protection, and many other naturally occurring features with intrinsic, historical, scientific, or commercial value.

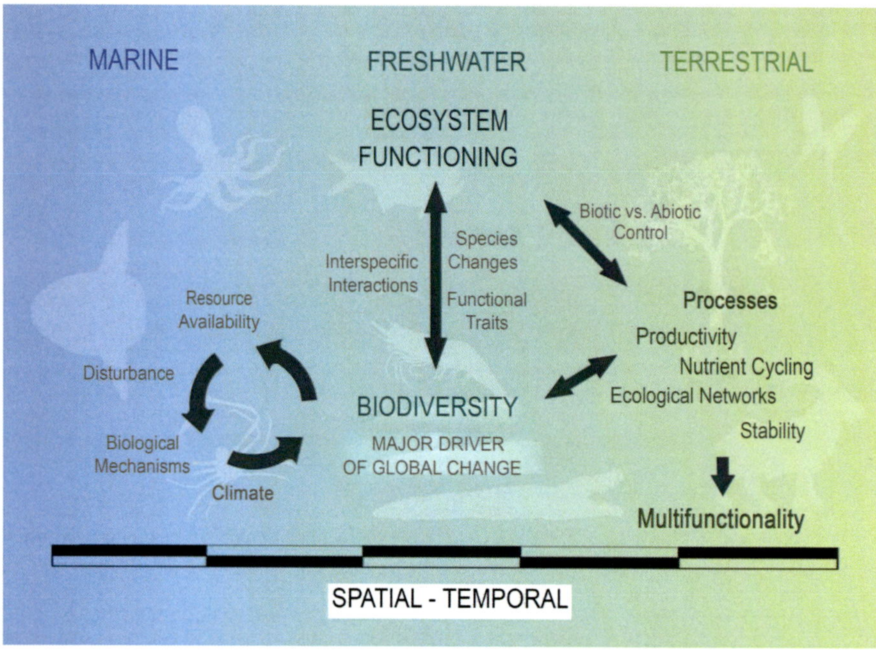

Figure 1.1. Levels of ecological study: Ecologists study within several biological levels of organization, which include organism, population, community, and ecosystem.

Source: Correia, A. M., & Lopes, L. F. (2023). Revisiting Biodiversity and Ecosystem Functioning through the Lens of Complex Adaptive Systems. *Diversity*. *15*(8), 895. https://doi.org/10.3390/d15080895.

Numerous subcategories of ecology exist as well, including animal, plant, and ecosystem ecology, which examine the variations and similarities between different plant species in a range of environments and climates. Furthermore, population ecology examines the similarities and differences between populations as well as how they replace one another over time, while physiological ecology, also known as Ecophysiology, studies how an individual organism responds to its surroundings.

Lastly, it's critical to understand that ecology is not the same as the natural sciences, environment, environmentalism, or natural history. Furthermore, although it is closely related to the fields of genetics, ethology, and evolutionary biology, it is distinct from them.

1.5.1. Organismal Ecology and Population Ecology

Studying an organism's ability to adapt to its surroundings and its relationships with other organisms is known as organismal ecology.

1.5.1.1. Organismal Ecology

When examining ecology at the organismal level, scientists are interested in the adaptations that allow organisms to live in particular environments. These adaptations can be behavioral, physiological, or morphological (relating to the study of form or structure). One example of a specialist butterfly is the Karner blue (*Lycaeides melissa samuelis*), whose females preferentially oviposit (lay eggs) on wild lupine. Because of this preferential adaptation, the Karner blue butterfly's survival depends largely on the existence of wild lupine plants (Figure 1.2.).

Figure 1.2. Karner blue butterfly: The Karner blue butterfly (*Lycaeides melissa samuelis*) is a rare butterfly that lives only in open areas with few trees or shrubs, such as pine barrens and oak savannas. It can only lay its eggs on lupine plants.

Source: https://bio.libretexts.org/ (https://bio.libretexts.org/Courses/Lumen_Learning/Fundamentals_of_Biology_I_%28Lumen%29/17%3A_Module_14-_Ecology_of_Living_Things/17.22%3A_Levels_of_Ecological_Research).

The larval caterpillars hatch, emerge, and feed exclusively on wild lupine for four to six weeks. After roughly four weeks, the caterpillars pupate (undergo metamorphosis) and become butterflies. The nectar of wild lupine flowers and other plant species is what adult butterflies eat. In addition to questions about egg-laying, a researcher interested in studying Karner blue butterflies at the organismal level might also ask about their preferred temperature (a physiological question) or the behavior of the caterpillars at different larval stages (a behavioral question) (Figure 1.3).

Figure 1.3. Wild lupine: The wild lupine (*Lupinus perennis*) is the host plant for the Karner blue butterfly.

Source: https://bio.libretexts.org/ (https://bio.libretexts.org/Courses/Lumen_Learning/Fundamentals_of_Biology_I_%28Lumen%29/17%3A_Module_14-_Ecology_of_Living_Things/17.22%3A_Levels_of_Ecological_Research).

1.5.1.2. Population Ecology

Remember

Organismal ecology focuses on the morphological, physiological, and behavioral adaptations that let an organism survive in a specific habitat.

A population consists of organisms living in the same area, breeding together, and belonging to the same species. Conspecifics are organisms that belong to the same species and form a population. A population can be distinguished in part by the boundaries that surround it, whether they are man-made or natural. Rivers, mountains, and deserts are examples of natural boundaries, whereas roads and mowed grass are examples of artificial boundaries. Population ecology is the study of how many organisms live in a given area and how and why that number varies over time. Because the Karner blue butterfly is considered federally endangered, population ecologists are especially interested in counting them. However, the abundance and distribution of wild lupine have a significant impact on the density and distribution of this species. Scholars may wish to investigate the causes of the wild lupine population decline and the consequences for Karner blue butterflies. Ecologists are aware, for instance, that wild lupines do best in open spaces devoid of trees and shrubs. Wildfires occur in natural settings on an irregular basis.

1.5.2. Community Ecology and Ecosystem Ecology

Community ecology is the study of species interactions and how ecosystems are impacted by biotic and abiotic factors.

1.5.2.1. Community Ecology

A biological community consists of the different species present in a specific area, typically in three dimensions, along with the interactions that take place within and between these species. Community ecologists study the mechanisms underlying these interactions as well as their effects. Questions concerning conspecific interactions frequently center on rivalry between individuals belonging to the same species over a finite resource. Scientists studying ethology also look at the relationships between different species; these individuals are known as heterospecific. Heterospecific interactions include things like herbivory, competition, predation, parasitism, and pollination. The regulation of population sizes and the influence of ecological and evolutionary processes on diversity can be achieved through these interactions.

The Karner blue butterfly's larvae, for instance, develop mutualistic bonds with ants. Mutualism is a type of long-term relationship that benefits both species and has developed through coevolution between them.

Figure 1.4. Karner blue butterfly caterpillar: Karner blue butterfly caterpillars form beneficial interactions with ants. This mutualistic relationship is an example of a community ecological study, which aims to examine the interactions between different species living in an area.

Source: https://bio.libretexts.org/

Each species must gain something from the other as a result of their relationship for mutualism to exist between individual organisms. Research indicates that when ants tend to the larvae (caterpillars) of Karner blue butterflies, there is an increased chance of survival. This may be because when ants tend to them, larvae spend less time in each life stage, giving them an advantage. A significant source of energy for the ants is a substance secreted by the larvae of the Karner blue butterfly that is rich in carbohydrates. The interaction benefits the ants as well as the Karner blue larvae.

1.5.2.2. Ecosystem Ecology

The study of ecosystems extends the study of populations, organisms, and communities. Together with the abiotic (nonliving) elements of a region, the biotic (living) components make up the ecosystem. Soil, water, and air are a few examples of abiotic elements. Ecosystem biologists investigate the storage of nutrients and energy, as well as their movement within and between organisms, as well as through the soil, water, and atmosphere.

The wild lupine and the Karner blue butterflies inhabit a desolate oak-pine forest. Natural disturbance and low nitrogen-rich soils are the defining features of this habitat. The distribution of the plants in this habitat is significantly influenced by the availability of nutrients. The significance of finite resources and the flow of resources, like nutrients, between the biotic and abiotic components of the ecosystem are topics that ecologists might investigate.

Did you get it?

1. **What is population ecology?**
2. **What is community ecology?**

1.6. RELEVANCE AND APPLICATION OF ECOLOGY

Learning Objectives

- Describe the applications of ecology

The relationship between living things and their surroundings is referred to as ecology. It is the study of how living things can use their surroundings and resources effectively now so that future generations can benefit from them as well. It is essential for food production, biodiversity preservation in a changing climate, and the preservation of clean air and water. It is crucial for resource allocation, environmental conservation, lowering pollution, and ozone layer depletion.

By preventing resource deprivation and scarcity, it can also prevent competition and the extinction or endangerment of specific organism species.

Knowing how each species interacts with others, what kind of habitat each one needs, how they affect one another, what population size assures its survival, and other details is essential to conserving and safeguarding nature and preventing the extinction of species. Ecology promotes peaceful coexistence among species, and with the right information, one can adopt a way of life and take precautions to preserve the ecology of life.

Compared to other more "dynamic" sciences like astronomy or economics, people rarely give ecology much thought. This is unfortunate because ecology teaches us about the inner workings of the world we live in. People have become more aware of the urgent news derived from ecological research in recent years, particularly concerning biodiversity loss and climate change. These are only a few of the numerous subjects that ecology studies. Prior to delving into those, it is imperative that we grasp the true essence of ecology.

Ecology is a large, multifaceted, and fairly complex science. To examine ecology and its significance, it is essential to first understand what is meant by the term "ecology.".

Ecology must consider a wide range of factors, including trophic relations, population growth, competition, biodiversity, and many more, to comprehend these relationships. Because of how complicated the system is to study, this science is broken down into multiple categories, each of which focuses on a different aspect of ecology. Microbial, behavioral, population, community, and ecosystem ecology are a few examples. They all provide us with different viewpoints on different kinds of organisms found throughout the world and aid in our comprehension of the relationships between them and the outcomes that result from those relationships. Global ecology is another field of study that looks at the entire biosphere and all living organisms within it. Since the climate can have a big impact on the biosphere, environmental factors like it are also studied here. It gets a little bit simpler to comprehend the goal of ecology after keeping this in mind.

It Helps with Conservation

Since humans are an indigenous species on Earth, ecology also studies them. Humans have a significant effect on the surrounding ecosystems. Given that we inhabit every part of the planet, we are intricately entwined with every system.

Humans and the ecosystems they live in may occasionally have beneficial interactions, but these are typically one-sided deals. It's no secret that people use ecosystems and resources for personal benefit. Ecology is another field that studies this in great detail.

Using ecological models and observations, we can demonstrate the extent of our harm to the surrounding ecosystem. Thankfully, we can also figure out how to lessen the harm and partially restore the ecosystem. The estimates and indicators of ecologists aid in instilling a sense of urgency in the general public and propel conservation efforts. It's difficult to envision where we would be without ecological intervention. We must heed the advice and conservation concepts offered by ecologists if we are to protect biodiversity and guarantee the planet's continued existence.

It allows for better resource allocation

How various creatures use resources is one of the more fascinating facets of the ecosystem that ecology discusses. It is very easy to determine which resources are excess and which are necessary for a species to survive and reproduce. This is a crucial aspect of the way ecology views the human species.

Because we rely on fossil fuels, humans have a larger carbon footprint than before. It has been demonstrated that, given the abundance of available alternative energy and power sources, this is not an absolute necessity. Ecology gives us a variety of viewpoints on this issue and enables us to decide how best to address it. It's critical that, to establish a long-lasting sustainable system, we as a species modify some of the demands placed on resources.

It helps with disease and pest control

As was already mentioned, microorganisms are a part of ecology. They play the same role in our ecosystem as any other kind of creature. Research on specific microbes is crucial for humans to stop diseases from spreading quickly. Studying everything from viruses to parasitic worms is necessary to determine the best ways to address the issues they cause.

An excellent illustration of how ecology addresses illness is the malaria situation. The Anopheles mosquito is the carrier of this specific disease. We must understand how Anopheles interacts with its surroundings if we are to stop the spread of malaria and possibly prevent it altogether. Its feeding, breeding, and propagation strategies are all crucial details that help us get closer to lessening their influence on human populations.

There are numerous ways to implement control measures. In the case of malaria, individuals have developed nets and other defense mechanisms to prevent mosquitoes from transmitting disease to people. A further strategy to stop them from spreading widely is to interfere with their habitats and breeding grounds. Mosquito populations are significantly reduced when freestanding surface water is reduced in a given area. These are only a few instances of how ecology manages certain pests and the diseases they spread.

It promotes eco-friendliness

In the end, ecology's research and knowledge encourage environmental friendliness. The ultimate objective is to bring every ecosystem into balance, and this takes a great deal of study and work to achieve. Every advancement ecology makes in a field fosters greater understanding among the populace, which in turn results in a more environmentally conscious way of looking at the world. People want to contribute to environmental preservation and are becoming more accepting of eco-friendly lifestyles.

This is accomplished by a variety of strategies that ecology supports. At the forefront of eco-friendly action are practices like using organic products and lowering one's carbon footprint. As soon as green options become available, many households begin to use them. There are many products available that, when used responsibly, can have a significant positive impact on the environment. People are starting to use organic materials for everything from their floors to their linens to their coffee cups, in place of plastic and non-recyclable products. Considering the other advantages that come with many organic products, some of these changes are not at all difficult to implement.

First of all, construction materials are more sustainable than before. Nowadays, cork—which comes from cork trees—is frequently utilized as flooring. It is a great environmentally friendly option because it is simple to recycle and biodegradable. There are also efforts underway to make other, more mainstream items eco-friendly. In addition to being more environmentally friendly, sleeping on sustainable bamboo sheets provides superior comfort and aeration compared to polyester sheets. Plastic cups are quickly disposed of because they are unpleasant to drink from and difficult to clean. A clay or ceramic cup can be personalized to your preferences and is simple to clean. Sustainable clothing is more cozy, as an example.

Every household can find plenty of environmentally friendly options, and when given the chance, people are more than happy to make changes. One of the main practical goals of ecology is this.

It enhances energy conservation

The energy crisis is one of the most urgent issues facing society today. Our energy-sourcing practices have a discernible negative impact on both the environment and human health. When finite resources like coal and fossil fuels are burned, harmful substances are released into the atmosphere. They contribute to the greenhouse gas effect and poison all ecosystems. Because of this, society needs to advance through the use of more sustainable energy sources.

Without an appropriate understanding of energy allocation and conservation, ecology makes it impossible to determine how unsustainable a particular energy source is. When it comes to using some energy sources, humans can be notably wasteful, but without the systems employed in ecological research, this would be impossible to measure.

While some practices, such as burning fossil fuels, are considered harmful because of their impact on populations, it can be difficult to quantify the harm caused by the cutting of trees . Ecology examines the ecosystems in which these trees exist and can demonstrate the severity of the effects that would result from their removal. To develop alternatives that are less damaging to both the environment and our species as a whole, we must remain aware of the ecological costs associated with particular fuel-consuming behaviors. Without this knowledge, humanity might continue on its current course and destroy the surrounding environment beyond repair.

1.6.1. Applications of Ecology

1.6.1.1. Agriculture

There are many uses for ecology in agriculture. We get our food and fiber from agriculture. Ecological principles are essential to agriculture. These four principles are productivity, pest control, irrigation requirements, and limiting factors. Ecology plays a role in land conservation. This helps in recovering the soil that is waterlogged and alkaline.

1.6.1.2. Forestry

Forestry is the study of forests. Pakistan and other developing nations are dealing with the issue of an increasing population. The need for food and shelter is increasing due to the population's rapid growth. As a result, man is felling trees for housing and agriculture. At that rate, there won't be any more leaf canopy if the forests are constantly cleared. This canopy of leaves shields the soil from torrential rain.

As a result, a portion of the soil will seep into the river and stream with the water. This soil will be thrown away. As a result, the soil's fertility will decline. Three facets of forestry exist.

(a) Reforestations: Deforestation is the process of clearing large tracts of forest to plant subsistence crops, raise cattle, or harvest lumber.

(b) Reforestation: Reforestation is the process of replanting trees in a forest. Reforestation is imperative to compensate for deforestation. For many species of conifers, reforestation is particularly crucial.

(c) Afforestation: Afforestation is the process of creating new forests where none previously existed.

Reforestation and afforestation are aided by ecology. Scientists are coming up with strategies to halt deforestation. overgrazing and desertification.

1.6.1.3. Wildlife Management

Wildlife includes all non-domesticated animals and non-cultivated plants. A vital role for wildlife exists in the food chain. In the absence of wildlife, the food chain is upset. Thus. Maintaining the ecosystem's balance in the context of wildlife becomes exceedingly challenging. Humanity is attempting to alter the surroundings. With time, the effects of these environmental changes become more and more noticeable. The unbalance in the natural world may endanger human existence. Every living thing depends on every other living thing. There is an equilibrium between the environment and all living things. Man has long been upsetting this equilibrium. Endangered species are those that are in danger of going extinct. Thousands of plants and animals are in danger of extinction today. The natural world is a finite resource. The goal of ecologists' work is to preserve endangered species.

1.6.1.4. Fishery and Aquaculture

Fishery is the breeding and raising of fish. Aquaculture is the culture of aquatic animals. In rivers, lakes, and oceans, fishing is nearly overfished. Consequently, fish populations in these bodies of water are declining. In rivers, ponds, and lakes, fish populations are maintained in part by ecological principles. Studies are conducted on fish habitat, feeding patterns, and reproductive physiology. Ponds are used for the culture of fish. They receive adequate nourishment as well as other ecological elements required for their development. In ponds, oxygen is a limiting factor. Thus, oxygen is provided through artificial means. Fish hatch in dedicated hatcheries. Fish populations have increased as a result of these actions.

1.6.1.5. Range Land Management

Rangeland is the term for the grassy areas where animals graze. Rangeland management is the application of science to rangelands in order to regularly produce the greatest number of grazing livestock. Rangeland is crucial for the growth of livestock. Sadly, he majority of rangeland has been converted to agricultural land by humans. There are thus fewer rangelands remaining in the nation. These rangelands must be kept safe from excessive grazing. A rangeland can become a desert due to overgrazing. Rangeland is preserved by making the best use of it. It also gives humans a steady supply of meat and milk. One of the uses of ecology is the upkeep of rangelands. Ecological research indicates the maximum number of grazers that a rangeland can support. It also explains the effects of rangeland destruction on the ecological cycle.

1.6.1.6. Water Shed Management

A watershed is the region from which precipitation, either as rain or snow, gathers to form a stream or river. Water shed management is the process of storing water for irrigation and using it to create electricity through the construction of dams and barrages. The foundation of our agriculture is made up of dams and barrages. All across the world, a lot of dams are being built. They affect the environment in numerous ways. Before building a dam, all of these ecological effects must be investigated. The majority of dams are having silting issues. The dam is quickly filling with silt. According to estimates, silting will render the majority of Pakistan's dams useless in ten years. It is ecology that halts the silting process. Tree planting in the area of the water shed can lessen silting.

1.6.1.7. Conservation of Soil

Plants receive all necessary nutrients from the soil. It offers water, oxygen and minerals to the plant. In the balanced state, each of these components is necessary. The soil has been severely impacted by water logging, alkalinity, and soil erosion. In a similar vein, overcropping has caused the soil's nutrient levels to drop. Ecology offers us the answers to all of these issues. Planting trees and maintaining a dense vegetation cover are effective ways to reduce soil erosion. Similar to this, some crops can thrive in alkaline soil as these crops lessen the soil's alkalinity. Ecology has developed numerous techniques to manage water logging. Crop rotation and the application of humus help keep the soil's nutrient balance intact.

1.6.1.8. Control of Pollution

Pollution is defined as the introduction of undesirable and hazardous chemicals into the environment. Our environment has been devastated by industrialization. There is an enormous amount of industrial waste produced by these industries. These wastes contaminate the surroundings. There are a lot of poisonous and carcinogenic (cancer-causing) substances in industrial waste. Heavy metals like chromium are produced in tanneries and automobiles. Our health is impacted by these metals.

Did you get it?

1. **What are the uses of ecology in agriculture?**
2. **What is a rangeland?**

CLOSER LOOK

ECOLOGICAL SUCCESSION

An important characteristic of all communities is that their composition and structure constantly change in response to the changing environmental conditions. This change is orderly and sequential, parallel with the changes in the physical environment. These changes lead finally to a community that is in near equilibrium with the environment and that is called a climax community. The gradual and fairly predictable change in the species composition of a given area is called ecological succession. During succession some species colonize an area and their populations become more numerous, whereas populations of other species decline and even disappear.

The entire sequence of communities that successively change in a given area are called sere(s). The individual transitional communities are termed seral stages or seral communities. In the successive seral stages there is a change in the diversity of species of organisms, an increase in the number of species and organisms as well as an increase in the total biomass.

The present-day communities in the world have come to be because of succession that has occurred over millions of years since life started on Earth. Actually succession and evolution would have been parallel processes at that time.

Succession is hence a process that starts where no living organisms are there – these could be areas where no living organisms ever existed, say bare rock; or in areas that somehow, lost all the living organisms that existed there. The former is called primary succession, while the latter is termed secondary succession.

Examples of areas where primary succession occurs are newly cooled lava, bare rock, and newly created ponds or reservoirs. The establishment of a new biotic community is generally slow. Before a biotic community of diverse organisms can become established, there must be soil. Depending mostly on the climate, it takes natural processes several hundred to several thousand years to produce fertile soil on bare rock.

Secondary succession begins in areas where natural biotic communities have been destroyed such as in abandoned farmlands, burned or cut forests, and lands that have been flooded. Since some soil or sediment is present, succession is faster than primary succession.

Description of ecological succession usually focuses on changes in vegetation. However, these vegetational changes in turn affect food and shelter for various types of animals. Thus, as succession proceeds, the numbers and types of animals and decomposers also change.

At any time during primary or secondary succession, natural or human induced disturbances (fire, deforestation, etc.), can convert a particular seral stage of succession to an earlier stage. Also, such disturbances create new conditions that encourage some species and discourage or eliminate other species.

Succession of Plants

Based on the nature of the habitat – whether it is water (or very wet areas) or it is on very dry areas – the succession of plants is called hydrarch or xerarch, respectively. Hydrarch succession takes place in wetter areas and the successional series progresses from hydric to mesic conditions. In contrast, xerarch succession takes place in dry areas and the series progresses from xeric to mesic conditions. Hence, both hydrarch and xerarch successions lead to medium water conditions (mesic) – neither too dry (xeric) nor too wet (hydric).

The species that invade a bare area are called pioneer species. In primary succession on rocks, these are usually lichens which can secrete acids to dissolve rock, helping in weathering and soil formation. These later pave the way to some very small plants like bryophytes, which can take hold in a small amount of soil. They are, with time, succeeded by bigger plants, and after several more stages, ultimately a stable climax forest community is formed. The climax community remains stable as long as the environment remains unchanged. With time, the xerophytic habitat is converted into a mesophytic one.

In primary succession in water, the pioneers are the small phytoplankton. They are replaced with time by rooted-submerged plants, rooted-floating angiosperms followed by free-floating plants, then reed swamp, marsh-meadow, scrub, and finally, trees. The climax again would be a forest. With time, the water body is converted into land.

In secondary succession, the species that invade depend on the condition of the soil, availability of water, the environment as also the seeds or other propagules present. Since soil is already there, the rate of succession is much faster and hence, climax is also reached more quickly.

What is important to understand is that succession, particularly primary succession, is a very slow process, taking maybe thousands of years for the climax to be reached. Another important fact is to understand that all succession whether taking place in water or on land, proceeds to a similar climax community – the mesic.

ROLE MODEL

ALEXANDER VON HUMBOLDT

Friedrich Wilhelm Heinrich Alexander von Humboldt, more simply called Alexander von Humboldt, was a notable Prussian geographer, explorer, and naturalist. He is widely recognized for his works on botanical geography which laid the foundation for biogeography. His last name is something many animal lovers may already be familiar with, as the Humboldt squid that lives in the Humboldt Current of South America, is named after him.

Early Life and Education

This naturalist and eager explorer was born on September 14, 1769, in Berlin, Germany. Alexander Georg von Humboldt, his father, was an army officer who died when he was only 9 years old. Alexander and his older brother, Wilhelm, were then raised by their mother, Maria Elisabeth Colomb. Alexander's family belonged to a prominent Pomeranian family, and so they were able to afford private tutors to provide the boys with a good education in mathematics, languages and classics. Less typically the brothers were also taught political history and economics.

As a child, Alexander had a hobby of collecting and labeling different plants, insects, and shells. He thus earned the affectionate nickname "little apothecary".

Intending on a political career, Humboldt took courses in finance for six months at the University of Frankfurt in 1787. Then, for a year he studied at Göttingen University. During this time Humboldt developed an interest in botany, geology and mineralogy. He met Georg Forster who had accompanied Captain Cook on his second voyage of discovery. Foster had a strong influence on Humboldt and they traveled across Europe and to England together.

In 1789 Humboldt wrote "Mineralogic Observations on Several Basalts on the River Rhine."

To further his interest in mineralogy and geology, Humboldt enrolled in the School of Mines in Freiberg, Saxony in 1790. A typical day would be spent underground in the mines in the morning, at classes in the afternoon and in the evening walking the surrounding countryside searching for interesting plants.

Humboldt worked as a government mines inspector in Franconia, Prussia from 1792 to 1797. During this time, He invented a safety lamp and founded a technical school for aspiring miners.

Explorations

His mother died when he was 27, in 1796 and this left him a good inheritance that would pay for his future explorations. A year after his mother's death, he left government service and planned his travels with the botanist Aime Bonpland. Humboldt had always wanted to travel and explore and he seized his chance. Together with Bonpland, they traveled to Madrid to obtain the special permission needed at that time to explore South America.

On the 5th of June 1799, Bonpland and Humboldt finally set sail aboard the ship Pizarro. They had a 6-day stop at Tenerife, where they explored the Teide volcano. On the 16th of July, they landed at Cumana, Venezuela. During their stay in South America, Bonpland and Humboldt studied the topography, flora, and fauna of the continent. Within a year, Humboldt had mapped more than 1,700 miles of the Orinoco River.

Bonpland and Humboldt continued their explorations of South America and they took a trip to the Andes, climbing to the top of Mt. Chimborazo, located in today's modern Ecuador. At that time, this summit was believed to be the highest peak in the world. They were not able to make it to the very top but ascended to over 18,000 feet. The exploration then continued to Lima, Peru, where Humboldt observed the transit of the planet Mercury (where Mercury appears as a tiny black dot moving across the disk of the Sun). He also noticed the fertilizing properties of guano.

While on South America's west coast, Humboldt discovered and measured the Peruvian Current. This current is also now known as the Humboldt Current.

Other notable experiences from his explorations in South America included being able to see an amazing annual meteor shower known as the Leonids on the nights of November 11 and 12 shortly after their arrival. Humboldt was the first to connect altitude sickness to a lack of oxygen, due to his many mountain climbs and his personal experiences at high altitudes. Bonpland and Humboldt captured and dissected electric eels from which they received many electric shocks themselves.

Other Travels and Achievements

After much exploration in South America, Bonpland and Humboldt visited Washington, D.C. They stayed for three weeks during which Humboldt was able to have several meetings with President Thomas Jefferson.

In 1804, Humboldt traveled to Paris and he spent his time writing 30 volumes about his different field studies. He stayed in Paris for 23 years, meeting and discussing with other bright minds of his age.

Eventually, the fortune from his mother's inheritance ran out. Humboldt then found a stable source of income and became one of the advisors of Prussia's King. He was invited to Russia in 1829 and on his extensive journey found diamonds in the gold mines of the Urals and collected data for an isothermal world map. In 1827, Humboldt visited in Berlin, giving public lectures. These lectures became so popular that he decided to write all his research in a work that he called the "Kosmos". The first volume was published in 1845 when he was 76 years old. Five volumes in total were published, the final volume posthumously.

Humboldt suffered a minor stroke on February 24, 1857. Two years later, his health began to decline and at the age of 89, he died on May 6, 1859. To this day, he is known as one of the most significant contributors to the earth sciences.

FOCUS ON CAREERS

ROLE OF ECOLOGIST

Ecologists study the intertwined relationships between humans, plants and animals as they interact with their physical environment. They research everything from soil microbial activity to humpback whale foraging and wastewater treatment.

Ecologists protect the health and sustainability of life on the planet. Their roles and responsibilities include managing natural resources, promoting biodiversity, protecting public health and improving the environment.

Duties of an Ecologist

Apart from studying living organisms, ecologists are also responsible for the study of the footprints humans have on nature. This study can help them suggest various solutions to the adverse effects of human activities on the environment. Here are a few duties of an ecologist:

- Classify plants, animals and other living organisms
- Conduct field surveys of the organisms and their habitats
- Research on theories in laboratories
- Present scientific data in a concise manner
- Monitor scientific investigations
- Record and analyze details of species and their environment
- Form plans to protect species that are on the verge of extinction
- Study the environmental impact on living organisms

- Recommend environmental legislation to government bodies
- Maintain equipment used for field research
- Work with software programs like geographic information systems
- Manage ecological projects
- Write proposals for funding opportunities
- Write and publish articles
- Present research at conferences and events
- Investigate the effects of changes in the ecosystem
- Devise sustainable practices
- Propose plans to conserve biodiversity
- Spread awareness among local communities about environmental issues

Fields of Study of an Ecologist

Ecologists study their field at the following five levels:

- **Organism:** It can range from a tiny insect to a giant blue whale.
- **Population:** It is a group of organisms of the same species, like bees, living in the same area at the same time.
- **Community**: It is a biological group of all the populations of varied species that exist in an area.
- **Ecosystem**: It comprises all the organisms that live in an area along with abiotic or physical factors that influence the community.
- **Biosphere:** It is the planet Earth where communities and populations of species live.

Fields of Specialization for an Ecologist

Ecologists can specialize in varied fields like:

- Marine biology
- Botany
- Toxicology
- Zoology
- Microbiology
- Soil science
- Chemistry
- Genetics
- Physics

How To Become An Ecologist?

If you wish to become an ecologist, you can follow these steps:

1. Complete School Education

Complete the higher secondary education (10+2) with physics, chemistry and biology from a board recognized by the Government of India. Secure an aggregate of 50% marks in the 12th standard's board examination, as this may be an important requirement to pursue a bachelor's degree. If these subjects are not part of your main curriculum, you may opt for them as additional subjects if your school allows for such an arrangement.

2. Finish a Bachelor's Degree

After completing higher secondary education and gaining a strong background in science, you can pursue a bachelor's degree in science or subjects like botany, biology, zoology or forestry. Depending on what area you want to specialize in, you may consider an additional bachelor's degree in environmental science, oceanography, geology or climatology. A bachelor's degree can help you better understand the various nuances of the field.

3. Gain a Master's Degree and Ph.D.

Once you have a relevant bachelor's degree, you may consider expanding your knowledge base by gaining a master's degree like a master's in science. A master's degree can also play an important role in getting admission to advanced ecology courses like a Ph.D. in ecology. A bachelor's degree, a master's degree, and a Ph.D. can help you show substantial educational experience to your employers during job interviews.

4. Pursue Additional Courses and Certificate Programs

If you plan to get a job immediately after a bachelor's degree, you may consider pursuing various diploma or certificate courses for specialized and comprehensive knowledge. Good ecology courses can help you understand topics like ecosystem dynamics, conservation, sustainability, biological diversity and other social and ecological systems. This exposure to in-depth learning can add to your resume and help you create a rewarding career in the field.

5. Work as An intern

You can consider working as an intern with industry experts who can help you acquire first-hand job experience. Internships can help you build subject awareness and you can use this knowledge when you start working as a professional in the field. Internships are a great way to perform a variety of tasks in different departments. With an internship, you can also build a network that may present job opportunities in the future.

6. Apply for Jobs

After completing your degree course, you can apply for the position of an ecologist. If you are applying as a fresher, create a resume to list your academic achievements. You can visit the websites of the organizations that work in the field of ecology and check out if there are any openings. Fill out the job application form and write a cover letter to feature your educational accomplishments.

Skills Required To Become An Ecologist

Here are a few important skills that you can develop before choosing ecology as your career path:

Interest in Nature

To become an ecologist, having a great interest in nature and its beings is important. An ecologist's passion for the environment and its conservation defines their dedication to the work. Loving their work can inspire them to achieve more.

Analytical Skills

Ecologists often deal with a lot of data and analyzing that data is one of their core duties. Possessing strong analytical skills can help an ecologist to efficiently compile and assess the many findings. These skills can also help them provide solutions to various problems.

Technical Skills

An ecologist uses scientific equipment and advanced computer software to record and process the research data. They also create many computer models that help them virtualize an ecosystem. Having the ability to quickly learn specialized computer systems can prove to be advantageous.

Communication Skills

Ecologists can work in a team and different scientists can have different theories.

Possessing good communication skills can help ecologists to clearly convey their points of view. Having interpersonal skills can also help them become good listeners.

Ability to be Patient and Pay Attention to Detail

Ecologists can sometimes spend years on research with indistinct results. Being patient can help them stay focused on their goals. Their ability to pay attention to minute details can also help them come up with solutions to various problems.

Writing Skills

One of the responsibilities of an ecologist is to write the findings of their research. They publish papers, share results with their colleagues and also write funding proposals. Knowing how to do that with clarity can prove to be advantageous for them.

Work Environment of an Ecologist

The organizations can ask ecologists to work in various fields like research or teaching. For instance, if an ecologist is working in the field, their job can require them to travel often and conduct field research. The fieldwork can take them to remote locations where the tasks can be mentally and physically challenging. Working outdoors can also involve uncomfortable situations such as extreme temperatures in deserts or heavy rainfalls in the rainforests.

An ecologist's profile can also be a desk job where they largely work on computers and laboratory equipment. Many universities also hire ecologists to teach their students subjects like chemistry, geography and earth sciences. In a teaching job, an ecologist interacts with young students and guides them towards a rewarding personal and professional life.

Where Can Ecologists Work?

Different organizations where ecologists can work are as follows:

- Government bodies
- Environmental groups
- Environmental consulting firms
- Local organizations
- Universities
- Conservation agencies
- Research institutes
- Laboratories

SUMMARY

- Ecology is the scientific study of how organisms interact with their surroundings. It includes studies of how living things relate to each other and how they interact with nonliving elements.
- Ecology is the study of interactions between living and nonliving elements in the environment, including relationships among individual organisms. Various scientists and ecologists have provided diverse definitions of ecology.
- The term "sustainable development" has gained global recognition in recent years due to increasing ecological awareness at local, national, and international levels among communities, movements, nation-states, and governments.
- Sustainable development is "development that meets the needs of the present without compromising the ability of future generations to meet their own needs," according to the Brundtland Commission's report.
- The enormous effects of human ecology are visible in the air, water, land, and energy life-support systems. The growing human population causes an imbalance in food production and a space crisis, which in turn causes environmental pollution and negatively changes our environment.
- Ecology is the study of interactions between living things and their environment. Researchers work at four levels: organism, population, community, and ecosystem, which can overlap or be separate.
- When examining ecology at the organism level, scientists are interested in the adaptations that allow people to live in particular environments. Adaptations can be behavioral, physiological, or morphological (related to form or structure).
- A population consists of organisms living in the same area and breeding together; they belong to the same species.
- A biological community is made up of the various species found in a given area, usually in three dimensions, as well as the interactions that occur both within and between these species.

MULTIPLE CHOICE QUESTIONS

1. **The natural residence of every organism is known as:**
   a. Biome
   b. Niche
   c. Habit
   d. Habitat
2. **The study of interactions between living organisms and the environment is called**
   a. Ecosystem
   b. Phytogeography
   c. Ecology

d. Phytosociology

3. Which one of the following factors is biotic?

a. Photoperiod

b. The CO_2 content of the soil

c. Texture and porosity of soil

d. Rainfall

4. The ratio between energy flow at different points in a food chain is known as

a. Ecological capacity

b. Ecological efficiency

c. Ecological assimilation

d. Ecological potential

5. The ability of a population to increase under ideal environmental conditions is called

a. Natality

b. Carrying capacity

c. Biotic potential

d. Absolute natality

REVIEW QUESTIONS

1. What is ecology with its objectives?
2. What is the definition of ecology?
3. What are the consequences of development on ecology and the environment?
4. Describe the level of study in ecology.
5. What are the applications of ecology?

Answers

1. (d) 2. (c) 3. (b) 4. (b) 5. (c)

REFERENCES

1. Bookchin, M., (2005). *The Ecology of Freedom*. AK Press: Stirling, UK.
2. Egerton, F. N., (2007). A history of the ecological sciences, part 23: Linnaeus and the economy of nature. *Bulletin of the Ecological Society of America, 88*(1), 72–88. http://www.esajournals.org/doi/abs/10.1890/0012 (accessed on 04 April 2024).
3. Ellison, A. M., (2006). What makes an ecological icon. *Bulletin of the Ecological Society of America, 87*(4), 380–386.

4. Odum, E. P., (2004). *Fundamentals of Ecology*. Cengage Learning: Boston, MA, USA.
5. Rotabi, K. S., (2007). Ecological theory origin from natural to social science or vice versa? A brief conceptual history for social work. *Advances in Social Work, 8*(1), 113–129.

CHAPTER 2

Ecosystem Ecology

LEARNING OBJECTIVES

After studying this chapter, you will be able to:

- Discuss the definition of an ecosystem and nutrient cycling
- Describe the trophic structure, ecology and productivity
- Explain the terms 'terrestrial' and 'aquatic' ecosystems
- Describe how energy flows through an ecosystem
- Discuss ecosystem functions and values

INTRODUCTORY EXAMPLE

Ecosystem ecology is the study of these and other questions about the living and nonliving components within the environment, how these factors interact, and how both natural and human-induced changes affect how they function.

Understanding how ecosystems work begins with an understanding of how sunlight is converted into usable energy, the importance of nutrient cycling, and the impact mankind has on the environment. Plants convert sunlight into usable forms of energy that are carbon-based. Primary and secondary production in populations can be used to determine energy flow in ecosystems. Studying the effects of atmospheric CO2 will have future implications for agricultural production and food quality.

A new focus in ecosystem ecology has been climate change. The world is being altered at an alarming pace from greater to lesser precipitation in some areas to changes in ecosystems from grasslands to deserts (desertification) or forests to grasslands (increased aridity). Ecosystem ecologists are now studying the causes and effects of climate change, hoping to one day minimize our impact on the planet and preserve natural ecosystems as we know them today.

INTRODUCTION

The study of ecosystems, known as ecosystem ecology, examines the interactions between living (biotic) and nonliving (abiotic) elements. It investigates how chemicals, soil, bedrock, plants, and animals interact within ecosystems.

Ecosystem ecology looks at the biological and physical components of an ecosystem and how they interact with one another. In the end, this clarifies how to preserve both commercially viable commodity production and high-quality water. Ecosystem ecology places a lot of emphasis on the ecological mechanisms and functional processes that preserve the composition and services provided by ecosystems. These consist of trophic interactions, decomposition, and primary productivity (the generation of biomass).

2.1. AN ECOSYSTEM AND NUTRIENT CYCLING

Learning Objectives

- Describe nutrient cycling

The structural and functional unit of ecology, the ecosystem is the place where living things interact with one another and their surroundings. Stated differently, an ecosystem is a series of relationships that exist between living things and their surroundings. The word "ecosystem" was initially used by A. G. English botanist Tansley in 1935.

An ecosystem is a functioning unit of nature that is made up of living things interacting with one another and the physical environment around them. The size of an ecosystem can range greatly, from a small pond to a large forest or sea. Many ecologists believe that the Earth's local ecosystems are combined to form the global ecosystem that is the biosphere. It is convenient to split this system into two main categories, the terrestrial and the aquatic, because it is too large and complex to be studied all at once. Examples of aquatic ecosystems include ponds, lakes, wetland areas, rivers, and estuaries; examples of terrestrial ecosystems include forests, grasslands, and deserts. An aquarium and crop fields are examples of man-made ecosystems.

2.1.1. Nutrient Cycling

Every element on Earth is continually recycled. The constituents of all living things, including oxygen, carbon, nitrogen, phosphorus, and sulfur, are major elements. The movement of these kinds of chemicals and elements between living things and their physical surroundings is referred to as a biogeochemical cycle. Chemicals that organisms consume are reabsorbed by the soil, air, and water through processes like respiration, excretion, and decomposition. These processes occur throughout the food chain. An element frequently forms compounds with other elements as it progresses through this cycle as a result of natural reactions in the atmosphere, hydrosphere, or lithosphere as well as metabolic processes in living tissues. The Biogeochemical Cycle refers to this type of periodic material exchange between living things and their nonliving surroundings.

Nutrient cycles are of two types:

(a) Gaseous; and
(b) Sedimentary.

Keyword

An ecosystem is a community of living organisms (plants, animals and microbes) in a particular area.

The gaseous type of nutrient cycle reservoir (e.g., The atmosphere contains elements such as the nitrogen and carbon cycles, and the sedimentary cycle (e.g., the reservoir is found in the crust of the Earth., phosphorus and sulfur cycle). The surroundings, for example, soil, humidity, pH, temperature, regulate how quickly nutrients are released into the atmosphere. The reservoir's purpose is to fill the deficit caused by an imbalance in the influx and efflux rates.

Did you know?

The term nutrient recycling appears in a 1964 paper on the food ecology of the wood stork: "While the periodic drying up and reflooding of the marshes creates special survival problems for organisms in the community, the fluctuating water levels favor rapid nutrient recycling and subsequent high rates of primary and secondary production".

2.1.1.1. Carbon Cycle

Through the process of photosynthesis, carbon is converted from carbohydrates into carbon dioxide, which is then absorbed by living things. The producers then transfer these organic materials (food) to the consumers (herbivores and carnivores). Eventually, this carbon is released back into the surrounding medium through respiration or the animals' and plants' natural decomposition. When fossil fuels are burned, carbon is also recycled (Figure 2.1).

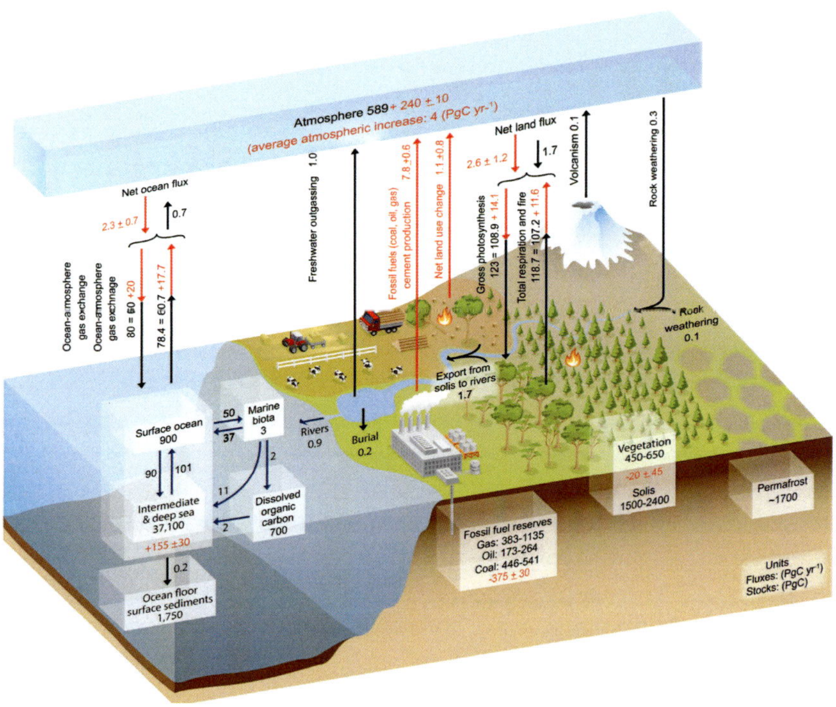

Figure 2.1. Carbon cycle.

Source: Rehman, Au, Lal, B. (2022). Gas Hydrate-Based CO_2 Capture: A Journey from Batch to Continuous. *Energies*. *15*(21), 8309. https://doi.org/10.3390/en15218309.

2.1.1.2. Nitrogen Cycle

As a necessary component of protein, nitrogen is needed by all living things, including humans. Although almost 79% of the nitrogen in our atmosphere is present, most living things are unable to directly utilize it. Similar to carbon dioxide, nitrogen undergoes a cycle of transitioning from gaseous to solid and back again through the actions of a diverse range of organisms. All living things depend critically on the cycling of nitrogen. The nitrogen cycle depends on five key processes, which are described in more detail below.

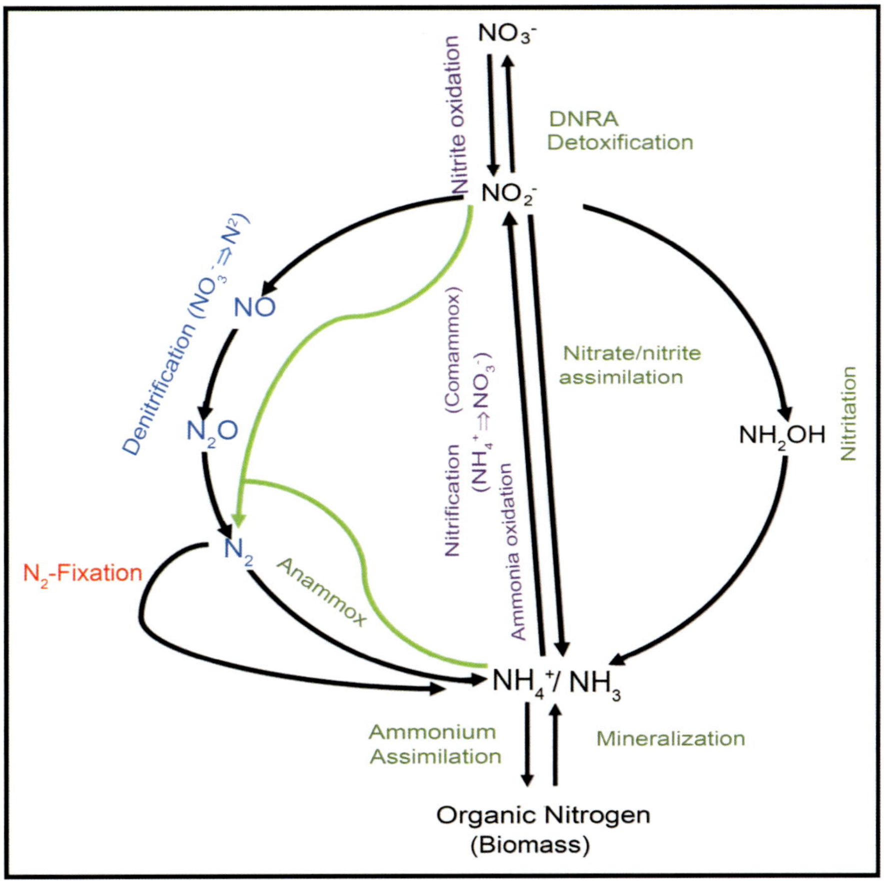

Figure 2.2. Nitrogen cycle.

Source: Martínez-Espinosa, R. M. (2020). Microorganisms and Their Metabolic Capabilities in the Context of the Biogeochemical Nitrogen Cycle at Extreme Environments. *International Journal of Molecular Sciences. 21*(12), 4228. https://doi.org/10.3390/ijms21124228.

2.1.1.3. Nitrogen Fixation

Through this process, gaseous nitrogen is converted into ammonia, which plants can utilize. There are three approaches to fixing atmospheric nitrogen.

(i) **Atmospheric fixation**: Lightning, combustion, and volcanic activity aid in the fixation of nitrogen.

(ii) **Industrial fixation**: At high temperatures (400° C) and high pressure (200 atm), molecular nitrogen is broken down into atomic nitrogen, which then combines with hydrogen to form ammonia.

(iii) **Bacterial fixation**: There are two types of bacteria – (i) Symbiotic bacteria, e.g., Rhizobium in the root nodules of leguminous plants. (ii) Free-living or symbiotic, e.g., 1. Nostoc 2. Azobacter 3. Cyanobacteria, can combine atmospheric or dissolved nitrogen with hydrogen to form ammonia.

Keyword

Cyanobacteria are a phylum of gram-negative bacteria that obtain energy via photosynthesis.

2.1.1.4. Nitrification

This is how the bacteria Nitrosomonas and Nitrococcus transform ammonia into nitrates or nitrites, respectively.

2.1.1.5. Assimilation

During this process, organic molecules like proteins, DNA, and RNA are created from nitrogen fixed by plants. Tissue from plants and animals is made of these molecules.

2.1.1.6. Ammonification

Urea and uric acid are examples of nitrogenous waste products produced by living things. The bacteria transform these waste materials and the lifeless remains of organisms back into inorganic ammonia. We refer to this procedure as ammonification. Bacteria that perform this reaction aid in this process.

2.1.1.7. Denitrification

Denitrification is the process by which nitrates are transformed back into gaseous nitrogen. Because they prefer an oxygen-free environment, denitrifying bacteria are found deep in the soil close to the water table. The opposite of nitrogen fixation is denitrification.

2.1.1.8. Water Cycle

Life requires water. Without water, no organism can survive. Precipitation, such as dew, rain, or snow, is the only water source on the planet. The term "water cycle" refers to the continuous movement of water in the biosphere as a result of direct evaporation and evapotranspiration, which return water vapor that was taken from the Earth's atmosphere (hydrological cycle). Earth is a watery planet in the solar system, with water covering roughly two-thirds of its surface. Nevertheless, only a very tiny portion of this is accessible to plants and animals. The distribution of water on Earth's surface is not uniform. The majority of water on Earth, approximately 95% of it, is chemically bonded to rocks and does not cycle. Approximately 97.3% of the remaining 5% is found in the ocean, and 21.1% is made up of polar ice caps. Therefore, the amount of fresh water that is present is only 0.6% in the form of soil, ground, and atmospheric water vapor. The water cycle is fueled by two factors: (1) solar radiation, and (2) gravity. The water cycle involves two primary processes: precipitation and evaporation. The sun causes water in lakes, ponds, rivers, and streams to evaporate. These two processes alternate with one another.

Through evaporation, the oceans lose an average of 84% of their water content. On the other hand, precipitation accounts for 77%. The ocean's evaporation deficit is balanced by the 7% of water runoff from lands via rivers. There is 23% precipitation and 16% evaporation on land (Figure 2.3).

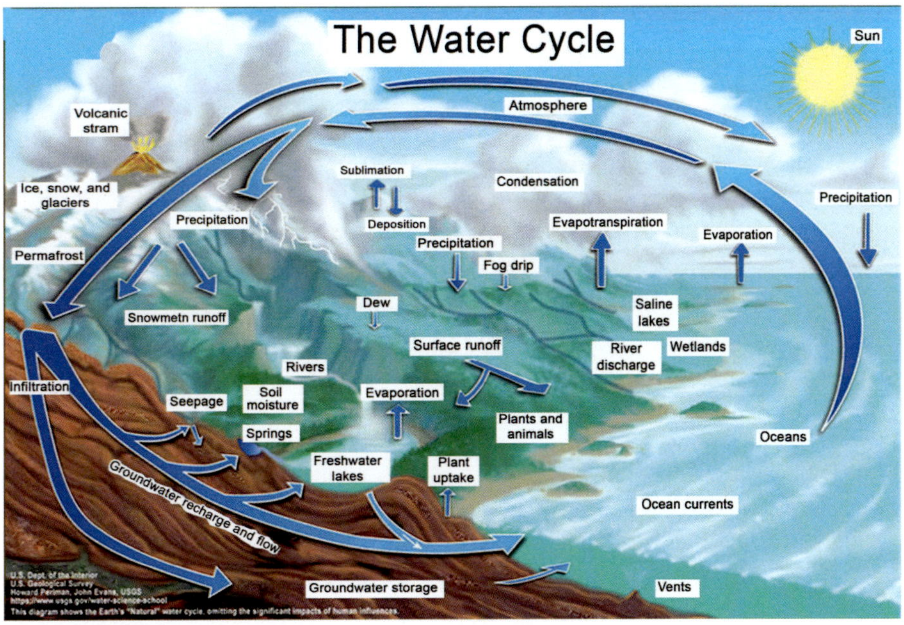

Figure 2.3. Water cycle.

Source: Wilfong, M., Pavao-Zuckerman, M. (2020). Rethinking Stormwater: Analysis Using the Hydrosocial Cycle. *Water*. *12*(5), 1273. https://doi.org/10.3390/w12051273.

2.1.1.9. Phosphorus Cycle

Nucleic acids, biological membranes, and cellular energy transfer systems are all heavily dependent on phosphorus. This element is also required in large amounts by many animals for the synthesis of teeth, bones, and shells. Phosphates, or natural phosphates, are found in rock, which serves as a natural source of phosphorus. Tiny amounts of these phosphates dissolve in soil solution as rocks weather, and plant roots absorb them. This element is obtained by herbivores and other animals from plants. Phosphate-solubilizing bacteria break down waste materials and dead organisms to release phosphorus. There is no respiratory release of phosphorus into the atmosphere, in contrast to the carbon cycle. The other two primary and significant distinctions between the carbon and phosphorus cycles are as follows: (1) the amount of phosphorus that is incorporated into the atmosphere by rainfall is significantly less than that of carbon, and (2) there is very little phosphorus gas exchange between organisms and their surroundings (Figure 2.4).

Keyword

The atmosphere is a layer of gas or layers of gases that envelop a planet and is held in place by the gravity of the planetary body.

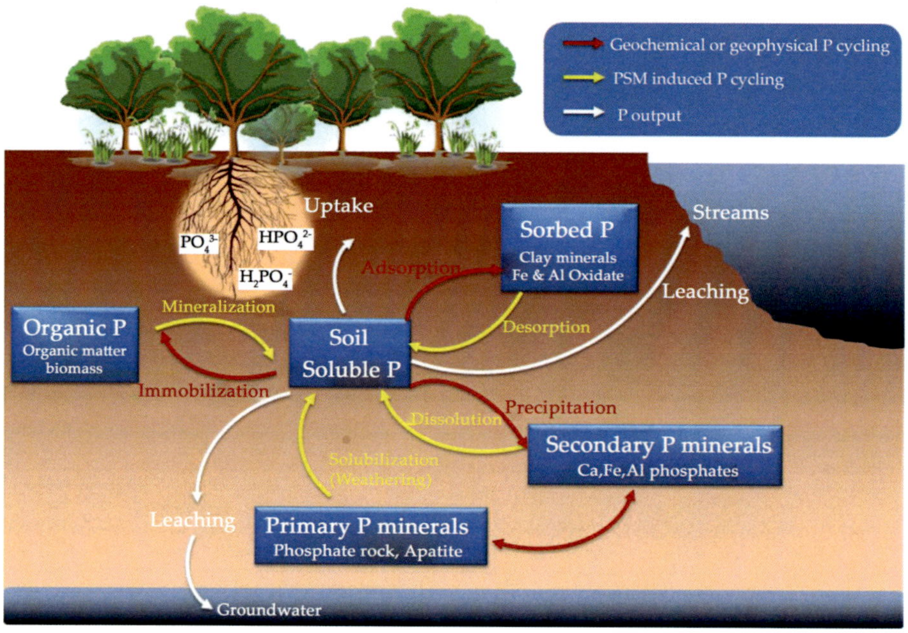

Figure 2.4. A simplified model of phosphorus cycling in a terrestrial ecosystem.

Source: Tian, J., Ge, F., Zhang, D., Deng, S., Liu, X. (2021). Roles of Phosphate Solubilizing Microorganisms from Managing Soil Phosphorus Deficiency to Mediating Biogeochemical P Cycle. *Biology*. *10*(2), 158. https://doi.org/10.3390/biology10020158.

Did you get it?

1. What two different types of nutrition cycles are there?
2. What are three approaches that fix atmospheric nitrogen?

2.2. TROPHIC STRUCTURE, ECOLOGICAL ENERGETICS, AND PRODUCTIVITY

Learning Objectives

- Describe ecological energetics
- Describe ecological productivity

In a tropical ecosystem, the organisms are arranged in tiers, with each tier denoting an organism that has a similar function and food source. Diagrams of trophic structure also show how energy moves from one trophic level to the next. We can determine the productivity of the estuary by arranging it according to a trophic structure. The estuary's productivity is essentially determined by its capacity to produce organic matter. An estuary that exhibits high diversity, high survival rates, minimal or no invasive species, and organisms that continuously carry out life processes is considered productive, in other words, a sustainable estuary.

The illustration of a trophic structure is provided below. The aquatic ecosystem is viewed from the bottom up by this trophic structure. The organisms at the top tier, or top predators, are the least energy efficient, and the organisms at the bottom tier, or primary producers, are the most energy efficient. Primary producers are more energy-efficient because they generate their own food, whereas predators and top fish need a variety of organisms to function. To put it another way, the energy requirements of predators are significantly higher than those of phytoplankton. An ecosystem driven by life processes and relationships is depicted in the figure below, which illustrates the trophic structure (Figure 2.5).

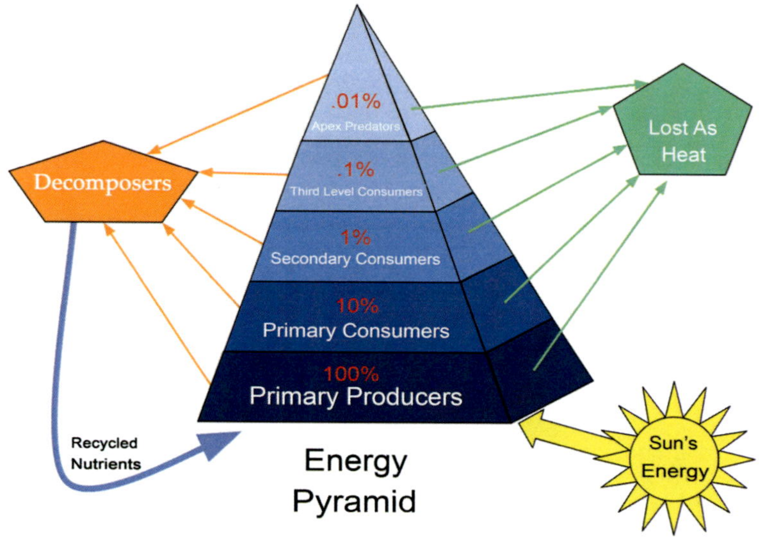

Figure 2.5. Trophic structure.

Source: By Swiggity. Swag.YOLO. Bro – Own work, CC BY-SA 4.0, https://commons.wikimedia.org/w/index.php?curid=40660207.

2.2.1. Ecological Energetics

Ecological energetics describes the energy flow in an ecosystem. Energy is the capacity for exertion. The sun emits radiation in addition to light. On Earth, some of this radiation lands. Light radiation strikes an object. The object warms up after absorbing light. As a result, heat energy is created from light energy. Heat is absorbed by land and sea, causing regions of the Earth to become hot or cold and creating clouds and air movement.

Keyword

Photosynthesis is the process by which plants use sunlight, water, and carbon dioxide to create oxygen and energy in the form of sugar.

The sun is the Earth's energy source. Green plants absorb a small portion of the radiant energy and convert it into chemical energy, specifically glucose. They provide this chemical energy to all other living things. Heat is a significant form of energy loss.

During *photosynthesis*, only a small amount of solar radiation is utilized. Living elements in the ecosystem derive their energy from this stored energy. Green plants absorb red and blue light, but only 1% to 5% of the absorbed energy is converted into food energy. Once an organism uses energy, it quickly loses it in the form of heat. Energy always flows in a single direction, following the laws of thermodynamics that govern energy movement in an ecosystem.

An estimated 57% of solar radiation is thought to be absorbed by the atmosphere. It is dispersed across space. Land or water absorbs about 36% of it. 8% of the fall lands on plants. Green plants absorb 80–85% of it, with the remaining portion being reflected. Photosynthesis uses about half of the energy absorbed i.e., It is kept in reserve as the potential energy of green plants' organic compounds.

2.2.2. Productivity

The term "ecological productivity" describes how plants initially fix solar energy and then use that energy to feed on dead biomass as detritivores, plant-eating herbivores, and animal-eating carnivores. A food web is this complex of energy fixation and utilization.

The productivity of green plants is known as primary productivity in ecology. Net primary productivity is lower than gross primary productivity because it considers energy losses needed for plant respiration. Gross primary productivity represents the total energy fixed by plants. If the net primary productivity of green plants in an ecosystem is positive, vegetation biomass will increase over time.

Herbivorous animals are associated with gross and net secondary productivities, while carnivores are linked to tertiary productivities. Ecological productivity in food webs is depicted as a pyramid structure. In general, plants contribute over 90% of the total productivity in the food web, followed mainly by herbivores and to a lesser extent by carnivores, accounting for less than 1%. Decomposer organisms eventually break down any dead plant or animal biomass; if ecological conditions hinder this process, the dead biomass may accumulate as peat or other nonliving organic matter.

The amount of productivity that is sustained by the world's ecosystems varies greatly due to variations in the availability of solar radiation, water, and nutrients. The least productive ecosystems are deserts, tundra, and the deep ocean, with an average energy fixation of less than 0.5×10^3 kilocalories per square meter per year (thousands of kcal/ m^2 /yr; one calorie is equivalent to one degree Celsius [34° F]); one kcal is equal to 1,000 calories. Generally speaking, the productivities of grasslands, montane and boreal forests, waters on the continental shelf, and rough agriculture range from 0 to 3.5×10^3 kcal/m^2/yr. The productivities of typical agricultural systems, moist prairies, shallow lakes, and moist forests range from 3 to 10×10^3 kcal/m^2/yr. Fertile marshes and estuaries, coral reefs, terrestrial vegetation on moist alluvial deposits, and intensive agriculture—which can have productivities of $10–25 \times 10^3$ kcal/m^2/yr—are the most productive ecosystems.

2.2.2.1. Types of Productivity

Productivity is basically how fast plants grow in an ecosystem. It's like the energy they get from sunlight. There are two types of productivity:

1. Primary Productivity

The term "primary productivity" describes the biomass produced by autotrophic organisms, like plants. The main process by which organic material is produced from inorganic substances like carbon dioxide and water is photosynthesis. Primary productivity is split into two categories:

- *Gross Primary Productivity:* Gross primary productivity is the amount of solar energy that is captured by the photosynthetic organism. Gross primary productivity includes all organic matter produced. This is dependent on both environmental conditions and photosynthetic activity.

- *Net Primary Productivity:* The gross productivity less the energy expended during respiration is used to estimate this.

NPP = GPP – Energy lost by respiration

It is the total energy that plants store. The animals that consume plants use this energy as food. It is calculated as the volume of organic matter generated in a community over a specific period. Over 170 billion tons of net primary productivity are produced annually throughout the biosphere.

2. Secondary Productivity

Secondary Productivity is influenced by heterotrophs, like animals. It is the energy that builds up at the level of the consumer. Unlike primary productivity, it is constantly shifting between different organisms. Organic materials are moved between different trophic levels, which causes this process to happen. It is also known as the rate at which the biomass of heterotrophs increases. Secondary production is influenced by a wide range of organisms, including bacteria, fungi, protists, and animals.

2.2.2.2. Unit of Productivity

Productivity is commonly measured in mass units per unit volume (or surface) per unit time.

Did you get it?

1. What is ecological energetics?
2. What are the two types of productivity?

2.3. TERRESTRIAL AND AQUATIC ECOSYSTEMS

Learning Objectives

- Describe the types of natural ecosystems
- Learn about terrestrial ecosystems
- Discuss aquatic ecosystems

All of the organisms, both living and nonliving, in a particular natural environment make up an ecosystem. Major elements of many ecosystems include rocks, soil, water, sunlight, animals, insects, microorganisms, and plants. Ecosystems of all kinds can be classified as either terrestrial or aquatic. Aquatic ecosystems are found in water, while terrestrial ecosystems are found on land. Terrestrial ecosystems that span a wide geographic region, like the tundra, can also be referred to as "biomes."

The various types of ecosystems are categorized according to several factors. We'll talk about the main kinds of ecosystems and try to figure out how these classifications are made. Understanding the various elements that set the ecosystems apart from one another is also crucial.

Ecosystems are typically divided into two categories, such as artificial and natural. Artificial ecosystems are natural areas that have been tampered with by humans. These are man-made towns, cities, lakes, and reservoirs. There are essentially two main categories into which natural ecosystems fall. They are the *terrestrial ecosystem* and the aquatic ecosystem.

2.3.1. Types of Natural Ecosystems

An ecosystem is a self-sufficient group of living organisms and the nonliving environment around them. An example of a natural ecosystem is displayed in the following chart (Figure 2.6).

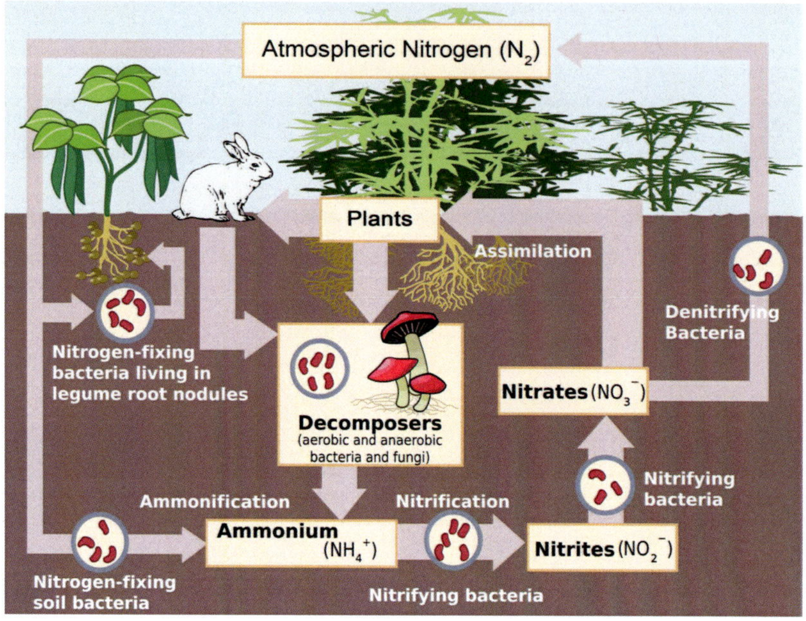

Figure 2.6. Chart of natural ecosystem.

Source: By U.S. Environmental Protection Agency – "Mid-Atlantic Integrated Assessment: Nitrogen," Public Domain, https://commons.wikimedia.org/w/index.php?curid=39327952.

Keyword

A terrestrial ecosystem is a land-based community of organisms and the interactions of biotic and abiotic components in a given area.

2.3.1.1. Biotic (Living Components)

Organisms like plants, animals, and microorganisms are examples of biotic components in ecosystems. The ecosystem's biotic components are made up of:

- Producers or Autotrophs
- Consumers or Heterotrophs
- Decomposers or Detritus

2.3.1.2. Abiotic (Nonliving Components)

Abiotic elements consist of the climate and its influencing factors such as temperature, light, humidity, precipitation, gases, wind, water, soil, salinity, topography, minerals, and habitat. For any ecosystem on Earth to operate, energy needs to flow, and water and nutrients need to cycle. Ecosystem function is supported by nonliving elements.

2.3.2. Terrestrial Ecosystems

The terrestrial ecosystem is the one that is restricted to landforms. The primary distinction between terrestrial and aquatic ecosystems is the relative scarcity of water in the former, and the consequent significance that water acquires in these environments as a result of its restricted availability. Since the environment on land is far cleaner than that of water, another factor in these ecosystems is the greater availability of light. The four primary categories of terrestrial ecosystems are those found in grasslands, mountains, deserts, and forests. Here, we will examine each of these in greater detail.

2.3.2.1. Forest Ecosystems

Because there is an abundance of flora, or plants, in these ecosystems, many different kinds of organisms coexist in a limited amount of area. This indicates that there is a high density of living things in these ecosystems. These ecosystems are categorized as tropical, temperate, or boreal based on the type of climate they experience, including taiga, temperate evergreen, temperate deciduous, and tropical evergreen and deciduous forests. More diverse flora and fauna can be found in tropical rainforest ecosystems than in any other area of the planet. Tall trees with lush, dense foliage grow in these warm, humid climates, and a variety of species can be found living from the forest floor to the canopy. Forest ecosystems in temperate zones can be deciduous, coniferous, or frequently a combination of the two, with some trees losing their leaves every fall and others having year-round evergreen status. Taiga, or boreal forests, are found in the far north, directly south of the Arctic. These forests are home to a large number of coniferous trees.

2.3.2.2. Grassland Ecosystems

Grasslands are areas primarily made up of grasses with a small amount of shrubs and trees. In these areas, the predominant organism types are herbivores, insectivores, and grazing animals. The three main categories of grasslands are steppes, savannas, and prairies. Though they can exist in colder climates as well, as is the case with the well-known Siberian steppe, grassland ecosystems are normally found in tropical or temperate regions. Semi-aridity is a climatic trait that all grasslands have in common. There may be flowers amidst the grasses, but trees are few or nonexistent. For animals that are grazing, grasslands offer the perfect habitat.

The term "savanna" refers to the seasonally dry tropical grasslands that are home to a wide variety of grazers and predators. Temperate grasslands without any big trees or shrubs are known as prairies. Three types of prairies exist: mixed grass, tall grass, and short grass prairies.

2.3.2.3. Desert Ecosystems

Low precipitation—typically less than 25 centimeters, or 10 inches, annually—is what unites desert ecosystems. Desert ecosystems make up nearly 17% of the planet's total land area. The high temperatures, intense sunlight, and scarcity of water in these ecosystems often result in relatively undeveloped fauna and flora. Shrubs, bushes, and a few grasses and trees make up the majority of the vegetation in these areas. These plants' stems and leaves are also designed to retain as much water as possible. The living things found in these areas include camels, reptiles, certain insects, and birds. Desert ecosystems can exist anywhere from the tropics to the Arctic, but they are usually windy regardless of latitude. Not all deserts are hot. While some deserts have sand dunes, others are primarily made of rock.

Remember

Another important feature about these ecosystems is that the animals that live here have thick fur coats for protection against cold and generally have a long hibernation period in the winters. The slopes at lower altitudes are generally covered with coniferous forests.

2.3.2.4. Tundra Ecosystems

Ecosystems in the tundra are characterized by a harsh environment, much like those in deserts. Permafrost is the term for the state in which the soil is permanently frozen in the wind-blown, treeless, snow-covered tundra. The variety and dispersion of habitats found in the mountain ecosystem make it the most diverse. In this ecosystem, a vast variety of plants and animals are present. Even so, only the treeless alpine vegetation can survive in these extremely high altitudes due to the extremely harsh conditions. Snowmelt in the short spring and summer creates shallow ponds that draw migratory waterfowl. This time of year can bring out lichens and tiny flowers. Although polar regions are most frequently associated with the term "tundra," alpine tundra—a term for communities that resemble tundra—can be found at high altitudes in lower latitudes.

2.3.3. Aquatic Ecosystems

Aquatic ecosystems are those that are found in bodies of water. Aquatic ecosystems are made up of communities of living things that rely on one another and the watery environment around them to survive. Freshwater ecosystems and marine ecosystems make up the two main categories of aquatic ecosystems.

2.3.3.1. Marine Ecosystem

The largest ecosystems are those found in the ocean. In addition to making up nearly 97% of all the water on Earth, they also occupy about 71% of the planet's surface. High concentrations of various minerals and salts in the water are typically present in marine ecosystems. Because they contain saltwater, which typically supports different types of species than freshwater, marine ecosystems differ from freshwater ecosystems.

2.3.3.2. Freshwater Ecosystem

Compared to marine ecosystems, freshwater ecosystems are much smaller in size, making up only 0.08% of the Earth's surface and 0.009% of all the water that exists on the planet. Freshwater ecosystems can be divided into three main categories: wetlands, lentic, and littoral.

Ponds and lakes with calm, slow-moving water are examples of lentic ecosystems. The water in littoral ecosystems flows quickly, much like rivers. Wetland systems are those in which the soil stays wet for an extended length of time. These ecosystems support a wide variety of amphibian and reptile species as well as approximately 41% of the world's fish species. Greater biodiversity is supported by the faster moving waters because they have more dissolved oxygen than the slower moving ones.

Remember

Marine ecosystems are the most abundant types of ecosystems in the world. They encompass not only the ocean floor and surface but also tidal zones, estuaries, salt marshes, and saltwater swamps, mangroves, and coral reefs.

2.3.3.3. Pond Ecosystem

These are typically contained and of a modest size. They typically consist of different kinds of insects, frogs, and plants. Fish are occasionally included, but this is less common since they are less mobile than insects and amphibians, and fish are typically artificially introduced to these habitats by humans.

2.3.3.4. River Ecosystem

Rivers are more likely to include fish in addition to the typical plant, amphibian, and insect life because they are always connected to the sea.

Because birds frequently hunt for small fish or insects in and near bodies of water, these kinds of ecosystems may also include birds.

Freshwater ecosystems are those that are restricted to freshwater environments, as the name makes obvious. Ponds, rivers, and other bodies of water that are not the sea are included in this (since the sea is naturally saltwater and cannot sustain freshwater life for very long). Freshwater ecosystems make up only 11.8% of the Earth's surface, making them the smallest of the three main classes of ecosystems. Freshwater systems are home to a variety of insects, plants, amphibians (including frogs, toads, and newts), and relatively small fish (larger fish are typically found in the sea). Plankton is the tiniest living organism found in these kinds of ecosystems; it is frequently consumed by fish and other small animals.

Did you get it?

1. **What are the types of natural ecosystems?**
2. **What is an aquatic ecosystem?**

2.4. ENERGY FLOW THROUGH AN ECOSYSTEM

Learning Objectives

- Describe how organisms acquire energy in a food web
- Discuss productivity within trophic levels
- Explain the term 'ecological efficiency'

In one way or another, energy is necessary for all living things. The majority of complex metabolic pathways require energy, particularly those that build large molecules from smaller ones (which frequently require adenosine triphosphate, or ATP). Life itself is an energy-driven process.

It is critical to comprehend how energy is obtained by organisms and how it is transferred between them via food webs and the individual food chains that make them up. Food webs show the direction of energy flow in ecosystems, showing how effectively organisms take in, use, and store energy for later use by other members of the food web.

2.4.1. How Organisms Acquire Energy in a Food Web

Three processes allow living things to obtain energy: photosynthesis, chemosynthesis, and heterotrophic consumption and digestion of other living or extinct organisms.

All organisms that can synthesize their own food, or more precisely, those that can use inorganic carbon as a carbon source, are classified as autotrophs. This includes both photosynthetic and chemosynthetic organisms. When it comes to energy sources, photosynthetic autotrophs (photoautotrophs) use sunlight, while chemosynthetic autotrophs (chemoautotrophs) use inorganic molecules. In every ecosystem, autotrophs play a crucial role. Energy would not be available to other living things and life itself would not be possible without these organisms.

An abundance of photoautotrophs, including algae, plants, and photosynthetic bacteria, provide the energy for most ecosystems on Earth. Grazing food webs are a common way to describe these ecosystems. Utilizing the sun's energy, photoautotrophs transform it into chemical energy in the form of ATP (and NADP). It takes ATP's stored energy to synthesize complex organic molecules like glucose.

Chemoautotrophs are primarily bacteria that are found in rare ecosystems where sunlight is not available, such as in those associated with dark caves or hydrothermal vents at the bottom of the ocean (Figure 2.7). Many chemoautotrophs in hydrothermal vents use hydrogen sulfide (H_2S), which is released from the vents as a source of chemical energy. This allows chemoautotrophs to synthesize complex organic molecules, such as glucose, for their own energy and in turn supplies energy to the rest of the ecosystem.

Remember

Living organisms would not be able to assemble macromolecules (proteins, lipids, nucleic acids, and complex carbohydrates) from their monomeric subunits without a constant energy input.

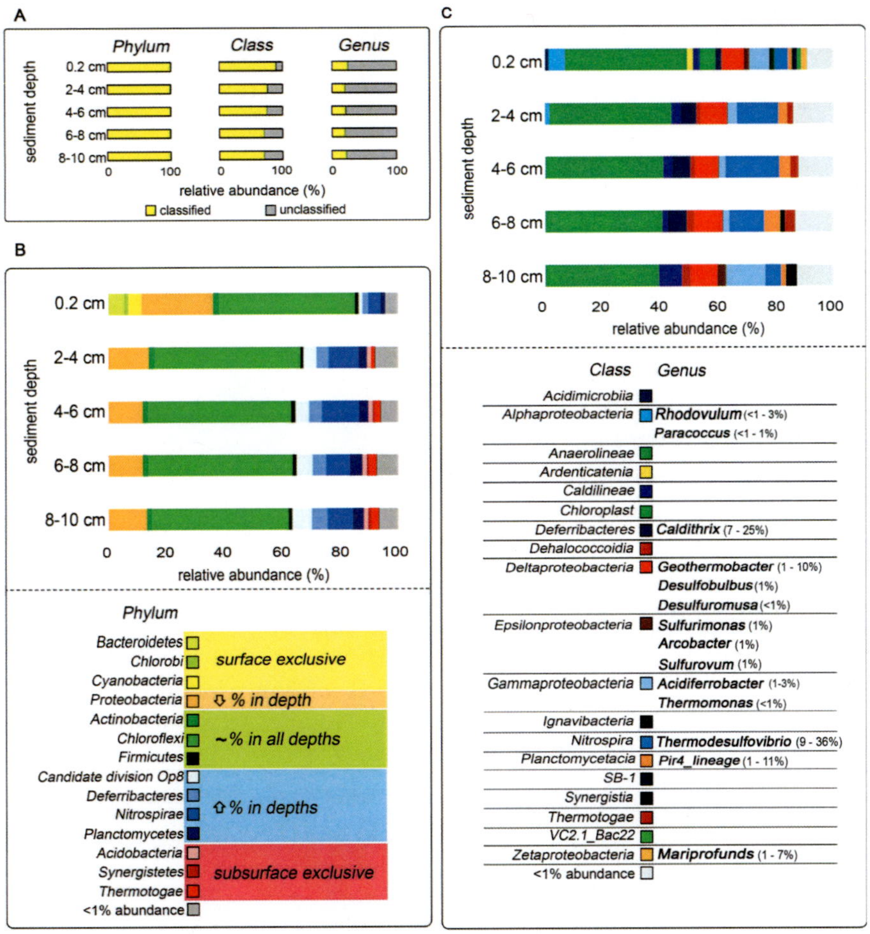

Figure 2.7. Examples of chemoautotrophic bacteria documented in a study in Soufrière Bay in the southwest of Dominica (Lesser Antilles).

Source: Gomez-Saez, G. V., Pop Ristova, P., Sievert, S. M., Elvert, M., Hinrichs, K.-U., & Bühring, S. I. (2017). Relative Importance of Chemoautotrophy for Primary Production in a Light Exposed Marine Shallow Hydrothermal System. *Front. Microbiol. 8*, 702. doi: 10.3389/fmicb.2017.00702.

Bacteria that are mainly found in rare ecosystems without sunlight, like those connected to dark caves or hydrothermal vents at the ocean's bottom, are known as chemoautotrophs. Hydrothermal vents release hydrogen sulfide (H_2S), which is used as a chemical energy source by a large number of chemoautotrophs. Because of this, chemoautotrophs are able to produce complex organic molecules like glucose for their own energy needs, which in turn provides energy to the ecosystem as a whole.

2.4.2. Productivity within Trophic Levels

An ecosystem's productivity is the proportion of energy entering the system that is incorporated into biomass at a given trophic level. The total mass of all living or previously living organisms within a trophic level, measured in a unit area at the time of measurement, is known as biomass. At every trophic level, ecosystems have distinct biomass amounts. In the English Channel ecosystem, for instance, the primary producers have a biomass of 4 g/m^2 (grams per square meter), whereas the primary consumers have a biomass of 21 g/m^2.

In any ecosystem, the productivity of the primary producers is particularly significant because, through photoautotrophy or chemoautotrophy, these organisms supply energy to other living things. Gross primary productivity is the pace at which energy from the sun is incorporated by photosynthetic primary producers. The energy flow compartment diagram in 2.8 shows an illustration of gross primary productivity in Silver Springs, Florida (Figure 2.8). The energy content and transfer between different ecosystem compartments are displayed in this study. The gross primary productivity (GPP) of the primary producers in this ecosystem was found to be 20,810 $kcal/m^2/yr$.

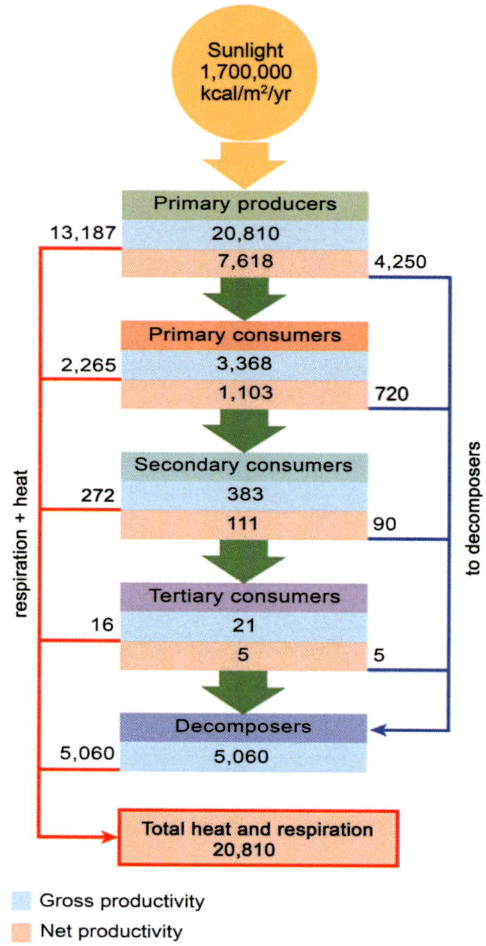

Figure 2.8. This conceptual model shows the flow of energy through a spring ecosystem in Silver Springs, Florida.

Source: https://courses.lumenlearning.com/wm-biology2/chapter/energy-flow-through-ecosystems/

Because all living things require a certain amount of energy for internal processes (such as respiration and the consequent release of metabolic heat), scientists frequently speak about an ecosystem's net primary productivity. The energy that is left in the primary producers after taking into consideration heat loss and respiration by the organisms is known as net primary productivity. The primary consumers at the following trophic level can then access the net productivity. 13,187 kcal/m^2/yr in our Silver Spring example were lost as heat or used for respiration, leaving 7,633 kcal/m^2/yr available for use by the primary consumers.

2.4.3. Ecological Efficiency

Figure 2.8 shows how the ecosystem loses a significant amount of energy as energy moves from primary producers through the different trophic levels. The primary cause of this loss is the second law of thermodynamics, which states that there is a tendency toward disorder (entropy) in the system whenever energy is changed from one form to another. This energy is lost when an organism consumes another organism in a biological system because it is expressed as metabolic heat. With 7618 kcal/m^2/yr of energy available to them from the primary producers, the primary consumers in the Silver Springs ecosystem example produced 1103 kcal/m^2/yr. Trophic level transfer efficiency (TLTE) is the measurement of the energy transfer efficiency between two successive trophic levels and is defined by the following formula.

$$\text{TLTE} = \frac{\text{production at present trophic level}}{\text{production at previous trophic level}} \times 100$$

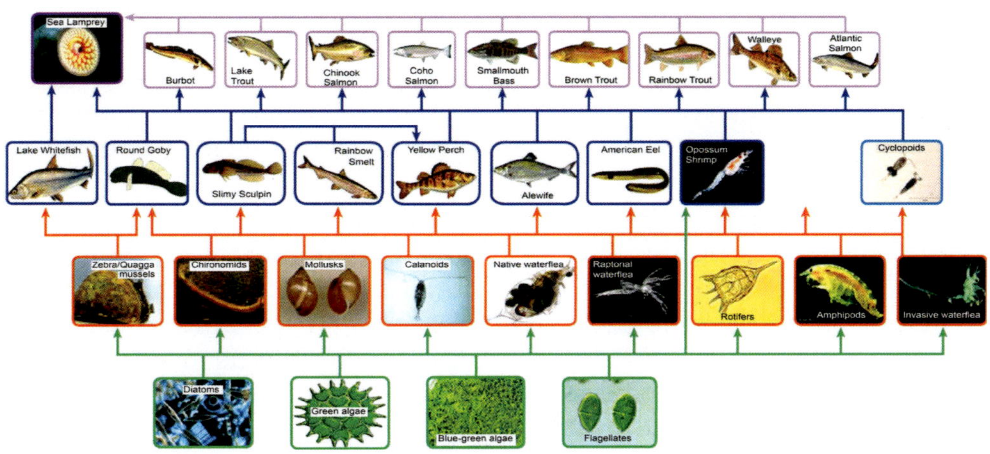

Figure 2.9. This food web shows the interactions between organisms across trophic levels in the Lake Ontario ecosystem. Primary producers are outlined in green, primary consumers in orange, secondary consumers in blue, and tertiary (apex) consumers in purple. Arrows point from an organism that is consumed to the organism that consumes.

Source: https://courses.lumenlearning.com/wm-biology2/chapter/energy-flow-through-ecosystems/ (credit: NOAA, GLERL)

Between the first two trophic levels in Silver Springs, the TLTE was roughly 14.8%. The main factor that typically restricts the length of food chains seen in a food web is the low efficiency of energy transfer between trophic levels. The truth is that not enough energy remains to support an additional trophic level after four to six energy transfers. Referring to Figure 2.9, the Lake Ontario example, there were only three energy transfers that took place between the apex consumer (Chinook salmon) and the primary producer (green algae).

Ecologists quantify energy flows within ecosystems using a variety of techniques. Depending on how complex the ecosystem is and how much access scientists have to observe it, certain transfers are easier or harder to quantify. Put differently, certain ecosystems are harder to study than others, and estimation is sometimes required when quantifying energy transfers. The net production efficiency is a crucial parameter for describing the energy flow within an ecosystem. Ecologists can measure how effectively organisms at a given trophic level integrate the energy they receive into biomass by calculating Net Production Efficiency (NPE), which can be done using the formula below.

$$\text{NPE} = \frac{\text{net consumer productivity}}{\text{assimilation}} \times 100$$

The energy content available to organisms at the next trophic level is known as net consumer productivity. The biomass, or energy content generated per unit area, of the current trophic level is known as assimilation. This is calculated by deducting the energy expended in respiration, incomplete food digestion, and waste production. The term "incomplete ingestion" describes how some consumers only consume a portion of their meal. For instance, an antelope killed by a lion will consume everything but the hide and bones. Because it lacks the bone marrow, which is packed with energy, the lion is unable to utilize all of the calories that its meal may offer.

The NPE, then, quantifies the effectiveness with which a trophic level converts the energy from its food into biomass to power the subsequent trophic level. Warm-blooded animals (endotherms) like birds and mammals use more of their energy for respiration and warmth than cold-blooded animals (ectotherms) like fish, amphibians, reptiles, and invertebrates. The additional heat produced by endotherms is a significant drawback for NPE even though it is advantageous for the organisms' activity in colder climates. As a result, in order to obtain the energy required for survival, many endotherms must eat more frequently than ectotherms. Ectotherms generally have NPE that is an order of magnitude (10x) greater than endotherms. For instance, the net present equity (NPE) of a caterpillar consuming leaves has been calculated to be 18%, while the NPE of a squirrel consuming acorns could be as low as 1.6%.

The world's food supply is significantly impacted by warm-blooded animals' inefficient use of energy. It is commonly acknowledged that the meat industry feeds livestock with vast quantities of crops, and that a significant portion of the energy from this feed is wasted due to the low NPE. For instance, growing cattle for beef consumption costs roughly $0.19, whereas producing 1000 dietary calories (kcal) from corn or soybeans costs about 1 cent. At about $0.16 per 1000 kcal, milk from cattle has the same energy content but is also more expensive. The low NPE of cattle is largely responsible for this discrepancy. Consequently, there is a global push to encourage the eating of non-meat and non-dairy foods to reduce the amount of energy used to feed animals for the meat industry.

2.4.3.1. Ecological Pyramids

Ecological pyramids, first described by Charles Elton's groundbreaking studies in the 1920s, are a useful tool for visualizing the structure of ecosystems. Ecological pyramids display the relative concentrations of different metrics across trophic levels, including biomass, energy, and the number of organisms.

Number pyramids can be upright or inverted, depending on the environment. A summertime grassland typically has a base of many plants, as depicted in Figure 2.10, and the number of organisms decreases at each trophic level. In contrast, the base of the pyramid in a temperate forest during the summer months has fewer trees than primary consumers, which are primarily insects. Due to their size and superior photosynthetic capacity, trees shade out other plants in this ecosystem in order to get sunlight. Primary producers in forests can support other trophic levels even in smaller quantities.

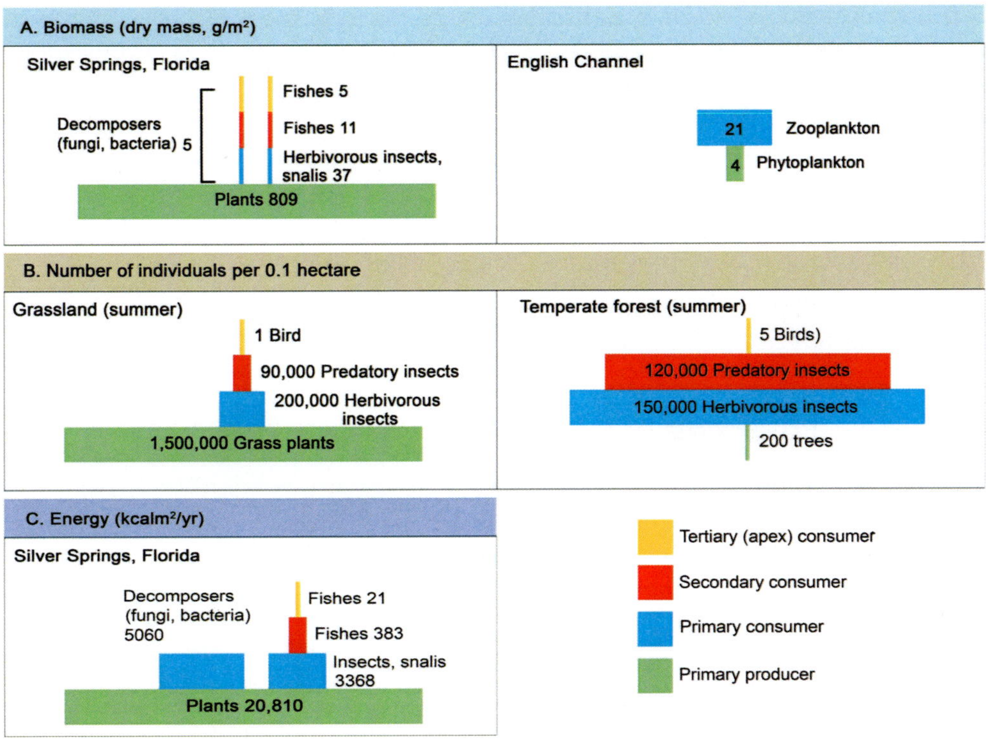

Figure 2.10. Ecological pyramids depict the (a) biomass, (b) number of organisms, and (c) energy in each trophic level.

Source: https://courses.lumenlearning.com/wm-biology2/chapter/energy-flow-through-ecosystems/.

Pyramids of biomass represent ecosystem structure in another way. At each trophic level, the amount of energy transformed into living tissue is represented by this pyramid. This data shows an upright biomass pyramid using the Silver Springs ecosystem as an example, while the English Channel example's pyramid is inverted. A significant portion of the biomass in the Silver Springs ecosystem is composed of plants or primary producers. But in the English Channel example, the zooplankton, which are the primary consumers, have a higher biomass than the phytoplankton. Similar to inverted numerical pyramids, this one is caused by a high phytoplankton turnover rate rather than a deficiency in primary producer productivity. Because the primary consumers eat the phytoplankton quickly, their biomass is never too high at any given moment. But because phytoplankton proliferate quickly, they are able to sustain the ecosystem as a whole.

It is also possible to demonstrate how energy moves through the trophic levels using pyramid ecosystem modeling. You'll see that these values match those in Figure 2.8's energy flow compartment diagram. Energy pyramids are always vertical, and an ecosystem cannot survive without adequate primary productivity. Ecological pyramids of all kinds are helpful in describing the structure of ecosystems. Nonetheless, the most reliable and accurate models of ecosystem structure when studying the flow of energy through an ecosystem are pyramids of energy (Figure 2.10).

Because a large organism, like a tree, can support many smaller organisms, organism pyramids can have an inverted or diamond-shaped structure. Similarly, because organisms reproduce quickly and can thus provide constant nutrition, a low biomass of organisms can support a larger biomass at the next trophic level. The laws of thermodynamics, however, require that energy pyramids remain upright at all times. Since energy cannot be created or destroyed, according to the first law of thermodynamics, each trophic level must obtain energy from the trophic level below it. Less energy is available at each higher trophic level because, according to the second law of thermodynamics, some energy is always lost as heat during the transfer of energy.

Did you get it?

1. **How do organisms acquire energy in a food web?**
2. **Who was the first ecologist to describe the ecological pyramid?**

2.5. ECOSYSTEM FUNCTIONS AND VALUES

Learning Objectives

- Learn about ecosystem services
- Explain some factors that complicate ecosystem management decisions
- Describe ecosystem values and their types

Ecosystem functions, or what the ecosystem does, are the physical, chemical, and biological processes or characteristics that support an ecosystem's ability to sustain itself. The provision of habitat for wildlife, the cycling of carbon, or the trapping of nutrients are a few examples of ecosystem functions. Therefore, the processes or functions that occur within ecosystems, such as wetlands, forests, or estuaries, can be used to describe them.

2.5.1. Ecosystem Services

Ecosystem services are the positive effects that ecosystem functions have on the environment or on humans. The provision of clean water, the maintenance of the food chain, the harvesting of plants or animals, and beautiful scenery are a few instances of ecosystem services. An ecosystem needs human interaction, or at the very least, human appreciation, to offer services to humans. Ecosystem functions are therefore value-neutral, but the services they provide are valuable to society.

According to the Millennium Ecosystem Assessment, the four main categories of ecosystem services are: provisioning services (like food, water, and timber), regulating services (like climate regulation and flood control), cultural services (like recreation and spiritual values), and supporting services (like nutrient cycling and pollination), which are all considered benefits humans receive from ecosystems.

- Provisioning services:

These are the tangible products humans directly obtain from ecosystems, including food (crops, fish), fresh water, timber, fuelwood, and minerals.

- Regulating services:

These are the benefits derived from the ecosystem's ability to regulate environmental processes, including climate regulation, water purification, flood control, and disease regulation.

- Cultural services:

These include the non-material benefits people gain from ecosystems, like aesthetic appreciation, spiritual enrichment, recreation, education, and cultural identity.

- Supporting services:

These are the fundamental processes that maintain ecosystem functions and enable the provision of other services, like nutrient cycling, soil formation, photosynthesis, and biodiversity maintenance.

Key points about the Millennium Ecosystem Assessment:

- It was a major UN-sponsored project that assessed the impacts of human activities on ecosystems and human well-being.
- It popularized the concept of "ecosystem services" and categorized them into the four mentioned categories.
- The assessment highlighted the importance of sustainable management practices to maintain the flow of ecosystem services for human benefit.

2.5.2. Factors that Complicate Ecosystem Management Decisions

The fact that different kinds of market failure are connected to the environment and natural resources makes decisions about ecosystem management more difficult. When a good's social costs and benefits are not fully reflected in the market, market failures occur. For instance, the costs of pollution that burning gasoline imposes on society are not fully reflected in the price of gasoline. The following are examples of market failures associated with ecosystems: (i) many ecosystems offer services that are considered public goods; (ii) many ecosystem services are impacted by externalities; and (iii) property rights associated with ecosystems and their services are frequently ill-defined.

Ecosystem services are frequently considered public goods, meaning that anyone can enjoy them without detracting from the enjoyment of others. An esthetic perspective, for instance, is a pure public benefit. It doesn't matter how many people appreciate the view—others can too. Certain services might be considered quasi-public goods, meaning that their use may reduce the enjoyment of others to a certain extent. For instance, anyone may be able to access a public recreation area. But, a crowded place can make people feel less satisfied with it. Public goods have a problem in that, despite their value, no one is motivated to pay to keep them maintained. Therefore, in order to produce the most beneficial quantity, collective action is necessary.

Externalities are uncompensated side effects of human activity that can have an impact on ecosystem services. For instance, there is a negative externality for the people downstream if runoff from agricultural land pollutes a stream. The issue with negative externalities is that they usually result in no compensation for the harm they cause to the people or ecosystems they affect. Ultimately, in the absence of a clear definition, property rights pertaining to natural resources may be exploited excessively, as there would be no motivation to preserve them. Uncontrolled fisheries, for instance, are a resource that is available to anybody who wishes to harvest fish. Since there is no single owner of the resource, unrestricted access may eventually result in significant declines in fish populations due to overfishing.

Ecosystem valuation measures the costs of market failures to society in terms of lost economic benefits, which can assist resource managers in mitigating their effects. The costs to society can then be used to hold those accountable or as a gauge for how worthwhile it is to take steps to lessen or eliminate negative environmental effects. For instance, by lessening the crowding in the crowded public recreation area, the benefits to the general public could be increased. This could be achieved by enlarging the space or reducing the number of guests. The higher economic benefits of lessening crowding can be compared to the costs of putting various options into practice.

If agricultural runoff pollutes a stream, the benefits of cleaning it up can be weighed against the costs of taking steps to lessen the runoff, or the amount of fines or taxes that should be imposed on the offenders. The advantages of lessening overfishing in open-access fisheries can be weighed against the costs of regulations or the financial impact on the commercial fishing sector if access is restricted.

2.5.3. Ecosystem Values

Ecosystem values quantify the importance and value of ecosystem services to human well-being. By calculating the price people are willing to pay to maintain or improve ecosystem services, economists can determine the value of these services to individuals. This isn't always the case, though, for a number of reasons.

Most importantly, many ecosystem services, such as a day spent watching wildlife or enjoying an ocean view, are not traded in markets, although some ecosystem services, like fish or lumber, are. Thus, many ecosystem services are provided without direct payment from humans. In addition, people's willingness to pay might not be evident because they are unfamiliar with buying such goods. However, this does not imply that ecosystems or the services they provide are worthless or that their worth cannot be measured in monetary terms.

Ecosystem services do not always need to be traded on open markets to be valued in US dollars. What is needed is an estimate of the amount of purchasing power (in dollars) that users are willing to forfeit to utilize the ecosystem's services, or the amount that users would have to pay to forfeit it if forced to make a decision that is comparable to one they would make in a market.

2.5.4. Types of Values

Ecosystem values are divided into various categories by economists. Use values and non-use, or passive use values, are the two primary categories. Non-use values are those that aren't connected to the actual use of an ecosystem or its services, or even the option to use them. In contrast, use values are determined by how the environment is actually used.

The value obtained from using an item or service—such as hiking, fishing, bird watching, or hunting—is thus referred to as use value. Indirect uses may also be included in use values. For instance, visitors to an Alaskan wilderness area benefit directly from its direct use. Indirect use values could be obtained by others from watching a television program about the region and its wildlife. Indirect use values can also be obtained by people from an input that goes into making something else that they use directly. For instance, recreational anglers who catch the fish that eat the lower-ranking aquatic organisms receive indirect use values from them.

The value that people attach to the possibility of enjoying something in the future, even if they aren't using it right now, is known as option value. It is a kind of use value as a result. For instance, someone might be willing to pay to preserve the Alaskan wilderness area in order to keep the option of visiting it at some point in the future.

Similarly, bequest value is the importance people attach to the knowledge that something will be available for enjoyment by future generations. Therefore, the willingness of people to pay to protect the environment for future generations is a key indicator of the legacy value. To ensure that future generations can enjoy the Alaskan wilderness area, for instance, someone might be willing to pay a price to preserve it.

Values that are unrelated to the actual use of an item or service, or even the option to use one, are known as non-use values, sometimes referred to as passive use values. The non-use value that people attach to the knowledge that something exists, even if they never see it or use it, is known as existence value. For instance, someone may be willing to pay to preserve the Alaskan wilderness area just because they appreciate its existence, even if they never plan to visit or even want to.

It is obvious that a single individual can gain multiple benefits from the same ecosystem. Therefore, the total economic value of a good or service is the sum of its relevant use and non-use values.

Did you get it?

1. **What are ecosystem services?**
2. **What are the types of values?**

CLOSER LOOK

RESEARCH INTO ECOSYSTEM DYNAMICS: ECOSYSTEM EXPERIMENTATION AND MODELING

The study of the changes in ecosystem structure caused by changes in the environment (disturbances) or by internal forces is called ecosystem dynamics. Ecosystems are characterized using a variety of research methodologies. Some ecologists study ecosystems using controlled experimental systems, while some study entire ecosystems in their natural state, and others use both approaches.

A holistic ecosystem model attempts to quantify the composition, interaction, and dynamics of entire ecosystems; it is the most representative of the ecosystem in its natural state. A food web is an example of a holistic ecosystem model. However, this type of study is limited by time and expense, as well as the fact that it is neither feasible nor ethical to do experiments on large natural ecosystems. To quantify all different species in an ecosystem and the dynamics in their habitat is difficult, especially when studying large habitats such as the Amazon Rainforest, which covers 1.4 billion acres (5.5 million km^2) of the Earth's surface.

For these reasons, scientists study ecosystems under more controlled conditions. Experimental systems usually involve either partitioning a part of a natural ecosystem that can be used for experiments, termed a mesocosm, or by re-creating an ecosystem entirely in an indoor or outdoor laboratory environment, which is referred to as a microcosm. A major limitation of these approaches is that removing individual organisms from their natural ecosystem or altering a natural ecosystem through partitioning may change the dynamics of the ecosystem. These changes are often due to differences in species numbers and diversity and also to environmental alterations caused by partitioning (mesocosm) or re-creating (microcosm) the natural habitat. Thus, these types of experiments are not totally predictive of changes that would occur in the ecosystem from which they were gathered.

As both of these approaches have their limitations, some ecologists suggest that results from these experimental systems should be used only in conjunction with holistic ecosystem studies to obtain the most representative data about ecosystem structure, function, and dynamics. Scientists use the data generated by these experimental studies to develop ecosystem models that demonstrate the structure and dynamics of ecosystems. Three basic types of ecosystem modeling are routinely used in research and ecosystem management: a conceptual model, an analytical model, and a simulation model. A conceptual model is an ecosystem model that consists of flow charts to show interactions of different compartments of the living and nonliving components of the ecosystem. A conceptual model describes ecosystem structure and dynamics and shows how environmental disturbances affect the ecosystem; however, its ability to predict the effects of these disturbances is limited. Analytical and simulation models, in contrast, are mathematical methods of describing ecosystems that are indeed capable of predicting the effects of potential environmental changes without direct experimentation, although with some limitations as to accuracy. An analytical model is an ecosystem model that is created using simple mathematical formulas to predict the effects of environmental disturbances on ecosystem structure and dynamics. A simulation model is an ecosystem model that is created using complex computer algorithms to holistically model ecosystems and to predict the effects of environmental disturbances on ecosystem structure and dynamics. Ideally, these models are accurate enough to determine which components of the ecosystem are particularly sensitive to disturbances, and they can serve as a guide to ecosystem managers (such as conservation ecologists or fisheries biologists) in the practical maintenance of ecosystem health.

Conceptual Models

Conceptual models are useful for describing ecosystem structure and dynamics and for demonstrating the relationships between different organisms in a community and their environment. Conceptual models are usually depicted graphically as flow charts. The organisms and their resources are grouped into specific compartments with arrows showing the relationship and transfer of energy or nutrients between them. Thus, these diagrams are sometimes called compartment models.

To model the cycling of mineral nutrients, organic and inorganic nutrients are subdivided into those that are bioavailable (ready to be incorporated into biological macromolecules) and those that are not. For example, in a terrestrial ecosystem near a deposit of coal, carbon will be available to the plants of this ecosystem as carbon dioxide gas in a short-term period, not from the carbon-rich coal itself. However, over a longer period, microorganisms capable of digesting coal will incorporate its carbon or release it as natural gas (methane, CH_4), changing this unavailable organic source into an available one. This conversion is greatly accelerated by the combustion of fossil fuels by humans, which releases large amounts of carbon dioxide into the atmosphere. This is thought to be a major factor in the rise of atmospheric carbon dioxide levels in the industrial age. The carbon dioxide released from burning fossil fuels is produced faster than photosynthetic organisms can use it. This process is intensified by the reduction of photosynthetic trees because of worldwide deforestation. Most scientists agree that high atmospheric carbon dioxide is a major cause of global climate change.

Analytical and Simulation Models

The major limitation of conceptual models is their inability to predict the consequences of changes in ecosystem species and/or environment. Ecosystems are dynamic entities and are subject to a variety of abiotic and biotic disturbances caused by natural forces and/or human activity. Ecosystems altered from their initial equilibrium state can often recover from such disturbances and return to a state of equilibrium. As most ecosystems are subject to periodic disturbances and are often in a state of change, they are usually either moving toward or away from their equilibrium state. There are many of these equilibrium states among the various components of an ecosystem, which affects the ecosystem overall. Furthermore, as humans can greatly and rapidly alter the species content and habitat of an ecosystem, the need for predictive models that enable understanding of how ecosystems respond to these changes becomes more crucial.

Analytical models often use simple, linear components of ecosystems, such as food chains, and are known to be complex mathematically; therefore, they require a significant amount of mathematical knowledge and expertise. Although analytical models have great potential, their simplification of complex ecosystems is thought to limit their accuracy. Simulation models that use computer programs are better able to deal with the complexities of ecosystem structure.

A recent development in simulation modeling uses supercomputers to create and run individual-based simulations, which account for the behavior of individual organisms and their effects on the ecosystem as a whole. These simulations are considered to be the most accurate and predictive of the complex responses of ecosystems to disturbances.

ROLE MODEL

STEPHEN R. CARPENTER

Stephen Russell Carpenter (born July 5, 1952) is an American lake ecologist who focuses on lake eutrophication, which is the over-enrichment of lake ecosystems leading to toxic blooms of microorganisms and fish kills.

Early Life

Born in Kansas City, Missouri, United States, his father, Richard, a chemist, became the Director of the National Academies' Board on Environmental Studies and Toxicology, so Carpenter was immersed in science at a young age. In his youth, Carpenter spent his summers on his grandfather's farm in Missouri. During this time, he and his relatives enjoyed fishing, hunting, and camping. "Hiking, camping, fishing, and hunting all come together in ecology," he says. "I was really excited when I discovered there was a way to get paid for being a scientist outdoors."

Education

His interest in ecology was sparked during his undergraduate program at Amherst College. After his sophomore year, Carpenter worked for the summer on a survey of tree cover in Glacier National Park. Carpenter performed undergraduate research in the Fort River of Massachusetts on the primary production of macrophytes under the instruction of Stuart Fisher, an aquatic ecosystem scientist. He received a B.A. in biology in 1974 and then entered the graduate programs in Botany and Oceanography and Limnology at the University of Wisconsin-Madison where he participated in the lab of Michael Adams to examine the role played by macrophytes in the phosphorus cycle of lake ecosystems. During his graduate years, he met his wife, Susan Moths, whom he married in the same year he finished his doctoral dissertation, 1979.

Career

He began a teaching career at the University of Notre Dame where he continued to work on lake research at the university's field station near Land O' Lakes, Wisconsin. Here he created a more broadly scoped study of lake ecosystems to include plants and animals and the food web. In 1982, he and Jim Kitchell began work on the Trophic Cascades Project, which involved the dynamics of the lake ecosystems. After 10 years spent at Notre Dame, he returned to the University of Wisconsin-Madison to a faculty position in the Center for Limnology and Department of Zoology. Madison had a strong limnology program allowing him to pursue other research including the accumulation of polychlorinated biphenyls in fish and invertebrates in Lake Michigan. He resumed work on the Madison lakes, including Lake Mendota, where his interest in the phosphorus cycle and eutrophication was renewed. His studies on the phosphorus cycle focused on nonpoint phosphorus pollution and how elevated phosphorus concentrations impacted the ecosystem of Lake Mendota. These investigations led Carpenter to devise strategies to manage the phosphorus cycle. By the mid-1990s, he began to study the economics of eutrophication, in which he compared the benefits factories and farms receive by causing eutrophication to the benefits of keeping a lake clean and clear to maximize the benefits on both sides. From 1999 to 2009, Carpenter led the North Temperate Lakes Long-Term Ecological Research at UW-Madison. In 2009, he became director of the UW-Madison Center for Limnology. In 2017, he retired as director in order to pursue research interests.

Carpenter was on the Science Committee for the Program on Ecosystem Change and Society and the board for the Stockholm Resilience Centre. As of 2021, he is the co-editor-in-chief of the journal *Ecosystems*. From 2000 to 2005, he was co-chair of the Scenarios Working Group of the Millennium Ecosystem Assessment. He led the North Temperate Lakes research site of the Long-Term Ecological Research Network program at the University of Wisconsin from 1999 to 2009. He is a former president of the Ecological Society of America. As of 2011, Carpenter has published 5 books and about 450 scientific papers, book chapters, reviewed reports, and commentaries.

Awards

Carpenter is the 2011 laureate of the Stockholm Water Prize for his research on how lake ecosystems are affected by their surrounding landscape and by human activities such as nutrient loading, fishing, and introductions of exotic species. His other awards include a Pew Fellowship in Conservation and Environment, the Naumann-Thienemann medal of the International Society of Limnology in 2007 for "research that has built bridges between ecological theory, ecosystem experiments, and management of complex limnological problems" and the work that "has elucidated the importance of the trophic cascade and regime shifts in the management of lakes", the G. Evelyn Hutchinson Award in 1999 from the Association for the Sciences of Limnology and Oceanography, the Robert H. MacArthur Award in 2000 from the Ecological Society of America, the Excellence in Ecology Prize (ECI Prize) in 2000 from the Ecology Institute for limnetic ecology, and the Ramon Margalef Prize in Ecology in 2018 from the Generalitat de Catalunya. In 2001, Carpenter was also awarded membership in the US National Academy of Sciences and Foreign Membership of the Royal Swedish Academy of Sciences and in 2006 he was elected to the American Academy of Arts and Sciences. In 2022 he was also awarded the Blue Planet Prize.

Research

Carpenter's research interest lies in whole-ecosystem experiments and adaptive ecosystem management in freshwaters. Specific topics include trophic cascades and their effects on production and nutrient cycling; contaminant cycles; freshwater fisheries; eutrophication; nonpoint pollution; ecological economics of freshwater; resilience of social-ecological systems; and early warnings of collapse in complex systems. "Eutrophication is a significant environmental problem that can impact humans on a recreational, economic, and even public health level," says Carpenter, "and it's likely to intensify in the coming decades due to increases in human population, demand for more food, land conversion, and fertilizer use."

SUMMARY

- Ecosystem ecology studies the interactions between living (biotic) and nonliving (abiotic) elements in ecosystems. It explores how chemicals, soil, bedrock, plants, and animals interact within ecosystems.
- The ecosystem, which is the structural and functional unit of ecology, is where living things interact with each other and their surroundings.

An ecosystem is a functioning unit of nature consisting of living things interacting with each other and the physical environment. Ecosystems vary in size, from small ponds to large forests or seas.

- Every element on Earth is constantly recycled. The components of all living things, like oxygen, carbon, nitrogen, phosphorus, and sulfur, are crucial elements. The transfer of these types of chemicals and elements between living things and their environment is known as a biogeochemical cycle.
- A tropical ecosystem's organisms are arranged in tiers, with each tier denoting an organism that has a similar function and food source.
- Ecological energetics describes the energy flow in an ecosystem. Energy is the ability to do work. The sun emits radiation along with light.
- The term "ecological productivity" refers to the process where plants capture solar energy and use it to sustain themselves through various trophic levels. This process forms a complex food web of energy fixation and utilization.
- Abiotic elements include the climate and its influencing factors, including temperature, light, humidity, precipitation, gases, wind, water, soil, salinity, topography, minerals, and habitat.
- The terrestrial ecosystem is the one limited to landforms. The primary distinction between terrestrial and aquatic ecosystems is the relative scarcity of water in the former and the consequent significance that water acquires in these environments as a result of its restricted availability.

MULTIPLE CHOICE QUESTIONS

1. **In an ecosystem, which one of the following types of entities occupies more than one trophic level?**
   a. Frog
   b. Phytoplankton
   c. Zooplankton
   d. Fish
2. **The rate at which food energy is assimilated at the trophic level of consumers is known as**
   a. Gross primary productivity
   b. Net primary productivity
   c. Secondary productivity
   d. None of the above
3. **Limitations of ecological pyramids involve all these statements, except**
   a. they assume a simple food chain and do not consider food webs
   b. in the ecological pyramids, saprotrophs are not given any place
   c. they do not consider the same species belonging to two or more trophic levels
   d. they do not represent relationships between organisms at different trophic levels
4. **The upper part of an aquatic ecosystem contains**
   a. Nekton
   b. Plankton
   c. Benthos
   d. Both a and b
5. **What type of food chain is dead animals → blowfly maggot → maggots → frog → snake?**
   a. Detrital food chain
   b. Decomposer food chain
   c. Predator food chain
   d. Grazing food chain
6. **"The pyramid of energy is always upright" states that**
   a. The energy conversion efficiency of herbivores is better than carnivores
   b. The energy conversion efficiency of carnivores is better than herbivores
   c. Producers have the lowest energy conversion efficiency
   d. Energy conversion efficiency is the same in all trophic levels
7. **Which statement is correct with respect to the food chain?**
   a. Every component of the food chain forms a trophic level

b. Inter-relation between different food chains is known as a food web
c. All the chains formed by nutritional relations are used to understand energy flow.
d. All of the above

8. This ecosystem has the maximum biomass

a. Grassland ecosystem
b. Pond ecosystem
c. Lake ecosystem
d. Forest ecosystem

9. Which of the following requires maximum energy?

a. Secondary consumer
b. Decomposer
c. Primary consumer
d. Primary producer

10. The bottom area where production is less than respiration in a pond ecosystem is termed as

a. Profundal zone
b. Tidal zone
c. Benthic zone
d. Limnetic zone

REVIEW QUESTIONS

1. What is the ecosystem?
2. What are the four main nutrient cycles in an ecosystem?
3. Write down the important features of a sedimentary cycle in an ecosystem.
4. What are the five main processes which are essential for the nitrogen cycle?
5. Describe the main types of terrestrial ecosystems.
6. Distinguish between ecosystem functions and services.
7. What are the two types of values?

Answers

1. (d)	2. (c)	3. (d)	4. (b)	5. (a)
6. (a)	7. (d)	8. (d)	9. (d)	10. (a)

REFERENCES

1. 1. Barot, S., Ugolini, A., & Brikci, F. B., (2007). Nutrient cycling efficiency explains the long-term effect of ecosystem engineers on primary production. *Functional Ecology, 21*, 1–10. doi: 10.1111/j.1365-2435.2006.01225.x.
2. 2. Berkes, F., Colding, J., & Folke, C., (2000). Rediscovery of traditional ecological knowledge as adaptive management. *Ecological Applications, 10*(5), 1251–1262. doi: 10.1890/1051-0761(2000)010[1251:roteka]2.0.co;2. ISSN 1051-0761.
3. 3. Chapin, F. S. III., (2011). Chapter 12: Temporal dynamics. In: Matson, P. A., Peter, M. V., & Melissa, C. C., (eds.), *Principles of Terrestrial Ecosystem Ecology* (2nd edn.). New York: Springer. ISBN 978-1-4419-9504-9. OCLC 755081405.
4. 4. Chapin, F. S. III., (2011). Chapter 8: Plant nutrient use. In: Matson, P. A., Peter, M. V., & Melissa, C. C., (eds.), *Principles of Terrestrial Ecosystem Ecology* (2nd edn.). New York: Springer. ISBN 978-1-4419-9504-9. OCLC 755081405.
5. 5. Gregorich, E. G., et al., (2001). Soil and environmental science dictionary. *Canadian Ecological Land Classification System* (p. 111). Canadian Society of Soil Science. CRC Press LLC. ISBN 0-8493-3115-3.
6. 6. Kellert, S. R., Mehta, J. N., Ebbin, S. A., & Lichtenfeld, L. L., (2000). Community natural resource management: Promise, rhetoric, and reality. *Society and Natural Resources, 13*(8), 705–715. doi: 10.1080/089419200750035575.
7. 7. Lavelle, P., Dugdale, R., Scholes, R., Berhe, A. A., Carpenter, E., Codispoti, L., et al., (2005). "12. Nutrient cycling" (PDF). *Millennium Ecosystem Assessment: Objectives, Focus, and Approach*. Island Press. ISBN 978-1-55963-228-7.
8. 8. Meffe, G., Nielsen, L., Knight, R., & Schenborn, D., (2013). *Ecosystem Management: Adaptive, Community-Based Conservation*. Island Press. ISBN 978-1-55963-824-1.
9. 9. Montes, F., & Cañellas, I., (2006). Modeling coarse woody debris dynamics in even-aged Scots pine forests. *Forest Ecology and Management, 221*(1–3), 220–232. doi: 10.1016/j.foreco.2005.10.019.
10. 10. Ochoa-Hueso, R., Delgado-Baquerizo, M., King, P. T. A., Benham, M., Arca, V., & Power, S. A., (2019). Ecosystem type and resource quality are more important than global change drivers in regulating early stages of litter decomposition. *Soil Biology and Biochemistry, 129*, 144–152. doi: 10.1016/j.soilbio.2018.11.009. S2CID 92606851.
11. 11. Podolinsky, S., (2004). Socialism and the unity of physical forces. *Organization & Environment, 17*(1), 61–75. doi: 10.1177/1086026603262092. S2CID 144332347.
12. 12. Romero, J., Lee, K., Pérez, M., Mateo, M. A., & Alcoverro, T., (2007). 9. Nutrient dynamics in seagrass ecosystems. In: Larkum, A. W. D., Orth, R. J., & Duarte, C. M., (eds.), *Seagrasses: Biology, Ecology and Conservation* (pp. 227–270). ISBN 9781402029424.
13. 13. Sutherland, W. J., (2003). Evidence-based conservation. *Conservation in Practice, 4*(3), 39–42. doi: 10.1111/j.1526-4629.2003.tb00068.x.
14. 14. Weiner, D. R., (2000). *Models of Nature: Ecology, Conservation and Cultural Revolution in Soviet Russia*. U.S.: University of Pittsburgh Press.

CHAPTER 3

Aquatic Ecology

LEARNING OBJECTIVES

After studying this chapter, you will be able to:

- Discuss freshwater ecology
- Learn about the inland water ecosystem
- Define the characteristics of a marine ecosystem

INTRODUCTORY EXAMPLE

Aquatic macroinvertebrates are insects in their nymph and larval stages, such as snails, worms, crayfish, and clams, that spend at least part of their lives in water. These insects play a significant role in freshwater ecosystems by recycling nutrients and providing food to higher trophic levels. Some aquatic macroinvertebrate species can tolerate wider fluctuations in pH, dissolved oxygen, and temperature, and can survive in a range of stream and water quality conditions. Others have very narrow tolerance ranges, and large fluctuations in pH and temperature can threaten their survival. Food sources for aquatic macroinvertebrates range from aquatic plants and algae to riparian vegetation, woody debris, and leaves. Diversity in food sources can promote a rich aquatic faunal community.

INTRODUCTION

According to the definition of an aquatic ecosystem, it is an environment centered on water in which living things interact with the environment's physical and chemical components. Aquatic organisms are those living things whose reproduction, food, shelter, and other basic needs are dependent on an aquatic environment. As a medium for material transfer and the catalyst for chemical reactions between the lithosphere, biosphere, and atmosphere, water is essential to the regulation of global ecosystem processes in aquatic systems. The distinct physicochemical characteristics of water mirror the caliber of the water body. An aquatic ecosystem's physicochemical properties dictate how well it works and how long it can support life forms. Sediments in aquatic ecosystems function similarly to soil in terrestrial ecosystems, offering substrate, nutrients, and a place for living aquatic resources to call home. Sediments play an important role as catalysts in the water quality dynamics and environmental food cycles.

An aquatic ecosystem's ability to function is directly or indirectly impacted by the quality of the sediment. The quality of sediment is determined by its numerous physicochemical characteristics. Similar to this, an aquatic environment's biotic mix dictates how well it works. They function as a trophic level and an energy source in the aquatic environment. At the trophic level, fish play a major ecological role in the entire food web. The most prevalent aquatic organisms are benthos, nekton, and plankton. In addition, wetlands, rivers, swamps, lakes, oceans, ponds, coral reefs, etc. are a few well-known instances of aquatic ecosystems.

3.1. FRESHWATER ECOLOGY

Learning Objectives

- Explain the classification of habitats
- Describe the abiotic factors that affect water conditions
- Describe freshwater communities and lentic waters
- Learn about still water animals
- Focus on the freshwater lentic communities and animals
- Describe freshwater plants and nutrients
- Discuss lotic communities
- Describe lotic communities and algae
- Describe lotic communities and animals
- Learn about freshwater communities and plankton
- Discuss pollution in freshwater ecosystems

Freshwater ecology is a specific subfield of the study of organisms and their surroundings (Figure 3.1). Unlike biology, ecology focuses on how organisms interact with and are influenced by their natural surroundings. Scientists specializing in freshwater ecology examine plants, animals, and water to understand freshwater systems. They study bodies of non-brackish water like lakes, ponds, and streams, distinct from marine ecologists who focus on saltwater ecosystems. Some work in freshwater wetlands. Data collected by freshwater ecologists can aid in plant and animal conservation and have implications for human health.

Scientists try to accurately understand how a body of fresh water goes about its daily life through the study of freshwater ecology. Every element affects the ecosystem's ability to function, from the large reptile or avian predators to the microbic organisms actively producing algae. An alien species that has been brought in, chemical runoff from a new industrial facility, or even increased use if the lake attracts more tourists are some examples of new factors that have the potential to drastically disturb and reorganize the ecosystem. Freshwater ecologists can estimate how new factors will affect the local environment by knowing how this body of water will behave under typical conditions.

Keyword

Conservation is the study of the loss of Earth's biological diversity and how to prevent it.

Figure 3.1. Freshwater ecology.

Source: Dr. William O'Connor. Our latest freshwater ecology surveys. https://ecofact.ie/aquatic_ecology_survey/.

Freshwater ecology can provide reasonably accurate information about how populations of plants or animals are surviving in their surroundings for conservation purposes. This can assist in determining whether government agencies have designated a declining species as protected or whether conservation efforts are helping an already endangered species to recover. Freshwater ecologists can provide a fairly clear picture of the direction a species is going in as well as identify important factors that determine its situation, even though the majority of their work is based on population graphs and probabilities rather than actual census taking.

Figure 3.2. Safeguarding sources of drinking water.

Source: www.usgs.gov/media/images/aquatic-ecology-pic Aquatic Ecology pic By Eastern Ecological Science Center.

Research on freshwater ecology also benefits humans. The water, which makes up the majority of the ecosystem, is continually examined and tested to find out vital information like its chemical makeup and potential risks. Freshwater ecologists' work can be used to test an existing water source for potential contamination or assess the viability of a new source of drinking water. Freshwater ecologists are benefiting not only the environment but also their own species by safeguarding sources of drinking water (Figure 3.2).

3.1.1. Classification of Habitats

There are two categories for the ecology of freshwater rivers: lowland and upland. Lowland habitats are found in relatively flat lowland areas and are characterized by warm, slow-moving rivers with water that is often colored by organic matter and sediment. Upland habitats are found in mountainous areas and are characterized by cold, clear, rocky, and fast-moving rivers. In freshwater ecology, dividing rivers and streams into upland and lowland categories is crucial because these river habitat types differ greatly and typically support quite different populations of fish and invertebrate species.

Keyword

Freshwater ecology is the study of the structural and functional interrelationships of organisms in fresh waters as they are affected by their dynamic physical, chemical, and biotic environments.

Swift-moving upland rivers and streams in freshwater ecosystems flow from hilly or elevated areas to wide alluvial plains, where they transform into lowland rivers. Additionally, there are other factors besides altitude that determine whether a river is lowland or upland. Course gradient and stream power are perhaps the most significant determinants. Rapidly falling altitude rivers have greater stream power, also known as the "force of water," and faster water flow. Thus, an incised course, a river bed primarily composed of bedrock and coarse sediments, a riffle and pool structure, and colder water temperatures are produced, which are the other features of an upland river.

Rivers with a very gradual altitude drop will have less force and a slower flow of water. The other features of a lowland river are thus produced, including a meandering course devoid of rapids, a river bed primarily composed of fine sediments, and warmer water temperatures. Lowland rivers typically contain higher concentrations of suspended sediment and organic matter; however, during seasonal low flow periods, some lowland rivers experience exceptionally clear water.

Upland rivers' bedrock and coarse sediment beds, along with their generally clear, cool, and swiftly flowing waters, attract fish species that have strict tolerances to temperature changes, high oxygen requirements, robust swimming abilities, and unique reproductive techniques to avoid their eggs or larvae being carried away. Invertebrate species with high oxygen requirements, restricted temperature tolerances, and ecologies centered on coarse sediments and the "gaps" or interstices between them are also encouraged by these traits (Figure 3.3).

Figure 3.3. Upland rivers-an example of the Factory Falls in the Pocono Mountains, Pennsylvania

Source: By Photo by and © 2006 Derek Ramsey (Ram-Man) – Self-photographed, GFDL 1.2, https://commons.wikimedia.org/w/index.php?curid=1099214.

3.1.2. Lowland

Lowland rivers are home to a variety of fish species that are better suited to withstand extremes of temperature and low oxygen levels due to their generally more turbid, warm, slow-moving waters, and fine sediment beds. These species have also evolved life histories and breeding strategies that take advantage of these characteristics. Along with supporting ecologies centered around fine sediments or alternative habitats like submerged woody debris ("snags") or submergent macrophytes ("water weed"), these traits also favor invertebrate species that are more tolerant of low oxygen levels and a wider range of temperatures.

The freshwater ecosystem is made up of the following four primary components of the living environment.

Ecological components and elements that organisms take in and use as a food source or for breathing. These substances are carried up the food chain by plants and are necessary for them.

- Autotrophic plants, or those that produce food on their own, obtain their energy from inorganic substances through photosynthesis and solar radiation. Since they create new energy and introduce it into the ecosystem, these plants—as well as some bacteria—are the main producers.
- The organisms that obtain their nourishment from other organisms are known as consumers. These could be secondary consumers who feed on the primary

consumers or primary consumers who feed on the plant material.

- To obtain energy, decomposers must break down dead organic material, or detritus. In the process, they release essential elements and compounds that plants need.

Abiotic factors are essentially nonliving elements that have an impact on the freshwater community's living organisms. The ability of newly emerging organisms to survive and procreate in an ecosystem that is empty and barren depends on the presence of conducive environmental conditions. These elements of the surroundings are abiotic. In an ecosystem with a diversity of species, the behaviors of those species can have an impact on the well-being of other species in the vicinity; these are known as biotic factors.

Remember

When a variety of species are present in such an ecosystem, the consequent actions of these species can affect the lives of fellow species in the area; these factors are deemed biotic factors.

A freshwater ecosystem heavily relies on solar light to illuminate the primary producers or plants. The amount of sunlight that reaches the ecosystem and its duration can vary depending on many factors that are summarized below:

- **Aspect** – The angle of incidence where light hits the water's surface. Because light is more direct during the day when the sun is high in the sky, more light can be absorbed by the water. Less light is absorbed at sunset due to the less intense light hitting the water's surface. Every season has a different aspect of the sun at different times of the day.
- **Cloud Cover** – The amount of cloud cover in a region will unavoidably impact the light's intensity and duration on freshwater ecosystems' water. Plant species need a specific window of time to receive light in order to perform photosynthesis.
- **Season** – An ecosystem's four seasons differ greatly from one another because the sun provides less heat and light in the winter and vice versa. As a result, the kinds of organisms that thrive in each season will differ as well.
- **Location** – The equator experiences roughly 12 hours of sunlight and darkness per day, compared to the extreme latitudes' 6 months of daylight and 6 months of darkness. Because of these variations, the kind of organisms that would inhabit freshwater ecosystems are significantly impacted by this kind of variance.
- **Altitude** – The average temperature decreases by one degree Celsius for every thousand meters above sea level. Altitude also influences how sunlight enters the freshwater ecosystem, which influences the kinds of organisms that live there.

- In the end, a variety of abiotic factors can influence which organisms survive and thrive in the freshwater ecosystem. In addition to providing heat and light for photosynthesis, the sun also maintains a temperature that is favorable for the growth of living things. The overall ecosystem and the creatures that live there can be directly impacted by the temperature of a freshwater environment. Enzymes function optimally at a specific temperature, and any departure from this "norm" will cause the organism to respire less than optimally. Since all aquatic life is ectothermal, its body temperature directly reacts to its surroundings.

Temperature influences a substance's density, and variations in water density translate into varying degrees of resistance for animals migrating through freshwater habitats.

3.1.3. Abiotic Factors – Water Conditions

It appears that the sun's heat and light play a significant part in creating the right environment. But life in the ecosystem is also inevitably impacted by the water's characteristics. Numerous factors will unavoidably disturb a still body of water, which will change the distribution of organisms in the water. Wind is thought to be the main cause of water disturbance; however, temperature variations can also result in convection currents, which spread the water's temperature evenly throughout the movement.

As water gives in to gravity and flows downstream, rivers inevitably experience water movement. Water movement is influenced by a number of comparatively constant factors, though it can also be caused by human activity. Depending on the cohesiveness of the water at the surface, surface tension can also have an impact on the organisms that live there by influencing the quantity of oxygen that reaches submerged organisms.

All of these elements have an impact on the lifestyle of the creatures that live in such a freshwater ecosystem. The chemical makeup of the surrounding air, soil, and water also influences the ecosystem's appearance on a more molecular level. Which organisms can survive in a given environment depends largely on the oxygen content of the surrounding air and water. Animals need oxygen for aerobic respiration, and the amount of oxygen present in a given space depends on a variety of factors, such as temperature and the number of living things present.

Since oxygen is necessary for a variety of chemical reactions and biological functions, changes in the oxygen concentration within an ecosystem will unavoidably have an impact on the ecosystem as a whole. The concentration of carbon dioxide is no different. In addition to being necessary for photosynthesis, CO_2 can also change other things, like the pH of water.

Freshwater ecology is typically separated into two categories: lotic (running) water and lentic (still) water. The organisms that are most likely to live there are also influenced by these two bodies of water.

3.1.4. Freshwater Plants and Nutrients

Plants need different nutrients, which are found in the nutrient-rich soil and nearby waters, in addition to maintaining an appropriate water concentration in their cells. Magnesium (Mg), nitrogen (N), phosphorus (P), and potassium (K) are among the macro elements that plants need in addition to carbon, hydrogen, and oxygen for photosynthesis. While some of these elements—most notably the gases—are easily found in the atmosphere, carbon dioxide is created when organic matter breaks down. Other elements are easily accessible in the soil, and as decomposing matter releases nutrients, the surrounding soil becomes more fertile. The photosynthetic processes in plants make oxygen available and establish a connection between the local concentrations of carbon dioxide and oxygen.

3.1.5. Freshwater Communities and Lentic Waters

Lentic (still water) communities can vary greatly; ranging from a small, temporary puddle to a huge lake, they can support various forms of life. Many of the enduring freshwater environments we see today have been shaped by long-term geological changes such as erosion, glacial shifts, volcanic activity, and to some extent, human involvement. These processes lead to the formation of uneven landscapes, creating depressions where water can accumulate and persist for extended periods. The size and depth of a still body of water significantly influence the characteristics of its ecosystem, and factors like erosion and glacial movement continuously alter them over time. The temperature's overall impact on a still water environment is a crucial factor. Unlike land, water takes longer to heat up under the sun, resulting in more gradual temperature changes, especially in larger water bodies. Various factors come into play to make this freshwater ecosystem suitable for life, shaping the environment to which organisms must adapt.

Keyword

Hypolimnion is the dense, bottom layer of water in a thermally-stratified lake. It is the layer that lies below the thermocline.

Just like osmosis, temperature will eventually become uniform within a substance, including a stagnant body of water. When sunlight hits the water, the surface will heat up and gradually create a temperature disparity between the surface and the rest of the water. This temperature discrepancy will depend on the surface area and depth of the water.

Two clearly distinct water layers form over time, separated by a significant temperature differential and offering distinct ecological niches for various organisms. This process, known as stratification, occurs when the surface and water bed have such distinct temperature differences that they are easily distinguishable from one another. The surface area, which is warmed water from direct sunlight contact, is called the epilimnion. The lower layer, known as the *hypolimnion*, is located beneath the water's surface. Because of its greater depth, it receives less heat from the sun, making the water below it colder.

The amount of light received by autotrophic organisms, which undergo photosynthesis, can be influenced by various factors. These factors, in turn, affect the organisms' photosynthesis and respiration, thus impacting their abundance. This subsequently affects other species that depend on them. The still water environment can receive organic material and sediment from deceased organisms in the vicinity, as well as water from hills and streams. Additionally, buoyant materials can obstruct the essential light needed by the primary producers in the ecosystem.

Loose sediment is disturbed when water moves because of the friction the water creates against the banks and the water bed. Lighter materials will stay suspended in the water, while heavier particles will gradually return to the bottom of the body of water based on their respective weights. Less light will be available for organisms below the surface as the lightest material rises to the surface.

Naturally, the effects of will cause organisms that depend on photosynthesis for sustenance to have less light, which will eventually mean that organisms that feed on these autotrophic organisms will have less access to their food source.

The surrounding oxygen content is another important factor influencing still water communities. Three factors primarily influence the concentration of oxygen.

- The surface area to which the body of water is exposed to the open air environment
- The circulation of water, chiefly due to temperature differentiations in different areas of the water body (convection currents)
- Oxygen is created as a result of respiration, consumption, and the oxygen consumed by animals and bacteria.

The temperature has an impact on the amount of available oxygen, which implies that the depth of the water will also have an impact. The organisms undergoing photosynthesis will then need the levels of carbon dioxide, which are directly related to the available oxygen. The ecosystem's creatures will be impacted by their availability. Their availability will also depend on how they relate to temperature. Apparently, under various circumstances, some of these factors change, and modifications to one of the factors typically have an impact on the others. This also applies to pH, where a rise in carbon dioxide causes the pH to decrease.

Hydrophytes are plants that live entirely or partially submerged in water. These hydrophytic plants exhibit a type of symbiosis whereby they offer a means of survival for surrounding organisms, including algae. This is so that organisms like bacteria and algae can survive in the environment, which is made possible by the hydrophytes. Herbivorous animals shield the plants from being eaten by feeding on this abundant layer of algae rather than the actual plants.

The detritus matter, or rich organic material on the waterbed, is also a source of food for the animals in this ecosystem. It is a region rich in organic material because the plants that thrive there offer both food and shelter, which can shield them from predators or serve as a place for their offspring to hatch in a closed protected area.

Plants known as hydroseres, which are partly or fully submerged in water along the banks, occupy the ecological niche next to the still water banks. While some of these hydroseres float on the surface with one side in contact with the water and the other with the open air, others are rooted in the water, though some of their leaves poke through the surface. Fundamentally, hydroseres have evolved to survive in such an environment thanks to their differing respiration rates and evolutionary adaptations from land plants. Because of this kind of evolutionary adaptation, plants have altered their physical structure to fit their surroundings, giving freshwater plants a very distinctive appearance.

Freshwater plants' absence of stiff structures is one illustration of these adaptations. This is because the water's density is significantly higher than that of an open environment, meaning that the plant is constantly "pushed" against during its life. This shields the plant from harm and makes such plants more resilient to approaching water tides.

Plants need a certain amount of buoyancy to allow them to contact the open air environment because they need a minimum concentration of gases in their diet, such as carbon dioxide. Among the adaptations could be:

- Air Spaces – Air spaces in the plant will decrease density and increase buoyancy.
- Broad Leaves – Broader leaves will spread their weight more evenly across the water surface allowing them to float.
- Waxy Cuticle – On the upper half to allow water to run off the surface to prevent the weight of the water from dragging the leaves under the surface.

Due to the abundance of water in their external environment—or, in some cases, the uptake of water from a wet environment—or the loss of water through their leaves in an open air environment, freshwater plants' transpiration process is altered in still water plants.

An example of transpiration problems for such plants is as follows:

- The plant lives in a marshy environment, where roots uptake water from soaked ground, allowing plenty of water to be taken up and transported up and across the plant.
- The difference in water concentration between the plant's leaves and the open air environment is so great that much of the water absorbed is lost to the external environment, meaning the plant loses water rapidly.
- Such a problem is solved by evolutionary adaptations. These adaptations essentially address the issue of rebalancing the critical deviations between the water that is absorbed and lost in a plant.

3.1.6. Still Water Animals

Throughout millions of years, aquatic animals have evolved to fill different ecological niches within the ecosystem. Three major regions of the freshwater environment can be clearly distinguished when examining the habitats of these specific organisms.

- **The Profundal Region** – An area of still water without sunlight is devoid of autotrophic organisms. **The profundal region's** more energetically dense regions provide the organic material that the animals in this zone eat.
- **The Pelagic Region** – The pelagic zone is located beneath the water's surface and is identified by the light that is accessible to it. The sea bed

and areas close to the shore are not included in the pelagic region.

- **The Benthic Region** – All freshwater environments that are in contact with land, except for shallow shore areas, are included in the benthic region. The benthic zone can support a wide variety of organisms because it contains sediments rich in nutrients and minerals that serve as food sources. Additionally, a portion of the benthic region can live in the euphotic zone, which is the portion of the water where light is present. This will give autotrophic organisms an ecological niche, which means that herbivores may be able to use them as a food source.

The area above and below the water's surface (epineuston and hyponeuston) is another unique niche for the animal community. The surrounding hydrosere vegetation provides food for the epineustic animals, who hunt small animals that fall into the water from the vegetation.

Nekton is a group of animals that live in both the pelagic and profundal zones below these surface-dwelling animals. They rise to the pelagic regions to feed on these epineustic animals. This nekton community includes fish, which are essential components of these freshwater ecosystems. Due to their frequent migration between fresh and saltwater, some of these fish are merely passing members of the community. Although they spend the majority of their lives in saltwater, anadromous fish spawn in freshwater. On the other hand, fish that are catadromous live primarily in freshwater environments. In every manner, the fish that are always present in the ecosystem serve as a bridge connecting the upper and lower reaches of the freshwater community.

3.1.7. Lotic Communities

Lotic communities—where "lotic" refers to flowing water—are another term for freshwater communities with running water. Water from various sources is introduced into the freshwater body to create lotic communities.

- **Precipitation** – Rainfall that directly enters the flowing water community will contribute to its presence to some extent.
- **Ground** Surface Water – Water from previous rainfall will seep into the flowing water community.
- **Subsurface Water** – Water absorbed into the soil can also enter.
- **Water Table** – Water for the flowing water community can be sourced from a "water table" deep beneath the surface.

The water in lotic communities flows at a specific speed, which is a key difference between lotic and lentic communities. This can significantly impact the types of organisms that inhabit the ecosystem and the specific ecological niches they occupy. Various factors influenced by flowing water can affect the organisms living in this particular environment.

- The movement of stones and minerals brought on by the water's volume and velocity causes the water bed to fluctuate continuously. Particles shifted downstream will increase in quantity and size due to the faster and larger volume of water present.
- Salmon at the base of waterfalls use standing waves to propel themselves upstream. Simultaneously, they generate tiny air pockets as a result of oxygen replacing the splashing

water, opening up a tiny micro-habitat that is ideal for specific organisms.

- The river's bed and bank are broken down by flowing water, which is what causes erosion and the long-term changes in the river's geography. This implies that, for example, hydroseres that had previously lived along a riverbank may now be separated from the flowing water, which will eventually cause the ecosystem as a whole to shift.

Some of the physical and chemical elements that support the running water community that organisms in their preferred ecological niches inhabit are listed below.

- **Temperature** – The temperature of running water communities varies more quickly but within a smaller range than the temperature of still water. Since the river's source has not been exposed to the warm air heated by the sun, the water there is typically colder in the summer than the water at the delta. In winter, the opposite happens: the water is warmer until it comes into contact with the colder air.
- **Light** – Generally speaking, overhanging shrubs that might be using a tributary water source, debris obstructing sunlight from reaching lower layers, and ripples in the water all contribute to a running water body's reduced ability to let light in. These are a few instances of how the amount of light that reaches the lotic community can vary, which can have an immediate impact on the rate at which the plants in the community perform photosynthesis.
- **Chemical Composition – Precipitation**, water seeping through vegetation, and sea spray are just a few of the numerous variables that can change the chemical makeup of the freshwater environment. All things considered, a variety of elements and compounds are needed by organisms for their everyday activities, and changes in these elements and compounds—or even their absence—have an immediate impact on the lives of these organisms.

Organic Matter – An ecosystem's modification may also be influenced by organic matter that was previously outside of the flowing water environment.

Keyword

Precipitation is any product of the condensation of atmospheric water vapor that falls from clouds due to gravitational pull.

3.1.8. Lotic Communities and Algae

Although some lotic communities have conditions that are similar to those of lentic communities, overall, the diversity of plant species in lotic communities is lower than in lentic communities (which are communities with still water). The majority of plants have evolved to withstand the forces and various environmental conditions that flowing water brings. Numerous species have been able to effectively occupy the lotic community as their ecological niche thanks to these adaptations.

Smaller species have discovered that the lotic community offers more hospitable conditions than typical plant species, particularly larger plants, which would find these conditions more harsh. This is because they can adapt better to the physical characteristics of the water. Because they can grow on a wide variety of surfaces and in a variety of locations, algae are an effective part of the running water ecosystem. Over time, the majority of these algae have evolved defense mechanisms against the water current that washes them away.

Keyword

Algae is an informal term for a large and diverse group of photosynthetic, eukaryotic organisms.

Algae come in a variety of species, and they can all proliferate and grow quickly. Due to competition for niches created by this outcome in the freshwater environment, algal colonies may heavily occupy a particular area at one point in time, only to be displaced weeks later by another species that can thrive in the more favorable conditions.

Figure 3.4. Algae.

Source: Cetz-Navarro, N. P., Espinoza-Avalos, J., Hernández-Arana, H. A., Carricart-Ganivet, J. P. (2013). Biological Responses of the Coral *Montastraea annularis* to the Removal of Filamentous Turf Algae. *PLoS ONE 8*(1), e54810. https://doi.org/10.1371/journal.pone.0054810.

Additionally, algae are the community's main producers, drawing fresh energy from the sun to power the ecosystem and give the main consumers an important source of food. With this knowledge, it becomes clear why short-term variations exist in algal populations and their locations within the lotic community (Figure 3.4).

3.1.9. Lotic Communities and Animals

The environment of flowing water provides a variety of microhabitats that mimic ideal circumstances for a wide range of animal species to flourish in the freshwater lotic community. Animals in this ecosystem have also undergone continuous evolutionary adaptations to better fit this running water environment, just like plants have. Some of these creatures are sessile, which means they are stationary and immobile. These creatures, which include some freshwater sponges and protozoans, are typically tiny. They either stay fastened to the bulk of a plant, the surface of a water bank, or a rock. Tentacles that splay out into the moving water and create a catchment area that can capture microscopic organisms (like plankton) floating downstream are typically how they get their food.

Despite their ability to adapt to avoid being washed downstream, these sessile animals are not regarded as essential members of the freshwater community. Their location may not be as ideal as it once was, as biotic and abiotic factors alter the ecosystem's landscape over time. Because they are immobile, they are powerless to change this. In light of this, certain animals have evolved adaptations that allow them to move through the water without being restricted to one location.

Throughout their evolutionary history, animals have acquired various adaptations that aid in their ability to adapt to their environment.

- **Suckers** – These suckers cling to something that helps them move in any direction and anchor them in place.
- **Hooks/Claws** – These sharp appendages can pierce through almost anything, giving the animal the ability to hold onto something or traverse surfaces.
- **Body flattening** – This adaptation can lessen the force of the water flowing downstream on the animal, allowing it to absorb less of the impact and lessen its ability to impede its movement. Additionally, this enables the animals to go into small spaces (like crevices beneath stones) that might offer a suitable living environment.
- **Streamlining** – Similar to transportation created by humans, animals with external appearance modifications that have been made to appear more streamlined provide less resistance to the flowing water when they try to move. Wings: Certain creatures possess modifications that enable them to soar, distancing themselves from the force.

3.1.10. Freshwater Communities and Plankton

Small organisms known as plankton are essential to the freshwater community because they are suspended in the water. They travel by currents created by wind or convection.

There are thousands of these organisms in almost every freshwater ecosystem's habitat, and because of their small size and ease of use, they can spread out over vast areas of water and multiply exponentially.

Plankton can be subdivided into two categories:

Remember

Zooplankton is primarily found in surface waters where food resources (phytoplankton or other zooplankton) are abundant.

- **Phytoplankton** – Phytoplankton are microscopic plants that use photosynthesis to get their energy. Nonetheless, this taxonomic group also includes certain bacterial species that are able to perform photosynthesis. Their significance to the ecosystem stems from their membership in the primary producing community and their role in recycling elements like carbon and sulfur that are needed by other communities.
- **Zooplankton** – In general, zooplankton is larger than phytoplankton and is primarily made up of crustaceans and rotifers. Because of the large populations that can exist in their environment, they are relatively unspecialized. A multitude of physiological adaptations have evolved over time to help with their buoyancy and to keep them safe from harm by enabling them to float in the water.

Since plankton are an essential component of the ecosystem, their distribution can be influenced by a variety of factors, which can have a negative impact on the ecosystem as a whole. Because they can transform light energy into chemical energy, phytoplankton are more prevalent in areas with high light intensity. Conversely, higher temperatures promote the growth and multiplication of both phytoplankton and zooplankton. Simple: The distribution and density of phytoplankton are also influenced by the amount of nutrients that are available in the surrounding environment.

3.1.11. Pollution in Freshwater Ecosystems

Similar to any ecosystem, the presence and activities of human society undeniably impact the way of life within a freshwater community. This is especially true in Western societies, where we extensively exploit land resources to support our way of living, consequently leading to repercussions on the planet's ecosystems.

- Many industries use water to cool their machinery. A discharge pipe empties this water into the river. This temperature rise may have an impact on the amount of oxygen that is readily available to living things, which may have an impact on respiration and ultimately their ability to survive. The ecosystem's ability to support life is impacted by this temperature shift.

- More flowing water can enter a freshwater ecosystem when nearby foliage is removed. Given this, extended periods of intense precipitation may cause the water levels to fluctuate greatly, which may also have a significant impact on the water's temperature and introduce new chemical agents into the stream.
- Recreational uses of waterways, like canoeing, are not without consequences. These people's litter can accumulate on the water's surface and obstruct the sunlight needed for photosynthesis, which is essential for the primary producers. All other organisms that depend on these primary producers for survival will be impacted if their way of life is so negatively affected that their population level falls.
- Chemicals released into the water, especially pesticides from farms or industrial waste, can have a significant molecular impact on the freshwater ecosystem. Lower concentrations of essential chemicals needed by the ecosystem's organisms are caused by higher concentrations of specific chemicals, some of which may be toxic. If so, the ecosystem's total biomass will be reduced since these organisms are unable to breathe and function at their best.

Did you get it?

1. **What are two categories for the ecology of freshwater rivers?**
2. **What is the importance of algae in an ecosystem?**

3.2. INLAND WATER ECOSYSTEM

Learning Objectives

- Learn about the origin of inland waters
- Discuss the environment

The inland water ecosystem refers to the collection of living organisms found in bodies of free water on the Earth's continents. These inland waters are essential components of the biosphere, containing a wide range of biological diversity, intricate biogeochemical pathways, and various energetic processes. Despite covering a relatively small area in terms of geography, inland waters play a significant role in shaping and maintaining the diversity, structure, and functioning of the biosphere when viewed from an ecological standpoint.

3.2.1. The Origin of Inland Waters

On continental landmasses, free water makes up a comparatively small portion of the total water content of the biosphere. About 97.6% of the water in the biosphere is found in the oceans, with the remaining 2.4% coming from water vapor, groundwater, and polar ice. Thus, continental free water—also known as inland water—makes up less than 1% of the total. Even with this low proportion, inland water is a vital component of the biosphere. It is found in many different forms and is home to a wide range of biological communities, which are very different from those found in terrestrial and marine ecosystems.

All inland waters come from the ocean, mostly from evaporation, and eventually return to it. This procedure is a component of the hydrologic cycle worldwide. The fact that more water evaporates from the ocean than is directly precipitated back into it is a key characteristic of this cycle. The remaining water vapor is precipitated over continental landmasses as rain, snow, or hail, from which it either drains into the sea or evaporates into the atmosphere (about 70%).

On land, there are two types of free water habitats: lotic (running water) and lentic (standing water). Lotic habitats consist of rivers, streams, and brooks, while lentic habitats include lakes, ponds, and marshes. These habitats are connected to drainage systems categorized into three major types: exorheic, endorheic, and arheic. Exorheic regions are open systems where surface waters eventually flow into the ocean through well-defined pathways involving streams and rivers that are sometimes blocked by permanent freshwater lakes. In contrast, endorheic regions are considered closed systems because their surface waters do not drain towards the sea. Instead, they flow towards inland termini where they either evaporate or seep away. These termini are usually permanent or temporary lakes that become salty due to the concentration of dissolved salts from rainwater or from the substrata in the drainage basin. Arheic systems, on the other hand, have unpredictable water flow in small quantities and follow random drainage patterns. In addition to rivers that originate outside the region (allogenic rivers) and areas supplied by underground water sources, most bodies of water in arheic regions are temporary.

Water under the ground is called inland water. groynes and other rocks, mostly limestone, have water and lots of groundwater in them. The world's water cycle relies on underground inland waters, some are important for life. there are 5 main types of inland waters based on if they're still or moving, fresh or salty, always there or not. There are three kinds of still waters: always fresh, sometimes fresh, always salty, and two kinds of moving waters: always there or not. Where these waters are found isn't the same everywhere. Still and moving waters that are always there are common in warm areas, while temporary waters are more in dry places. Salty lakes are usual in dry spots. But, water needs places to go, no matter what kind it is. Many earth processes made these features. Still waters are in basins made by wind, rivers, earth movements, glaciers, volcanoes, and weathering. People have made lake-like spots too. the lowest part of the land, shaped by water, is where moving waters start.

3.2.2. The Environment

3.2.2.1. Physical and Chemical Properties of Water

Numerous distinct physical and chemical characteristics of water have shaped life as it has evolved. In fact, the idea that the Earth is a biosphere depends on the unique physicochemical characteristics of water. These traits have had a major impact on how inland aquatic ecosystems are structured.

At the current global temperatures, the majority of inland waters are liquid. Water's unique thermal properties as a liquid reduce temperature swings. Its high specific heat, or I, is the first of these characteristics. The temperature of water must be raised with a comparatively large amount of heat. In addition, a significant amount of heat is needed to change water's state from a liquid to a gas (latent heat of evaporation) or from a solid to a liquid (latent heat of fusion). This ability of inland waters to absorb heat has multiple significant effects on the biosphere, one of which is their ability to moderate seasonal and diurnal (daily) temperature variations within aquatic ecosystems and, to a lesser extent, outside of them. The majority of the heat that enters inland waters comes from solar radiation. The actual amount that reaches inland waters at any given time is contingent upon many variables, including season, latitude, altitude, and cloud cover. Through reflection and backscattering, a sizable portion of the solar radiation that reaches the water's surface is lost. The residual portion goes into the water column, where its energy quickly decreases.

In inland waters, the transformation of light energy into heat has a number of important physical ramifications. The adjustments that the temperature makes to the density of water are particularly noteworthy. The density of pure water, which serves as a measure of heat content, is plotted against temperature to illustrate this relationship. Keep in mind that at 4° C, water has the highest density. This relationship is similar to that of fresh water, even though it pertains to pure water. Because freshwater lakes have slightly warmer, denser water at their surface, ice, which forms at 0° C, forms there first and keeps lakes from freezing solid. The biology of inland waters would be very different if this weren't the case. The relationship is somewhat different in saline waters, though, as higher dissolved salt concentrations lower the temperature of maximum density as well as the freezing point.

Water has many important chemical properties, one of which is its solvent capacity. This means that it is unparalleled in its ability to retain an extraordinarily broad range of substances in solution, including nonelectrolytes (substances like glucose that maintain their molecular structure and do not dissociate into ions), colloids (particulate matter small enough to remain suspended in solution), and electrolytes (salts that dissociate into ions in aqueous solution). Inland waters can contain a wide range of dissolved material combinations. However, certain significant patterns in the concentrations and kinds of solutes can be identified. The main inorganic solutes are the anions (negative ions) chloride, sulfate, and bicarbonate/carbonate, and the cations (positive ions) sodium, potassium, calcium, and magnesium. When combined concentrations of all of these ions (i.e., less than three grams of salt per liter, or salinity), are present (i.e., 3 parts per thousand, or 3 grams per kilogram), inland waters are typically considered fresh. The majority of freshwater is primarily composed of calcium, magnesium, and bicarbonate or carbonate ions, and their salinities are less than 0.5 grams per liter.

Every body of inland water is impacted by physicochemical phenomena, which form distinct interactions between and within the biotic and abiotic elements of the ecosystem. The biogeochemical cycles that carry nitrogen, phosphorus, carbon, and various micronutrients like iron, sulfur, and silica—the chemical elements necessary for life—are of special interest (biosphere: The organism and the environment: Resources of the biosphere: Nutrient cycling). Depending on the kind of inland water involved, different elements' outputs and inputs within a given aquatic ecosystem balance out to different degrees. In inland waters, however, all essential elements follow multiple, intricate, well-defined pathways that are frequently dependent on other biogeochemical cycles. Well-defined biogeochemical cycles are, in fact, a defining feature of all inland aquatic ecosystems, even the most basic transient bodies of highly salinized water.

The following are some of the most prominent general physicochemical characteristics of inland waters; however, it is crucial to note that these characteristics manifest in diverse ways amongst inland water types.

3.2.2.2. Permanent Bodies of Standing Fresh Water

Freshwater lakes that are permanent and deep contain about half of all inland waters. The largest of these lakes is Russia's Lake Baikal, which holds nearly 20% of all the freshwater found inland. The Great Lakes region of North America contains an additional 20%. Such waters are characterized by the development of vertical differences (vertical stratification) of several significant features, many of which also exhibit notable seasonal variation. Since light provides both heat and chemical energy for biological processes, it is by far the most significant variable feature. Most other spatiotemporal physicochemical differences within lakes are ultimately caused by variations in heat, specifically in the form of diurnal, seasonal, and vertical variations.

In deep freshwater lakes, different thermal patterns usually arise. The majority of these lakes are found in temperate biosphere regions, where seasonal variations in temperature and wind-induced mixing result in two periods of mixing—spring and autumn—per year in the lakes. This type of lake stratifies in the summer because the warmer epilimnion (surface layer) separates from the colder, lower layer (hypolimnion). Within the epilimnion, water circulates more actively than between the layers. The metalimnion, a region of abrupt temperature change, lies at the edge between these layers. When fall arrives, the water cools in the epilimnion and gets denser, sinking and combining with the hypolimnion. Wind provides the necessary effort to combine the two layers, and the lake fully circulates or overturns. Until the lake is shielded from additional wind action by surface ice, circulation will continue. After surface ice melts in the spring, the lake overturns once more, and by summer, stratification has returned. Monomixis is another type of thermal pattern where a single annual period of circulation alternates with a single event of thermal stratification.

Seasonal variations in vertical distribution are observed in numerous other physicochemical characteristics. The majority depend on and are strongly linked to seasonal variations in temperature. For instance, in summertime, dissolved oxygen concentrations in dimictic lakes may be higher in the epilimnion and lower in the hypolimnion. The case for dissolved carbon dioxide may be the opposite. However, there might not be any vertical differences other than during the summer. Because many aquatic animals cannot survive when oxygen concentrations fall below a certain level, changes in oxygen concentration are especially significant. The solubility of several significant substances, including phosphate, iron, and manganese, is also influenced by oxygen concentrations and as a result, these substances also exhibit vertical seasonal variation.

Salinity and water level are two characteristics of deep freshwater lakes that don't change with the seasons. In these kinds of lakes, salinity also lacks a vertical gradient. Shallow, permanent bodies of standing fresh water have few vertically stratified physicochemical features, though many of them change dramatically with the season. Thermal stratification and related phenomena, however, may form at shallower depths and last longer in permanent freshwater bodies situated in regions warmer than the temperate zone than they would in temperate lakes with comparable morphometry. The water density–temperature relationship, which states that at higher temperatures, water density changes quickly with only slight temperature increases, supports this.

3.2.2.3. Temporary Bodies of Standing Fresh Water

After rain, all of the biosphere's land surfaces temporarily become covered in freshwater bodies. Temporary bodies of standing fresh water play a significant biological role, even though their total volume is a small fraction of that of permanent freshwater lakes. They are among the most distinctive kinds of water bodies found in all of the biosphere's arid landscapes, which account for roughly one-third of the planet's total land area. In some areas and during specific periods, they serve as the most conspicuous aspect of the terrain (e.g., in the South African Highveld). There are numerous varieties that range in size from large, frequently flooded wetlands that last for several months to tiny, sporadic rainpools. These ephemeral bodies of water are so common that they go by a variety of names, including ones that are indigenous to the area, like vlei, claypan, pan, playa, and tinaja. Temporary waters are mostly caused by climatic and topographic factors, which also determine when and how often they occur. Generally speaking, though, their life span is shortened and their occurrence is less predictable in drier environments.

Since temporary water bodies are typically shallow, vertical variations in their physicochemical characteristics are less noticeable than in permanent, deeper water bodies. When vertical disparities do arise, they are typically momentary (refer to the comments regarding thermal stratification in hot regions, though). Compared to deeper, permanent waters, the physicochemical characteristics of transient bodies of water are far more susceptible to outside events. Consequently, hydrologic variability has greater effects than they do in deep, permanent freshwater lakes due to inputs, water levels, depths, sediment-water interchange, and wind. Nonetheless, salinity varies only seldom by 3 grams per liter. It can be concluded that, in general, aquatic ecosystems in transient freshwater bodies are considerably less protected from outside environmental events than those in permanent freshwater bodies.

3.2.2.4. Saline Lakes

Saltwater lakes (i.e., salinities exceeding three grams per liter) are common and can be found on all continents, including Antarctica. Saline lakes include many of the highest lakes, such as those in Tibet and on South America's Altiplano, as well as the world's largest lake, the Caspian Sea, and the lowest lake, the Dead Sea. Despite comprising about 45% of all inland water, salty inland water is mostly found in a small number of deep lakes, mainly the Caspian Sea. Due to two factors that are most frequently present in semiarid regions of the biosphere (roughly 27% of the total land area): a balance between precipitation and inflows, as well as the presence of endorheic drainage basins and water output, saline lakes are most common in these areas.

Salt lakes have a large total volume and are distributed widely geographically, but their significance as a key component of biospheric processes has generally been disregarded. In fact, the environmental significance of human impact did not become evident until after 1960, when the effects started to be noticed. The Aral Sea, a sizable salt lake in Central Asia, serves as an illustration of this. The lake level dropped, the salinity increased, and large stretches of the lake bed became visible after a large portion of the freshwater input was diverted before it reached the lake for irrigation purposes. The fishing industry crashed, islands that had been wildlife refuges turned into peninsulas, biological diversity and productivity decreased, biota vanished, massive amounts of salt blew onto nearby lands from the lake bed, groundwater salinity increased, and the local climate changed as a result. The consequences were equally disastrous for the local human population.

Temporary salt lakes and temporary bodies of standing fresh water react to environmental disturbances similarly. Permanent salt lakes exhibit the same kind of vertical differences in physicochemical attributes as permanent bodies of standing fresh water. All salt lakes, however, differ from all freshwater bodies due to their ionic composition and, obviously, significantly higher salinities. Salinities can reach values of up to 300 grams per liter, depending on the dominant ions that are present. Annual salinities and water levels may vary only slightly in permanent, deep salt lakes, but they may vary greatly in shallow, transient lakes, with salinities ranging from less than 50 grams per liter to more than 300 grams per liter in a single year, along with significant variations in water levels. Furthermore, compared to freshwater lakes, saltwater lakes are far more vulnerable to long-term climate changes because they all depend on a climatic balance. As a result, even big, permanent, deep salt lakes show noticeable variations in salinity and water level over time, which correspond to long-term changes in the climate. The diversion of water by humans frequently exacerbates these changes, as described.

Did you know?

Large saline lakes make up 44% of the volume and 23% of the area of lakes worldwide.

3.2.2.5. Running Water

Throughout the biosphere, there are streams, brooks, and rivers that are both permanent and ephemeral. Large, permanent rivers and streams are features of well-watered regions (temperate and humid tropical areas); transient streams are features of drier regions. Large, persistent allogenic rivers that originate in humid regions and flow into arid regions can exist in even the most arid of places—e.g., the North African Nile River.

Streams and rivers supply the crucial link in the global hydrologic cycle, i.e., the process that eventually returns all water that has precipitated onto land after evaporating from the sea. Even so, running waters make up less than 1% of all inland free waters, the majority of which are found in the Amazon River.

Running waters can be distinguished from standing waters by a number of physicochemical characteristics. The most noticeable ones are shallow depth, a generally linear morphology, and unidirectional water flow. The lotic environments' consistent low salinity is less evident, but still unique. All running waters are fresh, and they contain the same major ions as standing fresh waters, with very few exceptions. The aquatic environment differs greatly from the lentic environment due to these and other physicochemical characteristics. As a result, the majority of biological communities that start in lotic systems and the ecological processes that go along with them are restricted to this kind of habitat due to their extreme specialization. However, it's not always easy to distinguish between lentic and lotic environments. The duration that a specific volume of water stays in a particular area of an aquatic ecosystem—a notion that is unmistakably tied to flow rates—is the determining factor. Some large rivers with only a slight gradient have large floodplains with numerous interconnected bodies of lentic waters, as well as low discharge and flow rates. The broad reach of a large entity is comparable to this circumstance.

Running waters exhibit a greater diversity of physical form and a wider range of biological habitats than standing waters, despite not exhibiting the same salinity range. Small, ephemeral streams that only open up in deserts after sporadic downpours to massive, year-round tropical rivers that are so vast that their opposite banks cannot be seen are examples of running waters. Widespread flood plains could exist or not, the flow could be highly variable or roughly constant, ranging from high to almost nothing, and the substrate could be anything from fine mud to bare rock. Their primary stem, the hyporheic zone, contiguous terrestrial areas, the main stem of the river or stream, the biogeochemical pathways, and the relative ecological importance of the floodplain (if present) all differ greatly.

Did you get it?

1. **What is meant by the term 'marine ecosystem'?**
2. **What is a salt lake otherwise known as?**

3.3. MARINE ECOSYSTEMS

Learning Objectives

- Describe the origins of marine life
- Learn about the marine environment and marine biota
- Describe the patterns and processes influencing the structure of marine assemblages

A marine ecosystem is a complex of living things found in the ocean. Two-thirds of the Earth's surface is made up of marine waters. Certain locations have ocean depths greater than the height of Mount Everest; the Mariana and Tonga Trenches in the western Pacific Ocean, for instance, both have depths exceeding 10,000 meters (32,800 feet). Numerous different species that have evolved in response to different environmental factors can be found living in this ocean habitat.

3.3.1. Origins of Marine Life

Approximately 4.5 billion years ago, the Earth formed. Seas were created when the Earth experienced heavy rainfall that filled its large basins and condensed the water in the atmosphere as it cooled. Hydrogen, methane, ammonia, and water were among the inorganic elements present in the primordial atmosphere and waters. It is believed that when these materials were ignited by lightning strikes, they merged to create the first organic compounds. Cyanobacteria, also known as blue-green algae, are among the oldest known organisms. Australia has discovered evidence of these primitive photosynthetic prokaryotes in Precambrian marine sediments known as stromatolites, which date back about 3 billion years. Many different types of bacteria, algae, protozoa, and primitive metazoa evolved during the Precambrian (roughly 446 million to 542 million years ago) to take advantage of the world's early marine habitats, even though the diversity of life forms observed in modern oceans did not appear until much later. There was a significant radiation of life in the oceans between 542 million and 488 million years ago, during the Cambrian Period. Fossils of well-known creatures like cnidaria (e.g., echinoderms, jellyfish, etc.).

Over the last 600 million years, the world's oceans and seas have undergone substantial changes in shape. Plate tectonics theory states that the Earth's crust is composed of numerous dynamic plates. The Earth's mantle is home to two different kinds of plates: continental and oceanic. These plates can slide against, diverge from, or converge with one another. New crust is formed when two plates diverge, as magma from the mantle rises and cools; when two plates converge, one plate descends, i.e., is reabsorbed into the mantle while the other is subducted. There are instances of both processes in the marine environment. Large underwater mountain ranges like the Mid-Atlantic Ridge are known as oceanic ridges or rift areas, and these are where the oceanic crust is formed. Along subduction zones—which are typically identified by deep-water trenches like the Kuril Trench off the coast of Japan—excess crust is reabsorbed.

As sea levels fluctuate, the ocean's shape also changes. Sea levels are typically lower during ice ages because a larger percentage of the planet's waters are trapped in the polar ice caps. Sea levels rise during interglacial periods when the polar ice caps melt. Significant shifts in the distribution of marine environments, including coral reefs, are brought about by these sea level changes. For instance, the Great Barrier Reef did not exist as it does today during the last Pleistocene Ice Age; instead, it was located on the high-tide mark, a continental shelf.

3.3.2. The Marine Environment

3.3.2.1. Geography, Oceanography, and Topography

The marine life found in the oceans is not uniformly distributed. The different features of the marine environment determine the kinds of organisms that will live there and also create different habitats. Marine environments are affected by a variety of factors, including topographic complexity, proximity to land, water depth, and availability of light. The creatures that can live in a particular region of a marine ecosystem depend on the availability of light. Less light can enter the water the deeper it is, until eventually there is absolutely no light at all. The vast majority of the ocean is covered by this pitch-black region, known as the aphotic zone. The euphotic and disphotic zones can be identified within the lighted area, which is known as the photic zone. The layer nearer the surface where photosynthesis can take place is known as the euphotic zone. The disphotic zone is below; it receives enough light to support respiration at a rate higher than photosynthesis. The actual depth of these zones is determined by the local turbidity, cloud cover, and ocean surface conditions. Deeper than 80 to 100 meters is the general limit of the euphotic zone, and deeper than 80 to 700 meters is the disphotic zone. Although many organisms live in the aphotic zone and migrate vertically to the photic zone every night, marine organisms are most prevalent in the photic zone, particularly in the euphotic portion. Other organisms include certain species of sea cucumbers, brittle stars, and tripod fish.

Marine environments can be broadly classified as benthic environments, which are found on the bottom, and pelagic environments, which are found in water. The waters of the pelagic environment are separated into two provinces: the oceanic province, which comprises all open waters beyond the continental shelf, and the neritic province, which includes the water on the continental shelf. Because of the dissolved materials in riverine runoff, the neritic province has higher nutrient levels than the oceanic province. Photosynthesis takes place in the epipelagic zone, which is the uppermost part of the neritic and oceanic waters. It is similar to the photic zone in general. Within this region are three distinct areas: the mesopelagic, which spans 200–1,000 meters; the bathypelagic, which spans 1,000–4,000 meters; and the abyssalpelagic, which includes the lowest points of the ocean, from 4,000 meters down to the innards of the deep-sea trenches.

The benthic environment is also divided into different zones. The supralittoral is the high-tide mark and is usually not under water. The intertidal, or littoral, zone ranges from the high-tide mark (the maximum elevation of the tide) to the shallow, offshore waters. The sublittoral is the environment beyond the low-tide mark and is often used to refer to substrata of the continental shelf, which reaches depths of between 150 and 300 meters. Sediments of the continental shelf that influence marine organisms generally originate from the land, particularly in the form of riverine runoff, and include clay, silt, and sand. Beyond the continental shelf is the bathyal zone, which occurs at depths of 150 to 4,000 meters and includes the descending continental slope and rise. The abyssal zone (between 4,000 and 6,000 meters) represents a substantial portion of the oceans. The deepest region of the oceans (greater than 6,000 meters) is the hadal zone of the deep-sea trenches. Sediments of the deep sea primarily originate from a rain of dead marine organisms and their wastes.

3.3.2.2. Physical and Chemical Properties of Seawater

Seawater's chemical and physical characteristics change depending on its depth, latitude, proximity to land, and freshwater input. Pure water makes up the remaining 96.5% of seawater, with dissolved compounds making up the remaining 3.5%. Seawater's chemical makeup is influenced by a variety of processes, including rain, gas exchange with the atmosphere, volcanic activity, rock and sediment erosion, and the breakdown and metabolism of living things. Apart from carbon, nitrogen, and phosphorus are also vital nutrients for living things. Since they are not present in large amounts in seawater, they frequently act as constraints on the organic cycles of the ocean. Because marine organisms quickly absorb phosphorus and nitrogen, their concentrations are typically low in the photic zone. Due to organism decay, the highest concentrations of these nutrients are typically found below 500 meters. Additional elements of significance are calcium and silicon, which are necessary for the skeletons of numerous organisms, including fish and corals.

The ocean's chemical composition is influenced by that of the atmosphere as well. For instance, marine plant life releases oxygen into the atmosphere and absorbs carbon dioxide into the ocean. Contrary to previous beliefs, such as the idea that toxins could be safely disposed of in the ocean, the discharge of pollutants into the sea can also alter the ocean's chemical composition.

Organisms are greatly impacted by the physical and chemical characteristics of seawater, which vary significantly depending on the size of the organism. For instance, ciliates, which are tiny marine creatures measuring less than 1 millimeter (0.39 inch) in length, are viscous to seawater, but tuna, which are large marine creatures, are not.

Marine organisms can survive in the sea because they have evolved a wide range of special physiological and morphological traits. Because of antifreeze-producing proteins in their blood, notothenioid fishes in Antarctica can survive in waters as cold as −2° C (28° F). Numerous creatures can attain neutral buoyancy by releasing gas into internal chambers, as some fish do, or swim bladders, as cephalopods do. Other organisms use lipids, which are less dense than water, to accomplish this. Certain animals, particularly those found in the aphotic zone, produce light to draw in prey. To break up their body's outline and evade being seen by predators, animals in the disphotic zone, like hatchetfish, produce light using organs called photophores. Using specialized organs, many marine animals can detect sound or vibrations in the water over long distances. Whales communicate with one another through the use of melon, a sound-producing organ, and certain fishes have lateral line systems that they use to locate prey.

The degree to which organisms can withstand variations in salinity varies widely. For example, euryhaline organisms, which are found in estuaries—areas where rivers and seas converge—can withstand significant fluctuations in salinity levels, while stenohaline organisms cannot. Additionally, euryhaline organisms are highly resilient to temperature fluctuations. Using active pumping or salt retention, animals that migrate between fresh and salt water, like salmon and eels, can regulate their osmotic environment. In marine waters, body architecture varies widely. The sleek, elongated shape of the barracuda contrasts sharply with the body shape of the cnidarian by-the-wind sailor (*Velella velella)*, an animal that lives on the water's surface (pleuston) and sails with the help of a modified flotation chamber.

3.3.2.3. Ocean Currents

Many factors affect the movements of ocean waters: the Earth's rotation, which causes the Coriolis effect; atmospheric circulation patterns, which affect surface waters; and temperature and salinity gradients between the tropics and the polar regions, which are known as thermohaline circulation. You can read more on ocean currents for a thorough explanation of ocean circulation. The resulting circulation patterns vary in size from large-scale turbulences of less than one meter to those spanning vast regions, like the North Subtropical Gyre, which travels thousands of kilometers.

These patterns affect marine organisms of all sizes and have the power to define a species' range. For instance, only the Antarctic Circumpolar Current is home to krill (*Euphausia superba*). Large and small pelagic organisms' distribution patterns are also impacted. Larvae are transported over long distances by mainstream currents like the East Australian Current and Gulf Stream. Because of this, when fish and invertebrate larvae from the tropics are transported to high latitudes by these currents, cold temperate coral reefs get a tropical infusion. The Gulf Stream waters carry eels from their Caribbean spawning grounds to Europe and is essential to their successful recruitment. Strong biological activity is frequently seen in areas where the ocean is influenced by nearshore features like estuaries or where there is a vertical salinity gradient (halocline). Tiny organisms can concentrate in these settings, which gives other animals an abundance of food.

3.3.3. Marine Biota

The organisms that live in the benthic environment (benthos) or the pelagic environment (plankton and nekton) are collectively referred to as the marine biota. On the other hand, some organisms are benthic during one stage of their lives and pelagic during another. Both environments contain producers who create organic molecules. In the pelagic environment, the photic zone is produced by single- or multicellular plankton that have pigments used in photosynthetic processes. Microalgae are common benthic producers; other examples are large algae (e.g., kelp or *Macrocystis pyrifera*), and sea grass (e.g., Zostera).

3.3.3.1. Plankton

Plankton are the many, mostly microscopic organisms that live in the pelagic environment. Because they supply food for the nekton (e.g., fish and squid) as well as benthos (e.g., sea cucumbers and sponges), they are essential parts of food chains in all marine environments. Additionally, they have an impact on the biosphere globally because certain plankton's photosynthetic processes are crucial to maintaining the equilibrium of atmospheric components.

The word "plankton" comes from the Greek word "planktos", which means "to wander, to drift." This term perfectly describes how most plankton spend their lives—floating with the currents of the ocean. However, not all types of plankton lack the ability to control their motions; in fact, many rely on self-directed motions to survive. Plankton can range in size from microscopic organisms as small as 1 micrometer (0.000039 inch) to jellyfish, whose tentacles can reach a length of over 15 meters and whose gelatinous bell can reach up to 2 meters in width. But the majority of planktonic organisms, also known as plankters, are much smaller than 1 millimeter (0.039) in length. These bacteria are frequently photosynthetic and feed on the nutrients found in saltwater. Numerous different types of organisms, including algae, bacteria, protozoans, some animal larvae, and crustaceans, are considered to be part of the plankton. A large proportion of the plankton are protists—i.e., mostly single-celled, eukaryotic organisms. Plankton can be broadly classified into three groups: microbes, such as bacteria, zooplankton, which are animals or animal-like protists, and phytoplankton, which are plants or plant-like protists. The heterotrophic consumers in the marine community are zooplankton, while phytoplankton performs photosynthesis. Diatoms and dinoflagellates are two extremely diverse groups of photosynthetic protists that are essential parts of the plankton. They range in length from approximately 15 to 1,000 micrometers. The most prevalent type of phytoplankton is diatoms. Although many dinoflagellates are phototrophic, some also feed on algae or bacteria. Foraminifera, radiolaria, acantharia, ciliates, and flagellates are additional significant groups of protists. Numerous protists are significant consumers and zooplankton's food source.

Larger than 0.05 millimeters, zooplankton can be broadly classified into two groups: meroplankton, which only becomes plankton for a portion of its life cycle (usually the larval or juvenile stage), and holoplankton, which remains plankton throughout its entire existence. Many larval meroplankton, such as sea urchins, intertidal snails, crabs, lobsters, and fish, are very different from their adult counterparts in coastal, oceanic, and even freshwater environments. These larvae might display characteristics specific to the larval stage, like the amazing spiny armor that some crustacean larvae (e.g., Squilla), most likely employed to deter predators.

Copepods, cladocera, and euphausiids (krill), which resemble lobsters and are crucial to the marine environment because they provide food for fish and marine mammals, are examples of important holoplanktonic animals. Larvaceans, salps, and siphonophores are examples of gelatinous forms that feed on phytoplankton or other zooplankton. Certain copepods and euphausiids, which are omnivorous zooplankton, eat both phytoplankton and zooplankton; their feeding habits vary depending on the kind and quantity of prey that is available. Certain zooplankton species engage in intense grazing and predatory behaviors, which can lead to quantifiable decreases in phytoplankton or zooplankton biomass. For instance, when jellyfish are abundant in confined waters, they have the potential to eat so many fish larvae that the number of fish in the area is significantly decreased.

Predators, the jelly-like plankton are abundant. Nematocysts, which are stinging cells, and colloblasts, which are sticky cells, are how they secure their prey. Propelled by the wind along the surface, large numbers of the Portuguese man-of-war (*Physalia*), with its conspicuous gas bladder, the by-the-wind sailor (*Velella velella*), and the small blue disk-shaped *Porpita porpita* can be found strewn on the beach after strong onshore winds. Comb jellies are frequently abundant beneath the surface, along with salps, scyphomedusae, and siphonophores.

Compared to the variety of niches found in the benthic environment, it was previously believed that the pelagic environment offered fewer unique habitats. Plankton of varying sizes was thought to be the only factor that differentiated the pelagic realm due to its apparent uniformity. However, it has been found that biotic distributions are impacted by small-scale changes in the pelagic environment. Patchiness in biotic distributions is caused by plankton adhering to organic aggregates known as marine snow, which are formed by both living and dead matter.

Cell and mucus aggregates, drifting macroalgae, and other flotsam with sizes ranging from 0 to 1 centimeters (though these aggregates can be as small as 0.05 millimeters and as large as 100 centimeters) are examples of structures found in marine snow. Marine snow attracts a variety of microorganisms, phytoplankton, and zooplankton, as well as grazing copepods and predators that feed on the surface of these formations. Sometimes there is an abundance of marine snow, especially following plankton blooms. Large amounts of organic matter from the ocean's upper layers may descend to the bottom as marine snow, giving bottom dwellers a vital food source. Plankton also reacts to other structures found in the marine environment, such as large rafts formed by aggregates of phytoplankton cells found in tropical and temperate waters worldwide (e.g., Oscillatoria *Trichodesmium erythraeus* cells) and different kinds of seaweed (e.g., Phyllospora, Sargassum, and Macrocystis) that separate from the ocean floor and float.

3.3.3.2. Nekton

The most well-known marine organisms are frequently nekton, the ocean's active swimmers. In most marine food chains, nematodes rank as the top predators. Nekton and plankton are not always easily distinguished from one another. As previously noted, a lot of large marine animals, including tuna and marlin, spend their adult years as large, active members of the nekton after spending their larval years as plankton. Both macrozooplankton and micronekton are terms used to describe other organisms, such as krill.

Most nekton are made up of vertebrates, such as mollusks, crustaceans, fish, reptiles, and mammals. With about 16,000 species, fish are the most abundant class of nekton. Nekton can be found in marine waters at all latitudes and depths. In polar waters, whales, penguins, seals, and icefish are common. In the aphotic zone, lantern fish (family Myctophidae) coexist with other common species such as seven-gilled sharks, whalefish (family Cetomimidae), and gulpers

(Saccopharynx). The diversity of nekton is highest in tropical waters, especially those with a high concentration of fish species.

Neptunophora musculus, or blue whales, are the largest animals on Earth. They can reach lengths of 25 to 30 meters. These enormous creatures, along with other baleen whales (order Mysticeti), which are identified by the fine filtering plates in their mouths, consume plankton and micronekton, just like the world's largest fish, whale sharks (*Rhincodon typus*), which are typically 12 to 14 meters long, but can grow up to 17 meters. The largest carnivores that eat large prey are the great white sharks (*Carcharodon carcharias*), tiger sharks (*Galeocerdo cuvier*), giant groupers (*Epinephelus lanceolatus*), black marlin (*Makaira indica*), bluefin tuna (*Thunnus thynnus*), and toothed whales (order Odontoceti), such as killer whales (*Orcinus orca*). Nekton is the foundation of significant global fisheries. Large schools of tiny anchovies, herring, and sardines typically make up between 25% and 33% of the yearly ocean harvest. Squids are a valuable economic nekton as well. Cod, sole, and halibut are demersal fish (bottom-dwelling) that are valuable to the human food industry. They are typically found in the waters of the continental shelf. These regions are also important fishing grounds because pelagic nekton frequently thrives in upwelling areas where the waters are nutrient-rich.

3.3.3.3. Benthos

These organisms inhabit both deeper waters and the surface sediments of the continental shelf, which are rich in organisms, with a wide variety of species present. In shallow waters, beds of seagrass provide a rich habitat for polychaete worms, crustaceans, fishes, amphipods, etc. The majority of animal activities on the surface and in intertidal sediments are highly influenced by the tide. However, microscopic benthic diatoms are the only photosynthetic organisms found on many sediments in the photic zone.

Size is a useful factor in the classification of benthos. Organisms larger than one millimeter are known as macrobenthos. Deposit feeders are those that consume organic material found in sediments (e.g., holothurians, echinoids, and gastropods), while plankton consumers are predators (such as bivalves, ophiuroids, and crinoids) that eat other fauna in the benthic assemblage. Meiobenthos are organisms with a diameter of 0–1 millimeters. These larger microbes play a variety of roles in benthic food chains, including primary producers, decomposers, recyclers of nutrients, and predators. Examples of these roles include polychaetes, turbellarians, and foraminiferans. Organisms smaller than one millimeter, such as diatoms, bacteria, and ciliates, are referred to as microbenthos.

Bacteria break down organic matter aerobically in the vicinity of the sediment's surface, where oxygen is plentiful. On the other hand, deeper layers become oxygen-depleted due to oxygen consumption at this level, and marine sediments beneath the surface layer are anaerobic. The amount of organic matter and the oxygen permeability of the sediment are determined by the grain size, which also affects the thickness of the oxygenated layer. The dominance of anaerobic processes increases with decreasing oxygen concentration. The redox discontinuity layer, which is the layer that separates oxygen-rich and oxygen-poor layers, is visible as a gray layer between the black anaerobic layers.

Different strategies have evolved by organisms to deal with oxygen scarcity. Through metabolic processes, certain anaerobes release ammonia, hydrogen sulfide, and other harmful reduced ions. Sulfur is metabolized by the thiobiota, which is primarily composed of microorganisms. However, the majority of organisms that reside below the redox layer must produce an aerobic environment for themselves. Because the anoxic layer around them rapidly depletes the oxygen in the burrow, animals that burrow create a respiratory current along their burrow systems to oxygenate their living spaces. This influx of oxygen must be maintained continuously. Numerous bivalves (e.g., *Mya arenaria*) extend long siphons upward into oxygenated waters near the surface to breathe and feed while staying safe from predators deep in the sediment. A powerful "foot" is used by many large mollusks for digging, and in certain situations, it is also used to help them move away from predators like starfish. Consequently, burrow systems may become "irrigated," generating nutrient and oxygen fluxes that promote the growth of benthic producers (e.g., diatoms).

Not all benthic organisms reside in the sediment; certain benthic communities live on a rocky surface. Various types of algae, including red, green, and brown algae, are abundant and diverse in the sunlit area on rocky surfaces and play a crucial role as producers. In areas affected by tides, algae are most prevalent and largest near the point of low tide. Temporary algae like Ulva, Enteromorpha, and coralline algae cover a wide range of the tidal area. The assortment of algae species present in a specific location depends on its latitude and is also greatly influenced by wave exposure and the presence of grazers. For instance, Ascophyllum spores are unable to attach to rocks during even mild ocean movements, which is why this plant is primarily found on protected shores. the big plant, *Macrocystis pyrifera*, a giant kelp found on underwater rocks, can grow up to 1 meter every day. these plants, which can grow over 30 meters, are common in cool reefs. on cool rocky reefs, big algae and encrusting forms are found. on warm reefs with lots of corals, big algae are rare; sargassum and other algae are near the reef. on reefs, slow-moving animals are common. plant-eating snails and urchins are found in the water and can change where algae grow. barnacles are common on rocks. sponges, sea squirts, urchins, and sea anemones are found in deep water where it's dark. in caves and under big rocks, lots of animals are found.

The phylum Cnidaria contains organisms called reef-building coral polyps, or Scleractinia, which produce a calcareous substrate that supports a wide variety of organisms. In the Pacific and Indian oceans, there are about 700 different species of coral that fall into genera like Porites, Acropora, and Montipora. Coral reefs are home to some of the world's most intricate ecosystems. Zooxanthellae are single-celled, photosynthetic algae that coexist symbiotically in coral tissue and contribute to the formation of the solid calcium carbonate matrix that makes up the reef. Reef-building corals are found only in waters that are warmer than 18° C: the coral-algae complex needs both high light intensity and warm temperatures to secrete calcium carbonate. On top of volcanic rock, hundreds of meters of coral make up the entirety of many tropical islands.

3.3.3.4. Links Between the Pelagic Environments and the Benthos

Because the benthic and pelagic environments are interconnected in numerous ways, it is important to exercise caution when examining them separately. For example, animals that live on rocky or soft bottoms rely heavily on pelagic plankton as a food source. Detritus feeders feed on the buildup of particulate matter that falls from the water column, while suspension feeders, such as anemones and barnacles, filter both living and dead particles from the surrounding water. The rain of fallout from the pelagic environment to the ocean bottom is caused by dead plankton, marine snow, plankton feces, and crustacean molts. In certain weather patterns, such as El Niño, this fallout can be so strong that benthic animals on soft bottoms suffocate and perish. The rate at which plankton falls out varies based on seasonal production cycles. In the abiotic zone, where temperature and light fluctuations are minimal or nonexistent, this variance may give rise to seasonality. Marine sediments consist of plankton, and a variety of fossilized protistan plankton species, including coccoliths and foraminiferans, are used to date and identify rocks.

3.3.3.5. Organisms of the Deep-Sea Vents

When submarines began to explore the deep sea more frequently in the 1970s, producers were found in the aphotic zone. It is now recognized that deep-sea hydrothermal vents are rather common in tectonically active areas (e.g., spreading ridges). The vents provide organisms with a nonphotosynthetic supply of organic carbon. Deep-sea organisms such as vestimentiferan worms, big bivalve clams, and mussels depend on bacteria to oxidize sulfur (sulfide) and obtain chemical energy from the process. Rather than being photosynthetic, these organisms are called chemoautotrophic or chemosynthetic. Due to the symbiotic relationships that many of the species in the vent fauna have developed with chemoautotrophic bacteria, the megafauna is primarily responsible for the primary production within the vent assemblage. The situation is analogous to that found on coral reefs where individual coral polyps have symbiotic relationships with zooxanthellae. Around vents, there is a diverse community of free-living bacteria in addition to symbiotic ones. For instance, on any hard surface, Beggiatoas-like bacteria frequently form noticeable web-like mats.

3.3.4. Patterns and Processes Influencing the Structure of Marine Assemblages

3.3.4.1. Distribution and Dispersal

Physical and biological processes occur in both ecological time (tens of years) and geologic time (hundreds to millions of years) that impact the distribution patterns of marine organisms. Plate tectonics has an impact on the forms of the Earth's oceans, which in turn has an impact on the distributions of both extant and fossil marine organisms. According to the vicariance theory, biogeographic patterns are mostly determined by plate tectonics. For instance, Australia had few coral reefs and was once near the South Pole, some 90 million years ago. Australia has been approaching the Equator by a few millimeters every year since that time. Coral reef environments are gradually expanding southward due to this movement and regional oceanographic conditions. Additionally, dispersal might play a significant part in biogeographic patterns of abundance. Local oceanographic features, such as the direction and strength of currents and the biology of the organisms, have a significant impact on the significance of dispersal. The extinction of marine organisms and their distribution patterns can both be impacted by humans. For instance, catch limits determine the level of fishing in the Irish Sea. Bipartite life cycles are a feature of many marine organisms and can impact an organism's ability to disperse. The majority of animals found on both soft and hard substrata—including fish, lobsters, crabs, barnacles, polychaete worms, and sea urchins—spend their larval phase in the plankton, where they are widely distributed. One of the main factors influencing dispersal is the duration of the larval phase, which can range from a few minutes to hundreds of days. In contrast to many other kinds of reef fish, wrasses in the genus *Thalassoma* have a lengthy larval life. Additionally, populations of these fish are widely distributed throughout the reefs of remote volcanic islands in the Pacific. Algal spores are how algae disperse, and this process is influenced by their bipartite life cycle. While spores typically only travel a short distance from mature plants, storms and spores with limited swimming ability (Macrocystis spores have flagella) can spread spores farther.

3.3.4.2. Migrations of Marine Organisms

Many regions of the world have well-described plankton and nekton migrations throughout the water column. Day-to-day vertical migrations are typical. Certain fish, squid, and plankton species, for instance, spend the day below the photic zone, rising to the surface at dusk and sinking back below before dawn. The general consensus is that marine life migrates in reaction to light levels. Because plankton can avoid predators that use sight to locate their prey by spending the day in the low light or darkness beneath the photic zone, this behavior may be beneficial. Plankton can ascend to the surface waters when the sun sets, where there is more food available and they can safely feed while it is dark.

Larval forms can move vertically to aid in their horizontal transport along various currents. Currents can vary in direction based on depth, which makes this feasible (e.g., beneath thermoclines and haloclines), as is the situation in estuaries.

A wide variety of larger invertebrates (e.g., mysids, amphipods, and polychaete worms) move into the water column at night from beneath the cover of sediments and algae. When visual predators find it difficult to see them, it is believed that these animals swim to find mates or disperse to new areas. Only one sex may emerge at night in certain situations, and that sex is frequently more morphologically adapted for swimming. Fish frequently migrate horizontally over tens or even hundreds of kilometers, usually in response to patterns of feeding or reproduction. For the majority of the year, tropical coral trout (*Plectropomus* species) are scattered throughout a reef; however, when it comes time for spawning, the adults will congregate at specific spots. Many marine vertebrates have been seen to migrate across continents (more than 1,000 kilometers), and these movements are frequently connected to the needs of feeding and reproduction. The Atlantic Ocean is traveled by bluefin tuna (*Thunnus thynnus*) in a single year; they spawn in the Caribbean and then migrate to the Atlantic's high latitudes to feed on the abundant fish population. Large-scale migration is also experienced by sharks and turtles.

Fish that are diadromous—that is, they live their entire lives in both fresh and marine environments—display some of the most amazing migratory patterns. Anadromous fish, like Atlantic salmon (*Salmo salar*), have distinct migratory patterns because they spend the majority of their lives in the sea but migrate to freshwater to spawn. Once they spawn, the adults perish. Alevin, or newly hatched fish, are formed from spawned eggs and travel downstream through rivers in the direction of the sea. Fish that congregate near the ocean are called smolt, which are larger than juvenile parr. When the adult fish are ready to spawn, they follow many environmental cues, such as the water's chemistry, the Earth's magnetic field, and the Sun, to return to their natal river. The water chemistry of the fish's natal river is thought to be imprinted on them by the thyroid gland. Freshwater eels, including the European eel (*Anguilla anguilla*), migrate great distances from freshwater to the marine waters of the Sargasso Sea to spawn (a process known as catadromous migrations), where they eventually perish. The Gulf Stream carries eel larvae, also known as leptocephalus larvae, back toward Europe.

3.3.4.3. Dynamics of Populations and Assemblages

Numerous diverse processes impact both the composition of assemblages and the dynamics of individual species' marine populations (e.g., assemblages of populations of various species coexisting in the same region). Apart from aquatic mammals like whales, fish that give birth to live young (e.g., fish (Embiotocidae), as well as brooders (i.e., fauna that raise their progeny until they become larvae or juveniles), the majority of marine organisms produce a large number of offspring, only a small percentage of which survive. Plankton-related processes can significantly impact the quantity of juveniles that endure and are subsequently re-engaged, or transferred, into adult groups. Larval survival may be influenced by factors such as the quantity of food available at different times and locations, the presence of predators, and oceanographic characteristics that keep larvae close to ideal nursery sites. The size of adult populations and, consequently, the relative abundance of species in marine assemblages may ultimately depend on the quantity of organisms recruited to benthic and pelagic systems. However, a variety of factors can influence an organism's ability to survive after recruitment. Recruitment is consumed by predators, and the mortality rates of prey species can change over time. The way that colonies and successions occur can greatly affect benthic assemblages. When intertidal reefs are removed experimentally, for instance, the variety of organisms that occupy the exposed area frequently corresponds to the kinds of larvae that were present in the nearby waters at the time. If tube worms establish themselves first, they may become dominant; if not, algal spores may colonize the shore first and prevent the worms from settling. There's also a chance that organismal competition is involved. Long-term data from coral reefs collected over more than 25 years have shown that some corals (e.g., *Acropora cytherea*) competitively outgrow nearby corals. Many corals are physically disturbed by hurricanes, and during the reef's regrowth, competitively inferior species may coexist with those that are typically dominant. Sessile organisms' chemical defenses have the ability to inhibit the growth of settling organisms or increase their mortality rate. Larvae of ascidians (e.g., *Podoclavella*) frequently steer clear of sponges (e.g., *Mycale*); the larvae hardly ever mature into adults when this happens.

While the mechanisms governing species assemblages may be known, fluctuations in plankton composition make it challenging to forecast colonization patterns with high precision.

Did you get it?

1. **Elaborate on the physical and chemical properties of seawater.**
2. **What are plankton?**

CLOSER LOOK

HUMAN IMPACTS ON RIVERS AND STREAMS FROM DAMMING, CHANNELIZATION, AND FLOOD CONTROL MEASURES

Many rivers have been channelized, and riparian (streamside) vegetation has been removed to allow rapid boat travel, increased drainage, and agricultural and urban expansion. This has dramatic influences on the shoreline habitat. For example, a 25-km stretch (as the crow flies) of the Willamette River in western Oregon had 250 km of shoreline in 1854, but human activity decreased it to 64 km by 1967. There was a concurrent loss of at least 41% of the riparian wetlands during this time period. This removal of virtually all slow-moving portions and straightening of meanders is common in rivers in areas where humans live. Additional measures, such as creating projecting short structures that move flow from the edge to the middle of the channel (wing dikes), are common and further alter river discharge. Such alteration has been well documented on the Missouri River. In addition, the removal of large instream obstructions, such as logjams and stumps (often called snagging operations), to facilitate navigation is common and this destroys vital habitats for aquatic organisms. Many large river systems are stable substrate limited and much of the fish and invertebrate diversity and productivity is associated with accumulations of large woody debris. Even historic factors come into play in human alterations of rivers and streams; the more than 10,000 mill dams that were erected in the Northeast United States from the 1700s to early 1900s could have substantially altered stream geomorphology by burying wetlands in sediments and creating deeply incised stream channels throughout the region.

Flood control measures on many large rivers include levees to contain high discharge. When floods do occur, the levees constrain the water, making it move faster and deeper in the main channel instead of spreading out across the floodplain and flowing with a lower average velocity as it would naturally. If a flood does breach the levy suddenly, it causes considerable damage because of the rapid current velocity when the levee breaks. In addition, such levees constrain flows and act like dams to upstream regions that do not have levees. In the Mississippi River basin, mean annual flood damage has increased by 140% during the past 90 years. This increase is probably attributable to increases in the numbers of levees and the removal of riparian wetlands. Now, reestablishing floodplains and riparian wetlands on large rivers is beginning to be seen as a viable flood control strategy. Landowners often channelize smaller streams flowing across their property so they can develop or plant crops closer to the edge of the stream and drain their land more quickly. This channelization causes the water to move downstream at a higher velocity, increasing erosion in the channelized reach and upstream. Channelization also causes the stream to have a stronger flooding impact downstream, leading more landowners to channelize their stream banks. Removal of riparian vegetation creates an even worse situation because natural retention of sediment and slowing of floodwater does not occur, thus increasing the severity of floods.

Human alteration of stream and river hydrology has effects across all spatial scales. At the smallest scales, it removes habitat for aquatic organisms. At intermediate scales, alterations of stream hydrology are associated with increased erosion and more severe flooding. At the largest scale, human activities lead to global changes in the transport of materials by rivers. The economic impact of all these human influences is likely very great. Effective management requires an understanding of the dynamics of natural rivers.

Attempts at the restoration of hydrology and channel complexity have become common. Unfortunately, such restorations are rarely effective and often not based on appropriate design standards. Complete restoration of large rivers impacted by humans is unlikely, but partial rehabilitation may be possible.

A new trend of dam removal seems to be developing with the hope that some of the more harmful and less useful impoundments can be removed. At least 467 documented dam removals have occurred in the United States since 1912 for environmental, safety, economic, or other reasons. These removals have had some beneficial effects, but some of the degradation from dams to aquatic habitats is irreversible, and dam removal may even have some negative aspects such as allowing upstream movements of exotic species and the release of large quantities of sediments stored behind the dam. Removal effects on macro-invertebrates can even extend upstream. Analysis of a small dam removed from a Pennsylvania stream indicated that most effects of the removal downstream were gone once the sediments that were washed from the old reservoir had been flushed from the downstream channel, but some effects lingered a year afterward.

FOCUS ON CAREERS

ROLE OF AQUATIC ECOLOGISTS

Aquatic Ecology is a sub-discipline of ecology – the science of the environmental balance between forms of life, geography, geology, and soil type. Aquatic Ecologists examine freshwater areas such as marsh wetlands (salt flats and freshwater floodplains), lakes, and rivers.

What Does an Aquatic Ecologist Do?

Aquatic Ecologists research and work with areas of water and their associated dry land that depend on the working relationship between them. This will include the various forms of plant and animal life from the microscopic to the largest. These species work in an interdependent relationship. If one element has a sudden population increase or decrease, it will have a knock-on effect on the wider local ecology. They can further specialize in animal species or study the microbes that contribute to the environment. The job role differs from ecology in that they are concerned with aquatic (freshwater) ecologies rather than ecology in general. Their work can include studying the impact of invasive species or habitat loss, the result of pollution in water bodies and their impact on life forms. They can also have a good understanding of geology and soil analysis.

Where Does an Aquatic Ecologist Work?

As with Aquatic Biology, most with qualifications in this discipline will work for government organizations. Typically, they will work for State Parks or National Parks in conservation roles, for Federal organizations like the EPA or as researchers or policy-making and advisors. Many go on to become teachers in geography or emerging areas of environmental science in our High Schools; many more will get advanced degrees and teach at Colleges and Universities. Similarly, they may work as advisors or in public education in public roles such as zoos, wetland visitor centers or animal parks. Their expertise is vital in public engagement.

The greatest need is for environmental protection. Once considered areas ripe for drainage to turn over to agricultural use, today international law protects our vital wetlands. This country, like many others, must adhere to global laws on environmental regulation, protection and conservation as part of our global commitment to environmental issues.

Aquatic Ecology Jobs and Job Description

Aquatic ecologists study the interrelationships between flora, fauna, and the environment in lakes, streams, ponds, oceans, and other bodies of water. Given the vast amount of underwater environments found across the globe, most aquatic ecologists can expect to encounter the following tasks as part of their job:

- Stay abreast of research and current developments in the particular subfield of interest
- Provide knowledge and feedback on aquatic ecosystem interrelationships
- Collect field data on plant life, organisms and water quality in the field
- Examine human impact on the aquatic ecosystem in question
- Plan sustainable resource management practices
- Teach and implement sustainable practices for aquatic environments
- Study aquatic and terrestrial ecosystems, fish and wildlife habitat, population assessment
- Study how humans impact the aquatic ecosystem and vice versa
- Use biological modeling tools and software to create hypotheses and scenarios, model, and extrapolate results to real-world aquatic environments. Draw parallels between disparate environments where possible.
- Design fish and plant surveys

- Draw samples for organisms, water quality, substrate quality and composition
- Explain technical information in plain language to stakeholders
- Draft scientific and technical reports
- Implement aquatic harvest regulation systems and surveys
- Proficiently use software applications including statistical, spatial, population and habitat modeling, databases, neural networks/algorithms, and geographic information systems.

Senior aquatic ecologists are often promoted because of their past experiences, depth of knowledge, and management skills. In a senior role, aquatic ecologists will perform the following duties:

- Assess and offer feedback on scientific and technical reports
- Manage projects and delegate tasks to plan and lead scientific projects and subprojects
- Contribute to ecological assessments, ecological monitoring and survey programs, and environmental permitting
- Liaise with local, regional and national ecological and regulatory agencies
- Manage project budgets, benchmarks, procedures and reporting
- Write proposals for grant and fundraising purposes
- Manage existing and new projects such as surveys, aquatic components of environmental assessments, aquatic monitoring programs, data management, statistical analysis, and interpretation.
- Demonstrate leadership, coordinate staff across the operating unit.
- Prepare regulatory applications and approvals.
- Manage and communicate with clients, develop new projects and business leads.
- Interpret and apply relevant legislation, regulations, policies, procedures, and guidelines.
- Mentor staff.
- Support fieldwork and workgroups in a technical, interpersonal, and administrative capacity.
- Use technical theories, practices, and regulatory policies to resolve projects in line with goals and benchmarks.
- Make observations, collect samples, perform routine tests, and maintain documentation.
- Conduct 'Species at Risk' habitat assessments and species-specific surveys when required

What Is the Job Demand for Aquatic Ecology?

Employment demand is expected to be lower for the next ten years than most other sectors. The outlook for the period 2014–2024 is just 4%. It is a niche area with some low demand by state, but it is a transferable skill globally. Experts in this field may find more opportunities to work around the world than other comparable qualifications in ecology and environment.

Employment demand is expected to grow by 8% between 2020 and 2030.

What Are the Education Requirements to Become an Aquatic Ecologist?

There are no degrees in Aquatic Ecology in the US at present; this is a niche area requiring an approach from one of several different angles. Prospective Aquatic Ecologists should take a BS or BA in Ecology to get a strong enough foundation. Further, they should expect to enter post-graduate study to head towards a career in this direction. Environmental Biology, Microbiology, or Zoology are other options for approaching this career should the student wish to focus later on a key area.

As with other disciplines, a BA degree is suitable for practical work; students who wish to enter into fieldwork should study this type of degree. If desk-based research and lab testing are your preference, then a BS degree will be more suitable. A BS or BA will be suitable for the majority of jobs in this field, but should you seek long-term career development and project work, you should study for a master's degree. For academic lecturing and research, a Doctorate is vital.

What Kind of Societies and Professional Organizations Do Aquatic Ecologists Have?

Aquatic Ecology is a growing area related to both Ecology and Aquatic Biology. There are several prestigious groups related to this field.

- The National Stream and Aquatic Ecology Center (NSAEC) is part of the USDA Forest Service. Their aim is developing tools and methods for the promotion and conservation of these vital water-based ecologies.
- The British Ecological Society: Although dedicated to wider ecology, their understanding of the need for a dedicated aquatic ecology team has led to the setting up of this special interest group. Their aim is to promote research and bring together people inside and outside the discipline.

ROLE MODEL

G. EVELYN HUTCHINSON

George Evelyn Hutchinson was one of the top limnologists and ecologists of the 20th century, and likely the most influential of the century. His career spanned an era when ecology moved from a discipline that was mainly the province of natural historians to a modern experimental science. In large part, he and his students were responsible for these developments. Born in 1903 in Cambridge, England, Hutchinson was interested in aquatic entomology as a youth and authored his first publication at age 15. He obtained an MA from Emmanuel College of Cambridge University and worked in Naples, Italy, and South Africa before securing a position at Yale University. He remained at Yale until the end of his career and died in 1991.

Hutchinson's range of knowledge was immense, and his broad and innovative view of the world enriched his scientific endeavors. He was well-versed in literature, art, and the social sciences, and published works on religious art, psychoanalysis, and history, including some of the most widely read and cited ecological works of the century. His four volumes of the Treatise of Limnology are the most extensive treatment of limnological work ever published. His writings on diversity, complexity, and biogeochemistry inspired numerous investigations. Hutchinson organized a research team to study the Italian Lake Ianula in the 1960s; this multidisciplinary approach has since become a predominant mode of ecological research. It is reported that he was always able to find positive aspects of his students' ideas and encouraged them to develop creative thoughts into important scientific insights. As a consequence, many of Hutchinson's students and their students are among the most renowned ecologists today.

Hutchinson earned many major scientific awards in his career, including the National Medal of Science. He wrote popular scientific articles and books that were widely distributed. He was a staunch defender of intellectual activities and their importance in the modern world. Hutchinson's mastery of facts, skillful synthesis, a knack for asking interesting and important questions, evolutionary viewpoint, and cross-disciplinary approach make him an admirable role model for students of aquatic ecology.

CASE STUDY

RESTORATION OF AQUATIC ECOSYSTEMS: SCIENCE, TECHNOLOGY AND PUBLIC POLICY

Aquatic ecosystems perform numerous valuable environmental functions. They recycle nutrients, purify water, attenuate floods, augment and maintain streamflow, recharge groundwater, and provide habitat for wildlife and recreation for people. Rapid population increases in many parts of the United States—accompanied by intensified industrial, commercial, and residential development—have led to the pollution of surface waters by fertilizers, insecticides, motor oil, toxic landfill leachates, and feedlot waste. At the same time that water pollution and releases of nutrient-laden municipal sewage effluents have increased, water consumption has also increased, thus reducing the flows available for the dilution of wastes.

Increased sediment delivery resulting from urban construction, agriculture, and forestry has also resulted in greater turbidity and sedimentation in downstream channels, lakes, and reservoirs, with attendant losses of water storage and conveyance capacity, recreational and aesthetic values, and quantity and quality of habitat for fish and wildlife. Increased demands for drainage of wetlands have been accommodated by channelization, resulting in further loss of stream habitat. This has led to aquatic organisms becoming extinct or imperiled in increasing numbers and to the impairment of many beneficial water uses, including drinking, swimming, and fishing.

Although public and private decisions to manage aquatic ecosystems have enhanced water transportation, developed sources of hydroelectric power, reduced flood hazards, and provided water for municipal, industrial, and agricultural purposes, these activities have also altered the physical, chemical, and biological processes within aquatic ecosystems. This committee is convinced that U.S. public opinion strongly supports an increased level of attention to environmental protection. The nation's investment in different types of environmental programs has been considerable but piecemeal and has not always been effective. An accelerated effort toward environmental restoration and preservation is needed. The committee believes that a comprehensive and aggressive restoration component should be the centerpiece of such an effort.

The premise of this report is that ecological restoration of aquatic ecosystems is possible. Restoration means returning an ecosystem to a close approximation of its condition prior to disturbance. Accomplishing restoration means ensuring that ecosystem structure and function are recreated or repaired, and that natural dynamic ecosystem processes are operating effectively again. At times, however, restoration may be impractical or undesirable, as when a body of water that is naturally without fish is successfully transformed through stocking into a valuable trout fishery or when important urban developments have been situated on wetlands. In such cases, the committee recognizes that the economic value of these developments may preclude any attempt to restore preexisting natural systems at these locations. The committee also recognizes that preventive measures to protect aquatic ecosystems are important and that priority should be given to preventive measures that benefit more than one portion of the hydrologic cycle. Had environmental protection been adequate in the past, many expensive restoration projects would not be necessary today.

Naturally, restoration of aquatic ecosystems may be accomplished in stages, and particular ecosystem functions and characteristics— such as potable water— may be restored even when other ecosystem characteristics deviate from natural conditions. Thus, in certain situations, partial ecological restoration may be the operant management goal and may provide significant ecological benefits even though full restoration is not attained.

Therefore, since the loss and impairment of aquatic ecosystems are accompanied by the loss and damage of valuable environmental functions and amenities important to humans, and since restoration of aquatic ecosystems is possible, the committee concludes that a large-scale aquatic ecosystem restoration program in the United States should be implemented to regain and protect the physical, chemical, and biological integrity of surface water. Such a program should seek to:

- correct nonpoint source pollution problems.
- arrest the decline of wildlife populations; and
- restore all types of wildlife habitats with priority to endangered species habitats.

Failure to restore aquatic ecosystems promptly will result in sharply increased environmental costs later, in the extinction of species or ecosystem types, and in permanent ecological damage.

National Strategy

The committee recommends that a national aquatic ecosystem restoration strategy be developed for the United States. This comprehensive program should set specific national restoration goals for wetlands, rivers, streams, and lakes, and it should provide a national assessment process to monitor the achievement of those goals. The following recommendations are proposed as building blocks for the program and its guiding strategy. Details of the program design should be developed by federal and state agencies in collaboration with nongovernmental experts. A national strategy would include four elements:

1. National restoration goals and assessment strategies for each ecoregion (regions that have broad similarities of soil, relief, and dominant vegetation).
2. Principles for priority setting and decision making.
3. Policy and program redesign for federal and state agencies to emphasize restoration.

4. Innovation in financing and use of land and water markets.

Achieving these restoration goals will require planning, federal leadership, and federal funding, combined with financial resources and active involvement from all levels of government, as well as the involvement of nongovernmental organizations and businesses. Therefore, the federal government should initiate an interagency and intergovernmental process to develop the national aquatic ecosystem restoration strategy. The program should be developed and maintained under the firm leadership of a single responsible organization with the characteristics. Implementation of the program should include reliance on local and regional environmental restoration boards for program planning, synthesis, and leadership. Current appropriate federal programs should be reviewed to identify available opportunities for aquatic ecosystem restoration.

The Government

In light of existing budgetary constraints, innovative ways to finance restoration efforts are necessary. Therefore, it is imperative for the government to establish a National Aquatic Ecosystem Restoration Trust Fund. Private landowners and corporations should be provided with significant federal and state incentives to restore their aquatic ecosystems. Every effort should be made to utilize federal and other governmental funding to promote citizen participation in restoration. Citizen participation, whether through private citizen groups or public interest groups, has played a crucial role in initiating and sustaining restoration activities. Additionally, Congress should authorize states and local governments to exchange federal water development construction, maintenance, and major repair funds for financing aquatic ecosystem restoration programs.

The Food, Agriculture, Conservation, and Trade Act of 1990 empowered the U.S. Department of Agriculture (USDA) to engage in long-term contracts with farmers to remove former wetlands from agricultural production and facilitate their restoration as wetlands. However, the act restricts the number of acres eligible for the program to 200,000 per year, with a maximum of 1 million acres. Each acre of cropland withdrawn from production and restored as a wetland becomes ineligible for USDA program benefits. Therefore, the government should urge the USDA to explore opportunities for expanding the Agricultural Wetland Reserve Program in a manner that would lead to savings in USDA farm program expenditures; these saved funds could then be redirected to extend the wetland reserve program beyond 1 million acres.

Any redirection of federal policies and programs for aquatic ecosystem restoration should take into consideration the following:

- Use of a landscape perspective in restoration efforts.
- Use of adaptive planning and management (this refers to analysis of alternative strategies, reviewing new scientific data, and reanalyzing management decisions).
- Evaluating and ranking restoration alternatives based on an assessment of opportunity cost rather than on traditional benefit-cost analysis.

- Incorporating the definition of restoration as the return of an ecosystem to a close approximation of its condition prior to disturbance, in the mandates of all appropriate federal agencies.
- Reliance on nonfederal and federal units of government to coordinate restoration programs in local areas; and
- Initiating an interagency and intergovernmental process to develop a unified national strategy for aquatic ecosystem restoration.

Long-Term, Large-Scale, Coordinated Restoration-Planning, Evaluating, and Monitoring

Although restoration ecology applied to aquatic ecosystems is in a very early stage of development, the prospect for substantive improvements in damaged aquatic ecosystems is excellent. However, current federal and state environmental programs and policies are fragmented and do not adequately emphasize restoration based on the management of large, interconnected aquatic ecosystems. The diverse responsibilities of all layers of government affecting aquatic resources need to be better coordinated if large-scale restoration is to be accomplished efficiently and effectively. Because aquatic ecosystems are interconnected and interactive, effective restoration efforts should usually be conducted on a large enough scale to include all significant components of the watershed.

In addition, aquatic restoration efforts also need to be long-term to ensure that restoration project goals have been achieved and that restored ecosystems can endure stressful episodic natural events such as floods, droughts, storms, pestilence, freezing, heavy cyclical predation, invasion by exotics, and other perturbations. Because of limited resources, it is impossible in the short term to undertake all worthy aquatic ecosystem restoration projects. Criteria are thus needed to set priorities, select projects, and evaluate project designs. It is important to give priority to the repair of those systems that will be lost without intervention. A "triage" framework needs to be applied as a minimum initial step. In this approach, threatened systems would be divided into three categories: (1) those that will recover without intervention, (2) those that cannot be restored to a meaningful degree even with extensive intervention, and (3) those that can be significantly restored with appropriate action. Systems in the third group require further consideration. Selections from that group should be based on criteria such as the likelihood of success, opportunity cost, and technical review of the restoration plan. These criteria must be applied to the selection of projects because many restoration projects will not coincide with political boundaries.

Planning a restoration project must start with specifying the project's mission, goals, and objectives. Goals should be prioritized so that project designers and evaluators have a clear understanding of their relative importance. In addition to specifying goals, objectives, and performance indicators, project managers and designers need to propose a monitoring and assessment program that is appropriate in scale as well as in sampling frequency and intensity to measure the performance indicators accurately and reliably, and thereby assess progress toward the project's objectives, goals, and mission. Post-project evaluation will enable scientists to determine when and to what degree the system has become self-maintaining and whether or not the restoration attempt was effective.

Monitoring of a restoration effort should include both structural (state) and functional (process) attributes and should not be restricted to one level of biological organization. Monitoring of attributes at the population, community, ecosystem, and landscape levels is appropriate in a restoration effort.

Lakes

By far, the most widespread problem facing lakes and reservoirs is agricultural nonpoint runoff of silt and associated nutrients and pesticides. Lakes often do not cleanse or restore themselves. They are sinks for incoming contaminants that recycle and maintain the impaired conditions. Federal drinking water standards, for example, cannot be met except with great difficulty and expense unless degraded lakes and reservoirs are improved and then protected from further contamination.

A net gain over the next 20 years of 2 million acres of restored lakes, out of the current 4.3 million acres of degraded lakes, is an achievable goal. By the year 2000, it is recommended that a minimum of 1 million acres of lakes be restored. The costs for research, development, and technical guidance are federal responsibilities. The costs for actual restorations should be borne by federal and nonfederal sources, working through individual state lake programs. The committee realizes that the goals for the restoration of lakes should be realistic and tailored to individual regions of the country. Further development of project selection, goal setting, and evaluation techniques based on the concept of "ecoregions" should be encouraged and supported by the U.S. Environmental Protection Agency (EPA).

All states have degraded lakes, and each state should develop restoration plans and programs. States should consider establishing trust funds for environmental restoration and protection. The Clean Lakes Program (CLP) administered by the EPA has been the most reliable source of grant support for lake restoration efforts. This program should receive stable administrative support and increased funding from Congress. The 1991 appropriation for the CLP was $8 million. Although this amount will help to maintain or initiate a few lake restoration programs, it is inadequate for the large task of lake restoration facing the country. The program's mandate should be broadened to include all aspects of lake ecosystems, including habitat restoration, elimination of undesirable species, and restoration of native species.

Knowledge of the current ecological condition of the nation's lakes is grossly inadequate, and a national assessment of lakes is necessary to determine the severity and extent of damage and to measure changes in their status. The CLP should increase support for research and development of effective tools for restoration and should continue guiding states in developing lake restoration programs.

The federal government should support research and development for demonstration watershed-scale restorations that integrate lake, stream, and wetland components. Research could be coordinated under an interagency program, such as the Federal Coordinating Council for Science, Engineering, and Technology, to coordinate the selection, planning, and evaluation of demonstration projects. Although many techniques are available to restore lakes, further development is required to improve their efficiency and effectiveness. The research and development programs in lake restoration should take an experimental approach, emphasizing controlled manipulation of whole-lake ecosystems or large in-lake enclosures.

Research and development programs in applied limnology are needed to study

- improved techniques for littoral zone and aquatic macrophyte management.
- biomanipulation (food web management).
- contaminant cleanup in lakes, especially for mercury and polychlorinated biphenyls (PCBs).
- the relationships between loadings of stress-causing substances and responses of lakes.
- paleolimnological approaches to restoration; and
- prediction of lake trophic state from nutrient loading relationships.

Rivers and Streams

Given that healthy, vegetated riparian habitat and bottomlands are essential to the natural ecological functioning of associated streams and rivers, and are among the nation's rarest habitats due to prior devastation, riparian habitat and bottomland restoration should be made a high national priority along with the restoration of the stream or river channel itself. Because a river and its floodplain are intimately linked, they should be managed and restored as integral parts of an ecosystem. Remnant and undisturbed large river and floodplain ecosystems are rare and ecologically valuable. Therefore, reaches of certain large rivers and their floodplain ecosystems, such as portions of the Atchafalaya River and the Upper Mississippi River Fish and Wildlife Refuge, and at least 50 other large rivers (greater than approximately 120 miles in length) should be designated as "reference reaches" for use as restoration templates and should be protected as quickly as possible. Reference reaches should be designated and protected on representatives of all orders of streams and rivers in each of the nation's ecoregions. Highest priority should be given to protecting representative orders of rivers and streams not already protected as national wild and scenic rivers, or by being located in national or state parks.

Stream and river restoration should begin with improved land management practices that will allow natural restoration of the stream or river to occur. Therefore, the committee recommends the following:

- Erosion control programs in watersheds should be accelerated, not just to conserve soil, but also to restore streams and rivers.
- Grazing practices on federal lands should be reviewed and then changed to minimize damage to river-riparian ecosystems and to restore damaged rivers and streams.
- Erosion control by "soft engineering" approaches, such as bio-engineering techniques for bank stabilization and repair, should be considered first, in preference to "hard engineering" approaches, such as dams, levees, channelization, and riprap.
- Dikes or levees no longer needed or cost-effective should be razed to reestablish hydrological connections between riparian and floodplain habitats and associated rivers and streams.
- Classification systems for land use and wetlands should explicitly designate riparian environments and floodplains that retain their periodic connections to rivers.

The committee could not find a recent national assessment of the number of stream and river miles affected by channelization or leveeing, but the total is probably much greater than the number of miles of river dammed. Although water resources agencies track their own development projects, the only nationwide inventory of rivers and streams was conducted in the 1970s in response to the passage of the Wild and Scenic Rivers Act of 1968.

Therefore, the committee believes there is a need for a comprehensive up-to-date nationwide assessment of rivers, comparable to the National Wetland Inventory. It would be very useful to know how many miles of free-flowing, unchannelized rivers remain in the United States, and where these reaches are located.

The Food, Agriculture, Conservation, and Trade Act of 1990 and Section 404 of the Clean Water Act of 1977 now encourage the restoration and protection of wetlands. These laws should be expanded to provide for the protection and restoration of large active floodplains and riparian zones that are key components of riverine ecosystems. In addition, the Conservation Reserve Program, the Environmental Easement Program, and short-term agricultural set-aside programs should be amended to ensure that riparian zones and floodplains of all kinds are eligible for inclusion along with wetlands.

Opportunities to allocate water to in-stream uses arise (1) when land with water rights is sold or transferred, (2) when municipalities and irrigators decrease water withdrawals through conservation, and (3) when operating permits for dams are scheduled for renewal. Although the prior appropriations system (the basis of water law in the West) initially did not permit in-stream flow rights, many Western states now recognize in-stream flow water rights. Therefore, states that have not established a water right for in-stream uses should do so. Flow that becomes available as a result of water conservation or lapse of permits should not automatically be reassigned to a consumptive use or withdrawal. Instead, consideration should be given to assigning the flow to in-stream uses. In addition, operating plans for dams should consider the annual water regime required by riverine fish and wildlife.

Federal agencies should be requested to update channelization estimates and to estimate miles of bank stabilization work already performed. The agencies should provide average and mean costs per mile for construction and maintenance of these conventional river management strategies so that unit costs are available for comparison of different strategies. Government agencies should also conduct post-project evaluations of fluvial modifications, enhancement, improvement, channelization, and restoration projects to determine whether these projects achieve the benefits (e.g., flood protection, fish and wildlife enhancement) for which they were designed at costs that were projected.

The committee also recommends that a national river and stream restoration target of 400,000 miles of river-riparian ecosystems be restored within the next 20 years. This target represents only about 12% of the total 3.2 million miles of U.S. rivers and streams and is recommended because it is comparable to the miles of streams and rivers affected by point source and urban runoff (EPA, 1990).

Education and Training

To accomplish the preceding tasks, the nation will require resource management professionals with multidisciplinary training. Restoration of aquatic ecosystems requires an integrated, broad-based approach; those trained to help restore these systems must have an interdisciplinary education. Although specialization will still be necessary, professionals will need the ability to coordinate work that draws on aquatic biology and fisheries, chemistry, hydrology, ecology, fluvial geomorphology, hydraulic engineering, social sciences, and wildlife management.

Some well-intentioned restoration projects have failed because fluvial and biological processes were not adequately considered in their design and implementation. The public has become increasingly aware of the need for restoration of river-riparian ecosystems, and numerous public and private agencies and citizen groups are likely to initiate further stream and river restoration projects. These organizations, if properly guided and supported, can be a valuable impetus for effective aquatic ecosystem restoration and, in some cases, a valuable source of volunteer labor to accomplish restoration.

A new emphasis on resource stewardship and restoration cannot succeed without public understanding and support. Thus, educational programs aimed at raising the level of public knowledge and comprehension of aquatic ecosystem restoration rationales, goals, and methods should receive adequate government funding.

The committee believes that hydrological advisory services should be operated by states or federal agencies to provide technical assistance to groups interested in stream and river restoration. Universities with experts in natural resources or hydrology and water resources institutes, based at universities in every state, also should contribute technical assistance required for the restoration of aquatic ecosystems through free or at-cost expert hydrological and biological advisory services.

SUMMARY

- An aquatic ecosystem's ability to function is influenced, either directly or indirectly, by the quality of the sediment. This quality is determined by various physicochemical characteristics.

A specific subfield within the broader study of organisms and their environments is freshwater ecology. Ecology, in contrast to biology, is the study of organisms and how they interact with and are influenced by their natural surroundings, or ecosystem.

- Two categories exist for freshwater river ecology: lowland and upland. Lowland habitats are in flat areas with warm, slow rivers colored by organic matter and sediment. Upland habitats are in mountainous regions with cold, clear, rocky, fast rivers.
- Upland rivers and streams are swift-moving waterways in freshwater ecosystems that drain hilly or elevated terrain, frequently onto wide alluvial plains (where they become lowland rivers).
- Lowland rivers are home to a variety of fish species that are better suited to withstand extremes of temperature and low oxygen levels due to their generally more turbid, warm, slow-moving waters, and fine sediment beds.
- Hydrophytes are plants that live entirely or partially submerged in water. These hydrophytic plants exhibit a form of symbiosis where they provide a means of survival for nearby organisms, such as algae.
- Plants require various nutrients from nutrient-rich soil and nearby waters, along with maintaining proper water levels in their cells. Macro elements like magnesium (Mg), nitrogen (N), phosphorus (P), and potassium (K) are essential, along with carbon, hydrogen, and oxygen for photosynthesis.
- Inland water ecosystems consist of organisms in continental bodies of free water, playing a crucial role in the biosphere with diverse biological life and complex biogeochemical processes.

MULTIPLE CHOICE QUESTIONS

1. How many types of aquatic ecosystems are there?

a. One

b. Two

c. Three

d. Four

2. Where do plants and animals live in aquatic ecosystems?

a. Water

b. Land

c. Air

d. Fire

3. Where can we find both running water as well as stagnant water?

a. Marine ecosystems

b. Wetlands

c. Coral reefs

d. Freshwater ecosystems

4. In which of the following we can see fluctuation in the water level dramatically in different seasons?

a. Coral reefs

b. Brackish water

c. Wetlands

d. Deep oceans

5. Which is the simplest aquatic ecosystem?

a. Pond

b. Stream

c. Lake

d. Marine

Answer to Multiple Choice Questions

1. (b) 2. (a) 3. (d) 4. (c) 5. (a)

REVIEW QUESTIONS

1. Explain the classification of habitats.
2. How do abiotic factors affect the water conditions?
3. Discuss still water animals.
4. What are the features of freshwater lentic communities and animals?
5. Give an overview of lotic communities.
6. Elaborate the physical and chemical properties of water.
7. Demonstrate the origins of marine life.

REFERENCES

1. Bakun, A., (2001). "School-mix feedback": A different way to think about low frequency variability in large mobile fish populations. In: McKinnell, S. M., Brodeur, R. D., Hanawa, K., Hollowed, A. B., Polovina, J. J., & Zhang, C. I., (eds.). Pacific climate variability and marine ecosystem impacts from the tropics to the arctic. *Progress in Oceanography*. In press.
2. Bianchi, G., Gislason, H., Graham, K., Hill, L., Jin, X., Koranteng, K., Manickchand-Heileman, S., et al., (2000). *Impact of Fishing on Size Composition and Diversity of Demersal Fish Communities. ICES Journal of Marine Science*, Volume 57, Issue 3, June 2000, Pages 558–571, https://doi.org/10.1006/jmsc.2000.0727
3. Christensen, V., (2000). *Indicators for Marine Ecosystems Affected by Fisheries. Marine and Freshwater Research 51(5) 447 - 450*
4. Claessen, D., De Roos, A. M., & Persson, L., (2000). *Dwarfs and Giants: Cannibalism and Competition in Size-Structured Populations. The American Naturalist 155(2):219-237 155(2):219-237 DOI:10.1086/303315*
5. Cury, P., & Cayré, P., (2001). *Hunting Became a Secondary Activity 2000 Years Ago; Marine Fishing Did the Same in 2021. Fish and Fisheries 2(2):162 – 169 2(2):162 – 169 DOI:10.1046/j.1467-2960.2001.00044.x*
6. Ritter, D. F., Kochel, R. C., & Miller, J. R., (2002). *Process Geomorphology*. McGraw-Hill, Boston.
7. Shin, Y. J., & Cury, P., (2001). *Exploring Fish Community Dynamics Through Size-Dependent Trophic Interactions Using a Spatialized Individual-Based Model. Aquatic Living Resources 14(2):65-80 14(2):65-80*
8. USGS, (2013). *Water Quality Tests of Big Thompson River in Moraine Park*. Retrieved from: http://co.water.usgs.gov/ (accessed on 04 April 2024).

CHAPTER

4

Forest Ecology

LEARNING OBJECTIVES

After studying this chapter, you will be able to:

- Discuss forest ecosystems
- Clarify the water cycle in the forest ecosystem

INTRODUCTORY EXAMPLE

There is still a plethora of life in boreal forests that have evolved to survive extremely cold temperatures all year round, such as caribou reindeer and other creatures that can travel great distances during the winter. Boreal forests, which are abundant in conifers and deciduous trees, swath large portions of Russia, Alaska, and Canada. Furthermore, boreal forests are significant as carbon sinks. Like all forests, their ability to absorb carbon dioxide, a major cause of climate change and global warming, removes it from the atmosphere and contributes to the planet's overall health.

Despite the fact that these forests are all unique, they are all unfortunately connected by a common threat. Due to deforestation and degradation, we lose 18.7 million acres of forests annually, mostly in the tropics. It's the same as 27 soccer fields in a minute. The reasons may vary—some are illegally harvested for timber, while others are cleared for ranching and agriculture—but the devastation and degradation of the forests are the same.

INTRODUCTION

The study of how organisms interact with one another and the surroundings of a forest ecosystem is known as forest ecology. For people who care about the preservation of forests, this particular area of study is crucial. A sustainable plan cannot be implemented in the absence of reliable data about the health of the forest. An expert in this field may be employed by a park service, university, or even a private company. The goal of forest ecology is to provide answers to various questions. Research topics could include the value of old growth and new growth forests, the degree of threat posed by invasive plant and animal species, and the effects of human activity on the surrounding ecosystem. Depending on the size of the forest, the ecologist may occasionally need to take data from multiple zones or biomes into account. Even though these disparate regions might not seem connected, they could all be indicators of a forest's general health. It takes a lot of time to record the various species that are present in the forest ecosystem. These data points will frequently be contrasted with information gathered in prior years. This comparison ought to give you a decent idea of the long-term patterns in the field. A decrease in the number of species or an increase in the number of species could be especially concerning.

Diversification of species may be beneficial to some, but as anyone working in forest ecology knows, it has to be the right kind of diversification. In a matter of decades, invasive plant and animal species have the potential to completely alter almost every species present in the forest ecosystem by decimating native populations. Some people blame humans for the introduction of certain species, while others may view this as the normal course of nature. This may result in an extremely strange outcome.

The next step is to ascertain whether the changes are the consequence of unnatural processes or a natural maturation in the forest ecology after the species have been cataloged and trends have been examined. This can be inferred in part from the amount of non-native species and knowledge of the changes that occur in forests over time. For example, as canopies get denser, there will be less light available for ferns and other ground cover plants to grow. This is anticipated.

The ecologist might be asked to devise a strategy to bring the forest's health back if the changes are deemed unnatural or unsustainable. There might not be a solution in some situations because the species and processes involved are too established to be changed. In other situations, methods for reestablishing a more organic forest ecology might exist.

4.1. FOREST ECOSYSTEMS

Learning Objectives

- Describe the types of forest ecosystem
- Illustrate the forest ecosystem structure
- Describe the soil and its types
- Give an overview of the microclimate

Trees are large, usually long-lived perennial plants with large woody stems and extensive root systems that dominate biological and ecological conditions and processes in local or regional segments of landscapes known as forests. Due to their size and longevity, trees can outcompete other plant species by depriving them of resources, such as light and soil. As a result, trees play a crucial role in regulating key ecological processes, determining habitats for animals, microbes, and other plant species, and influencing the abundance of these organisms in the forest. Within forest ecosystems, trees also influence the hydrological cycle, soil development processes, microclimate, and biological characteristics of streams. Additionally, large plants with woody stems that do not fit the strict definition of trees, such as bamboo or tree ferns, can also exert dominance in forests.

A structure can be found in both local ecosystems and regional landscapes, where living and dead plants, animals, and microbes are arranged in both vertical and horizontal patterns. In terrestrial ecosystems, the physical environment is made up of soil, geological substrates, and atmosphere. Sediments or geological substrates may or may not have an impact on the ecological processes in aquatic ecosystems, where water serves as the medium. In order for the system of living and dead organisms to function and for life as we know it to be possible, physical (light) or chemical energy must combine with chemical elements (nutrients) from the soil, geological substrate, water, or atmosphere. This process results in the complex, high-energy organic molecules that make up living things. In forests, photosynthesis provides nearly all of the energy needed to create these molecules. The proper mixes of various nutrient elements circulate throughout the ecosystem (nutrient cycling), which in turn is greatly influenced by temperature and moisture, and thus by climate, as well as by the activity of soil animals and microbes. This process provides the chemical building blocks for their synthesis.

A complex assemblage of various plant life forms, including lichens and bryophytes (e.g., mosses), different kinds of shrubs and herbs, trees, and climbers. A variety of microorganisms and animals that utilize the various living plants for sustenance and energy are connected to them. A distinct group of animals and microbes uses the energy from dead plant and animal matter to break down organic matter and release its nutrients so that new generations of living plants and microbes can absorb them. Ecological systems are complex because of the interactions that occur between living things and between living things and dead organic matter and the physical environment.

Keyword

Vegetation is a general term for the plant life of a region and the ground cover it provides. It can also refer to the growth process of a plant.

Interactions among the elements of an ecosystem. The relationship between the structural elements of an ecosystem is crucial, as it is for all systems. For instance, plants influence soil development, but soils also affect plants. Climate influences the development of soil and vegetation, but plants can also create microclimates in their immediate environments. Moreover, vegetation can occasionally affect the characteristics of regional climates, and globally, vegetation significantly affects the global climate. Plants are impacted by animals and microbes, but the animals and microbes that have an impact on them are mostly determined by the plants.

- *Modify over time.* In the human race, "taxes and death are the only certainties.". Change is the only constant in ecosystems. Ecosystems experience "renewal" when human or non-human processes or events disturb them, much as individual animals are born, grow, mature, and eventually die. When an ecosystem "matures," it is eventually disturbed again, and the cycle of renewal begins again. More often than not, this is a necessary condition for the long-term stability of ecosystems as well as the preservation of historical ranges of variation in their structure, function, complexity (including biological diversity), and component interactions. This change over time is as fundamental to ecosystems as any other characteristic.

A forest ecosystem is a section of the landscape, ranging in size from a few hectares or less for a local stand to an entire continent, where trees predominate in terms of structure, function, complexity, interactions, and patterns of change over time. The study of these landscape units dominated by trees is called forest ecology. While landscape-level forest ecosystems typically include streams, rivers, lakes, and areas of non-forested terrestrial ecosystems, stand-level forest ecosystems are terrestrial ecosystems. However, these other types of ecosystems' overall characteristics are greatly impacted by their placement within a landscape that is dominated by trees. Forest ecology is not a science that can be used to make moral judgments. No stage of the development of a forest ecosystem, no structure, no function, no complexity, and no pattern of interactions between ecosystem components can be considered "better" or "worse" than any other. There is no difference between an old and young forest. High productivity forests are neither superior nor inferior to low productivity ones. There is no difference between a forest with high species or structural diversity and one with low levels of these metrics.

The examination of forest ecology will cover why forest ecosystems vary and change over time. Following this, we will discuss the ecological principles supporting sustainability and stewardship, which are societal goals influenced by human preferences but reliant on an ecological framework for success.

4.1.1. Types of Forest Ecosystems

According to Elizabeth Mygatt, "World's Forests Continue to Shrink," all types of forests cover about thirty percent of the earth's surface. Mygatt continues, "The forest ecosystem is essential to preserving a healthy planet because it regulates the water cycle, stabilizes soils, helps to level the climate by absorbing and storing carbon dioxide, gives animals a home, and produces wood, food, and medicine. There are many distinct types of forests in the world.

4.1.1.1. Temperate Forest Ecosystem

The ecosystem of temperate forests is crucial to life as we know it. In areas where summer and winter bring significant temperature variations, temperate forests are found. Tropical rainforests are found in areas where the weather is consistent throughout the year. Deciduous and evergreen trees predominate in temperate forests almost universally. Trees that lose their leaves in the winter are called deciduous trees.

Evergreens trees that retain their leaves throughout the year—include pine trees. Forests can be one or the other, or they can be both at once. A temperate rainforest is one of the four types of forests. These can be found in the US states of California, Oregon, and Washington.

The tallest trees in the world, sequoias and redwoods, make up these forests. If there is a forest in a given area depends on how much rain falls there. Generally speaking, a forest will grow if there is enough rainfall to sustain the trees. If not, the area will turn into grasslands.

4.1.1.2. The Tropical Rainforest Ecosystem

Earth's tropical rainforests are among its most significant regions. Thousands of species of animals and plants call these unique ecosystems home. In contrast to what the general public believes, rainforests are made up of tall trees that shade the sun from below in addition to densely packed plants. Smaller plants cannot grow above this ceiling. Interesting plants can be found in any area where sunlight can reach the surface.

Rainforests get their name from receiving an average of 80 inches of rain annually. Can you guess where this comes from? Throughout the year, there is little variation in the temperature. It's hot and humid all year round. Brazil is in South America, and it is home to the well-known Amazon rainforest. The Neotropics are the name given to this specific forest. There are other sizable blocks in West and Central Africa.

4.1.1.3. Insects of the Tropical Rainforest

The world's most infamous and feared spider lives in the jungle. One of the most unsettling creatures you will ever see is a tarantula. To protect themselves and to kill prey, the majority of tarantula species have poisonous fangs. Certain substances pose a threat to human life, while others are innocuous. In the rainforest, army ants are merely one type of ant. The reason they march through the jungle in a long, thick line is why they are called army ants. They don't quit until the young larvae are in the pupil stage. The ants resume their march after the queen lays her eggs.

The forest is full of lovely butterflies, but these insects weren't always so pretty. Initially, caterpillars—which can be a little unattractive—become butterflies. They undergo metamorphosis, or the transformation into a butterfly. Not so fortunate are centipedes. Instead of changing into butterflies, they wander the forest in search of food. Some centipedes kill their prey with poison.

4.1.1.4. Tropical Rainforest Birds

The world's most beautiful birds can be found in the rainforest. The forest tops come alive with a plethora of colors that dart between the trees. Because of their attractiveness, many species of tropical birds are kept as pets.

In the rainforest, there are hundreds of species of parrots. Among them is the scarlet macaw. Not only is it the longest, but its entire length, from head to tail tip, measures three feet. Upon consuming a toxic fruit, these macaws consume a unique variety of clay that counteracts the toxicity.

Toucans are fascinating birds as well. They can reach fruit they can't reach with their large beaks. According to scientists, the rainforest is home to 33 different species of toucans. Not all tropical birds have good looks. The hoatzin, without its beautiful tail, resembles a peacock more.

Hoatzins are not good flyers; they frequently crash land. The brown kiwi is a flightless bird with a long feathered beak and a look more akin to a rodent. Kiwis don't live in trees; they live on the ground. They have unique claws that they use for defense, digging, and running.

4.1.1.5. Tropical Rainforest Mammals

There are other animals besides birds that soar through the rainforests. The jungle is home to several species of flying mammals. The tropical rainforests are teeming with surprises, from the harmless fruit bat to the unusual flying squirrel.

Its wings have a maximum spreading width of five feet. These bats do not reside in caves, in contrast to bats found in other parts of the world. During the day, they prefer to hang in trees. One tree may contain hundreds or even thousands of bats.

The Amazon jungle in South America is home to vampire bats. This is most likely where the well-known legends of bloodsucking bats originated. Indeed, the blood of their victims is consumed by these bats. They have consumed human blood in addition to their usual prey, farm animals. However, vampire bats consume very little liquid.

4.1.1.6. Tropical Rainforest Reptiles

Reptiles abound in the world's tropical rainforests. Since reptiles are cold-blooded, their body temperature is influenced by their surroundings. It is crucial that they continue to live in warm climates. Reptiles like snakes are common in rainforests. Of all the families, the mamba family is the most toxic. With their razor-sharp fangs, they inject venom into their victim to kill it.

Anacondas make up another snake family. They are some of the longest creatures in the world, reaching up to 30 feet in length. Anacondas wrap themselves around their prey and constrict. They swallow their prey whole and sleep while the food is digesting. Chameleons are interesting lizards that can change color.

This enables them to blend in with their surroundings. Not only is this a great disguise from predators, but it is also an easy way to sneak up on their prey. Chameleons only eat insects. Geckos are also fascinating creatures. The flying gecko can glide from tree to tree to escape from predators and their grip on surfaces is incredibly strong.

4.1.1.7. Tropical Rainforest Primates

Primates include all monkeys and their relatives, including humans. Tropical rainforests around the world are home to a diverse range of monkey species. There are two categories of monkeys: New World and Old World. South and Central America are home to New World monkeys. For example, spider monkeys reside in the rainforests of the Andes Mountains.

Due to their diet mainly consisting of fruit and nuts, spider monkeys are categorized as frugivores. Howler monkeys also inhabit these areas. These primates derive their name from a distinctive sac that amplifies their vocalizations.

Old World monkeys are found exclusively in Africa and Asia. One such species is the colobus monkey. As they primarily feed on leaves, these monkeys are referred to as folivores. While some primates live in larger groups of up to 200 individuals, they reside in smaller groups of 15 or fewer. Gorillas, orangutans, bonobos, humans, and chimpanzees are collectively known as great apes . Gorillas are able to climb trees, despite spending the majority of their time on the ground.

4.1.1.8. Boreal or Taiga Forests

Boreal forests, a vast expanse in the Northern Hemisphere, consist of a continuous belt of deciduous and coniferous trees. In North America, Alaska and much of northern Canada are enveloped by this forest. This expansive forest ecosystem is globally renowned.

This forest ecosystem is the largest land-based ecosystem in North America, making up about 35% of Canada's total land area. It is also home to a sizable portion of Canada's biodiversity and is well known for its role as a global carbon sink.

The boreal forest is crucial because it serves as North America's "great lung," "breathing in" carbon dioxide and "exhaling" oxygen into the atmosphere, despite its relatively unknown status. To put it succinctly, the boreal forest accomplishes the same tasks as the Amazon rainforest, but with a much smaller variety of plants and animals.

The largest and smallest mammal species on the continent of North America, the wood bison and pygmy shrews, are found in this forest ecosystem. Great lakes, northern rivers, enormous bogs, fens, and other organic wetlands are just a few of the many features of the Boreal forest. The Boreal region boasts an abundance of diverse wildlife, such as woodland caribou and lynx, whooping cranes and wood bison, northern owls, woodpeckers with three instead of four toes, and vibrant wood warblers. More than 5,000 species of strikingly colorful fungi can be found in the Boreal, which are far more popular in Scandinavia and Siberia than they are in North America. Among the Boreal forest communities, the most valuable and biologically diverse are the old-growth forests, which are vital to numerous Boreal species.

Did you know?

Taiga covers 17 million square kilometers (6.6 million square miles) or 11.5% of the Earth's land area, second only to deserts and xeric shrublands. The largest areas are located in Russia and Canada. In Sweden, taiga is associated with the Norrland terrain.

4.1.2. Forest Ecosystem Structure

An ecosystem (or biotic community) with a preponderance of trees is called a forest. Ecology is the study of living things' interactions with their surroundings and ecological systems. Understanding the factors that influence the distribution patterns and abundance of various organisms in global forests is the main goal of forest ecology. The observations that forests dominate the natural landscape over a significant portion of the world and that forests are home to a large proportion of the world's species make forest ecology relevant to understanding biodiversity. Furthermore, many forest ecosystems are heavily altered and modified by human societies due to the value that humans place on forests and the products and services they offer, particularly wood, which frequently disturbs the natural ecological patterns. Additionally, a lot of forests grow in areas with soil and climates that support intensive agricultural production, and turning forests into farms drastically alters the biota's composition. Therefore, in order to forecast and mitigate the effects of forest utilization and conversion on *biodiversity*, a thorough understanding of forest ecology is required.

Biodiversity is the variety of life on Earth. It is a measure of variation at the genetic, species, and ecosystem levels. Many ecosystems obtain materials and energy from the sun, transfer these resources among ecosystem components, and then release the materials and energy back into the surrounding environment. Plants, animals, and decomposers are the fundamental biological elements that make up ecosystems. To add carbon to the ecosystem, plants absorb solar radiation. Certain ecosystems, like deep-sea hydrothermal vents, lack plants and are dominated by bacteria that oxidize hydrogen sulfide (H_2S) to produce organic matter. Dead organic matter is broken down by decomposer microorganisms, or microbes, which release CO_2 into the atmosphere and nutrients that other microbes and plants can use. Large collections of decomposing organic matter would sequester the nutrients needed to support plant growth if there was no decomposition. Animals are essential parts of ecosystems because they move materials and energy, and they have a significant impact on the number and activity of soil microbes and plants. An ecosystem's primary abiotic elements are water, the atmosphere (which provides carbon and nitrogen), and soil minerals (which provide additional nutrients needed by organisms).

The main ecosystem fluxes and pools, as well as the variables controlling these fluxes, are depicted in an ecosystem model. Differentiating between nutrients, water, and energy is the relative importance of ecosystem inputs and outputs in comparison to each other. In inside-out recycling, for instance, plants mostly take up carbon from the atmosphere, and the majority of the carbon they breathe out is taken up again by the atmosphere. Because of this, the cycling of carbon through ecosystems is highly open, with significant inputs into and losses from the system. However, because ecosystems store relatively large amounts of carbon, the activities of microbes and animals are somewhat protected from changes in plant uptake of carbon. Ecosystems have a relatively open water cycle, with precipitation being the main source of entry and evaporation, transpiration, and drainage to streams and groundwater the main ways that water leaves. Because most ecosystems cannot store as much water in their plants and soil as they can store carbon, organism activity is directly correlated with water inputs. Mineral elements like nitrogen and phosphorus are recycled relatively tightly within ecosystems, with small annual inputs and losses, in contrast to carbon and water.

Ecosystems vary greatly in their pool sizes and cycling rates. Compared to tundra or deserts, tropical forests have significantly larger plant pools of nutrients and carbon. On the other hand, large pools of soil carbon, not plant carbon, are found in peat bogs (Figure 4.1).

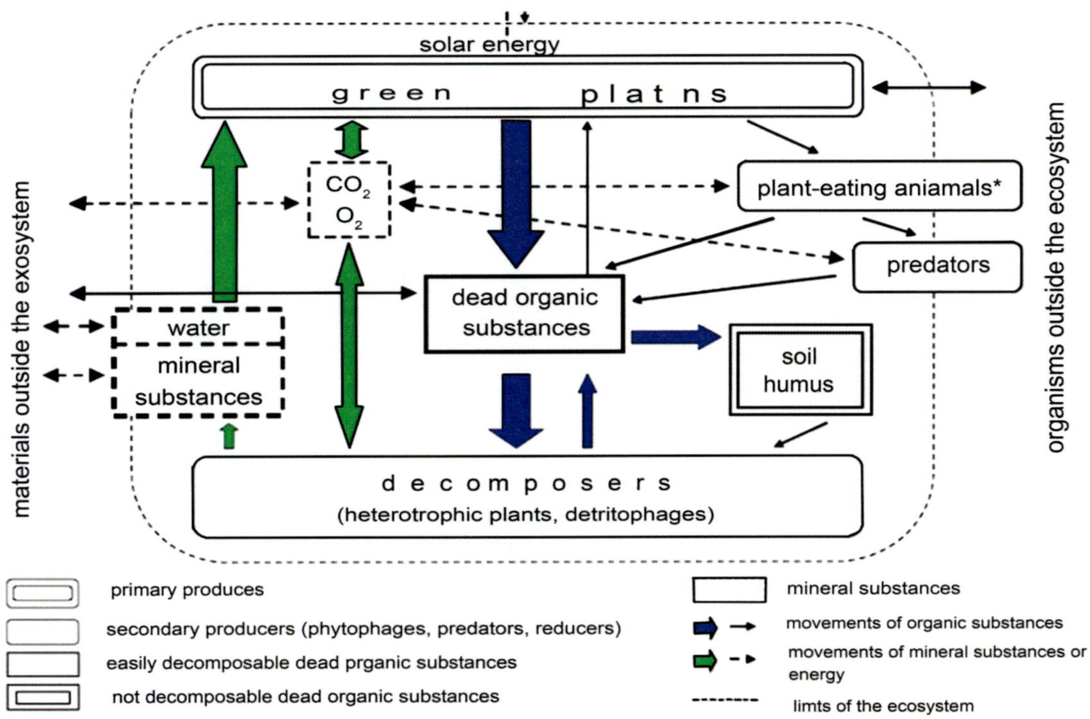

Figure 4.1. Model of ecosystem structure and functioning.

Source: Forest Ecology Textbook Ji&#xFFFD;í Kulhavý.

- The soil or other organic or geological substrate that serves as the roots for trees and other plants.
- An atmosphere, a regional climate that is affected locally by aspect and slope, a microclimate brought about by the trees' increased humidity, decreased wind speed, and shade.

- Organisms, among them plants. The arrangement of living forest plants of various life forms, both horizontal and vertical, is largely reflected in the structure of the forests. Living forms include lichens, mosses, liverworts, and grasses; forbs, or non-grass herbs; shrubs and climbers; and trees of all kinds, both deciduous and evergreen, coniferous (gymnosperm) and broadleaved (angiosperm), short- and long-lived, short and tall, nutrient-demanding or tolerant of inadequate nutrition, etc. Apart from the flora, there are dead trees in different stages of decay and collapse, dead tree stems on the ground, and an organic layer on the soil surface, or the forest floor, which is composed of different combinations of dead fine roots, ground litter fall (leaves, branches, reproductive structures), decaying wood, dead fungi, bacteria, and animals, the feces of decay-resistant soil animals, and various other highly decomposed organic matter. These living and dead plant components of ecosystem structure are linked to a diverse range of animals and microbes.

Keyword

Organic matter refers to the large source of carbon-based compounds found within natural and engineered, terrestrial, and aquatic environments.

4.1.3. Soil

The loose mineral material at the Earth's surface known as soil has been shaped over time by physical, chemical, and biological processes to the point where it is now able to support ecosystem functions and plant life. Certain tree species, like certain pines, have adaptations that allow them to root and grow slowly on fractured rock as long as there is enough moisture. As a result, some types of forests can grow even in areas without soil. Some, like alders, can fix nitrogen from the atmosphere, which allows them to grow on physically and chemically unweathered but fragmented rock that is devoid of organic matter and nitrogen. They get the other mineral nutrients they require by generating acidity, which breaks down the unweathered rock fragments and releases the nutrients they contain (e.g., iron, potassium, calcium, magnesium, sulfur, phosphorus, and other "micronutrients"). If moisture is not a limiting factor, many trees can grow in an accumulated layer of organic matter covering the rock; acids released from the organic matter's decomposition break down the rock, releasing mineral nutrients. On the other hand, the majority of trees and other plant species thrive in rich, well-developed soil that contains a combination of minerals and *organic matter.*

4.1.3.1. Types of Soil

The loose layer of earth that covers the entire planet is referred to as soil. It is the outcome of rocks breaking down into their parts, which are subsequently influenced by a variety of factors, such as the physical actions of wind and water as well as chemical interactions with salts. Three states of matter are represented by soil as a whole: air and liquid are present in the pore spaces between soil particles, and the soil itself is solid. The earth is home to a wide variety of soil types due to the various mineral and organic matter compositions and environmental stresses that each soil has experienced.

One important part of the Earth's ecosystem is soil. The processes that occur in the soil have a profound effect on the world's ecosystems, affecting everything from water pollution and rainforest destruction to ozone depletion and global warming.

Based on the size of the particles that make up the soil, six primary types of soil are typically discussed in agricultural and basic soil science classes.

Sandy soil is one of the types of soil and the type with the biggest particles. Sandy soil typically consists of large spaces between the particles that allow water and minerals to easily flow through, as well as granules of rock and minerals. It also contains a fair amount of grit.

Another type of soil is clay soil, but the particles in clay are incredibly small. As a result, there is not much room for drainage between the individual particles in clay. Because of this, clay soil is poor for growing plants because it prevents water from escaping and makes it difficult for root systems to pierce the clay layer. Since it takes many, many years for rock particles to break down to the small enough size to form clay, clay soils are typically much older than sandy soils.

Silty soil has rich nutrients and excellent drainage, making it one of the most fertile types of soil. Although it contains more minerals and nutrients than sandy soil, its composition is otherwise very similar. Its size is slightly smaller. You can plant almost anything in silty soil, which is typically quite dark and aromatic.

In reality, loamy soil is a combination of several different types of soil, including varying proportions of silty, sandy, and clay soil. Loamy soil is nutrient-rich, well-drained to prevent water buildup and root rot, and retains water effectively due to the sand's grittiness. The best soil for gardening is loamy soil, and most supply stores carry a variety of loam.

Two other types of soil are found all over the world, but they aren't very good for easy growing: chalky and peaty soils. Chalky soils typically contain mixed-in stones of different sizes and are very alkaline. Chalky soil is not the best for planting because it dries out quickly and prevents plants from absorbing vital minerals. Peaty soils have a high organic matter content and typically contain a lot of dead plants. However, the high acidity of the soil prevents the organic matter from decomposing completely. Although peaty soil doesn't have a lot of nutrients, it can be a great soil for planting if it is treated properly.

4.1.4. Microclimate

A microclimate is a tiny, noticeably distinct climate found in a larger region. For example, in a garden, a sunny, wind-sheltered area will be much warmer than the rest of the garden for the majority of the year. Because of a number of factors that cause their weather to differ from the more general weather in the area, valleys and hills typically have their own climates;

in this case, the microclimate would be very small, but it can also get much larger.

Remember

It's important to understand microclimates as they represent the physical conditions experienced by organisms.

Gardeners and architects are often interested in microclimates because they can be highly significant areas to work with. Instead of designing a house that will clash with the land, an architect who considers the unique climate at a home site can design a more energy-efficient home by utilizing the natural features to keep the home warm in the winter and cool in the summer. These spaces can be used by gardeners to cultivate plants that aren't meant to grow in their local climates and to design water-efficient landscapes.

Because microclimates can occasionally support rare or unusual species in an area where these species are not normally found, biologists are also frequently interested in them. Rare plants often find habitats in which to flourish, sometimes drawing attention to them when people recognize plants that are not typically found in a given area.

Several factors influence the conditions in these areas. Both the land's slope and the direction that it faces can be significant. The conditions can also be affected by wind, drainage, exposure, shade, and other factors like the proximity of a body of water or the presence of an urban area. Urban areas are known for reflecting heat from surrounding buildings, sidewalks, and pavements, making them significantly warmer than the open land around them.

Particularly for those who are unfamiliar with the area, it can take some time to identify microclimates. Though neighbors can occasionally aid, it usually takes a year or two to recognize a particular area of a garden or neighborhood as having a distinct climate. For example, it's not unusual for the ones on either side of a street to be noticeably different, or for certain neighborhoods to have higher or lower temperatures than others.

Did you get it?

1. **What is a boreal forest?**
2. **What are the types of soil?**

4.2. WATER CYCLE IN FOREST ECOSYSTEM

Learning Objectives

- Understand the term hydrology
- Describe energy and water
- Illustrate the cycles of a forest ecosystem
- Discuss deadwood, wood debris and buried wood: ecological partitioning
- Explain the strategies of forest management

The water cycle, also known as the hydrologic cycle or H2O cycle, is the continuous movement of water on Earth's surface. While the total amount of water on Earth remains relatively constant, its distribution among ice, fresh, salty, and atmospheric reservoirs varies due to different weather conditions. Evaporation, condensation, precipitation, infiltration, runoff, and subsurface flow are some of the physical processes that transfer water from one reservoir to another, such as from a river to the ocean or from the ocean to the atmosphere. Water goes through several phases while doing this: liquid, solid (ice), and gas (vapor). Temperature fluctuations are a result of energy exchange in the water cycle. For example, as water evaporates, it absorbs energy from its surroundings and causes the atmosphere to cool. It condenses, releasing energy and warming the surrounding air. This exchange of heat affects the climate. Freshwater is restored to the land by the cycle's evaporative phase, which also purifies the water. Minerals are transported around the world by liquid water and ice.

All ecosystems, but especially forests, which typically grow in climates where precipitation provides more water than the vegetation can use or the soils can store, depend on the hydrological cycle. The extra water helps streams flow, which meets urban demands and irrigation needs far from the precipitation source. One important component of *the hydrological cycle* is vegetation. Water from the surface of vegetation and the litter layer is captured and evaporated before it reaches the soil. The density and depth of root channels and organic residue incorporated into the soil influence how quickly water seeps into the soil, runs off the surface, or percolates through to the water table.

4.2.1. Hydrology

Hydrology is the name of the scientific branch of physical geography that studies the water cycle. It is focused on the global origin, distribution, and characteristics of water. Consequently, in many scientific and educational materials, the water cycle is also referred to as the hydrologic cycle. Meteorology and oceanography, taken together, provide a broad overview of a set of global physical processes involving water that are also central to the field of hydrology. Geologists use the movement of groundwater within the planet's subterranean features to explain another aspect of the physical processes. Measurable data and understanding regarding the water cycle are of interest to hydrologists. Measuring the volume of water used in the transitional phases that take place when the water switches from one cycle process to another is also crucial. Therefore, in order to quantify the movement of water, hydrology is a broad science that integrates data from many different sciences. The foundational instruments of hydrology are derived from auxiliary scientific methods that have their roots in physics and mathematics.

Keyword

The hydrological cycle is the continuous movement of water between the Earth's surface and the atmosphere.

Thus, one of the interdisciplinary sciences that forms the foundation for the development and management of water resources is hydrology.

4.2.2. Energy and Water

The Earth System cannot function without solar energy and water. The global hydrologic cycle and energy budget are two significant mechanisms for redistributing both, as they are not distributed equally across the planet. These processes are inseparable due to their close interdependence. The hydrological cycle is propelled by solar energy, which is responsible for the vertical movement of water from Earth to the atmosphere through transpiration (the loss of water from plants) and evapotranspiration (the total of surface evaporation). On the other hand, evapotranspiration is a crucial process in Earth's energy budget since it contributes to 75% of the turbulent energy transfer from Earth to the atmosphere. The hydrological cycle affects all biotic processes, dissolves nutrients, and transfers them between and within ecosystems to regulate Earth's biogeochemical cycles. These nutrients give organisms the resources they need to grow. The flow of substances suspended and dissolved in water connects different ecosystems in a landscape (Figure 4.2).

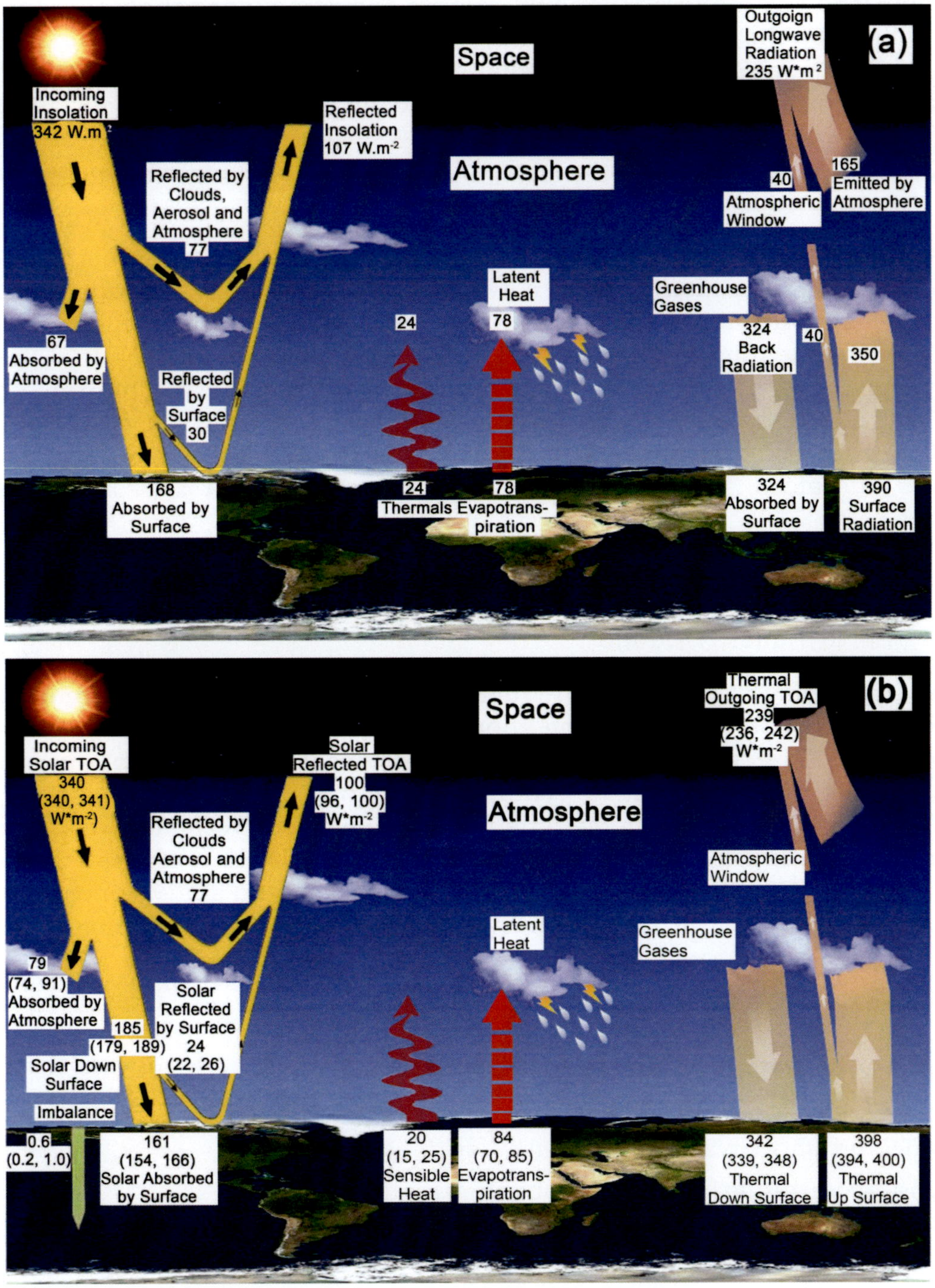

Figure 4.2. Estimate of the Earth's annual and global mean energy balance.

Source: Liou, Y.-A., Kar, S. K. (2014). Evapotranspiration Estimation with Remote Sensing and Various Surface Energy Balance Algorithms—A Review. *Energies*. 7(5), 2821–2849. https://doi.org/10.3390/en7052821.

Over time, the Earth and atmosphere release the same amount of outgoing longwave radiation to balance the amount of incoming solar radiation that they absorb. The Earth's surface absorbs roughly half of the solar radiation that enters the planet. Through evapotranspiration, longwave radiation absorbed by clouds and greenhouse gases, and warming of the air in contact with the surface (thermals), this energy is transferred to the atmosphere. Longwave energy is then radiated both out into space and back to Earth by the atmosphere. Units are $W.m^{-2}$ (IPCC).

The Sun can, with high accuracy, be described as a black body emitting energy at 5800 K (5530 °C) with a maximum at a wavelength of 0.5 µm. Averaged over the year and all surfaces of the Earth, this amounts to 342 W/m^2. As this shortwave radiation hits the Earth's atmosphere, 77 W/m^2 (23%) is directly reflected back into space, 67 W/m^2 (20%) is absorbed by molecules in the atmosphere, heating it. The remaining part is partly reflected (30 W/m^2 or 9%) and partly absorbed (168 W/m^2 or 49%) by the Earth's surface. The total annual energy flux from the Sun that is absorbed by the surface of the Earth is, therefore, $168 \times 365 \times 24 \times 3600 \times 4 \times \pi \times (6371 \times 10^3)^2 = 2.7 \times 10^{24}$ J.

The gross photosynthesis on Earth is 220 Pg (C) yr^{-1}. The heat of combustion of 1 g in the form of glucose (C) is 38.9 kJ, and the gross photosynthesis corresponds to 7.84×10^{21} J. The solar energy trapped in photosynthesis is thus a negligible component (3‰) in the Earth's energy budget.

Short-wave (0–3.4 µm) and long-wave (4–80 µm) radiation components are both included in the radiation balance. The short-wave portion of solar radiation, comprising approximately 95% of the radiation, comprises roughly equal amounts of near-infrared radiation (NIR) and visible light (0–4 µm; PAR, or photosynthetically active radiation). Short-wave radiation can either be reflected or absorbed by plants or other surfaces. The absorbed radiation can either evaporate water (latent heat) or heat the surface and convert it into sensible heat. According to Stefan-Boltzmann's law, long-wave radiation from all surfaces at absolute zero is emitted at a rate that is proportional to the fourth power of temperature (Figure 4.3).

The total radiation incident on any surface is the sum of

(1) direct short-wave radiation from the sun.
(2) diffuse short-wave radiation from the sky.
(3) reflected short-wave from nearby surfaces.
(4) long-wave radiation from atmospheric emission; and
(5) long-wave emitted from nearby surfaces.

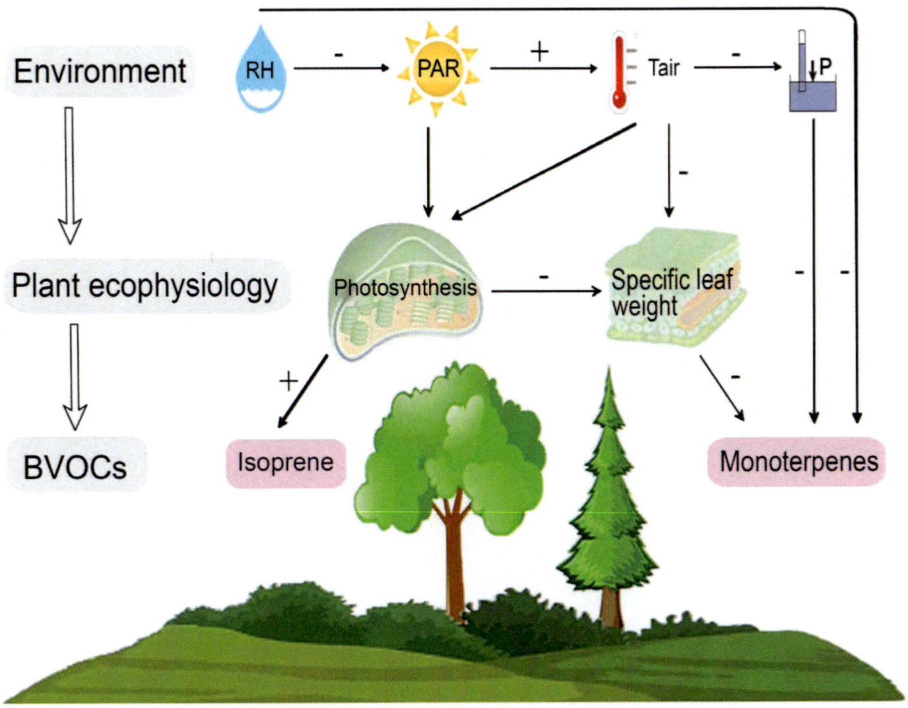

Figure 4.3. Energy exchange between vegetation and the environment involves a number of processes.

Source: Duan, C., Wu, Z., Liao, H., Ren, Y.(2023). Interaction Processes of Environment and Plant Ecophysiology with BVOC Emissions from Dominant Greening Trees. *Forests*. *14*(3), 523. https://doi.org/10.3390/f14030523.

Direct, scattered, and reflected sunlight—all of which contain some short-wave components crucial to photosynthesis—are how solar radiation enters plant canopies. Cloud reflection can enhance incident short-wave radiation at the ground surface by up to 30% on partly cloudy days. Less than 10% of incident short-wave radiation is scattered by the atmosphere on clear days, and no shadows are cast when short-wave radiation is diffused and reduced on cloudy days. As a function of their absolute (Kelvin) temperature, plants and other surfaces emit thermal radiation in addition to absorbing and reflecting short- and long-wave radiation. The majority of the heat load that plants experience is radiated back out; the remainder is removed by transpiration-based evaporative cooling and convection-based heat transfer (advection). *Plant tissue* and the soil both temporarily store heat that is subsequently released back into space.

Remember

Plant tissue **is a group of similar cells that work together to perform a specific function for the plant.**

4.2.3. Cycles of Forest Ecosystems

4.2.3.1. Energy Cycle

The photosynthesis process is fueled by the sun, which initiates the food pyramid. All life processes in trees and other green plants start with photosynthesis, which produces plant biomass as a byproduct. As plants transform light into plant biomass, which is subsequently ingested by other organisms, the sun's energy is continuously changing forms. Energy fixation, a crucial function carried out by plants, is the foundation of all life.

Remember

The water cycle is an extremely important process because it enables the availability of water for all living organisms and regulates weather patterns on our planet.

4.2.3.2. Water Cycle

Water flows in cycles. The water cycle gathers, cleans, and distributes water throughout the planet. Earth's life would not be possible without the water cycle. This water cycle includes plants and trees. Plants lose water through the pores in their leaf structures through a process called transpiration, which is regulated evaporation. When in growth, a mature tree can transpire hundreds of gallons of water every day.

4.2.3.3. Nutrient Cycle

As we already know, nutrients like phosphorus and nitrogen are essential for the healthy growth and reproduction of trees. Trees continuously take up nutrients from the soil and give them back throughout their life cycles. On a regular basis, nutrients are cycled from the living environment (rocks, soil, water, and air) to the nonliving environment.

4.2.4. Deadwood, Wood Debris, and Buried Wood: Ecological Partitioning

Numerous well-established, crucial ecological functions are provided by tree mortality and the ensuing physical, chemical, and biological decomposition in forest systems. The conceptual model employed by forest ecologists depicts how organisms use deadwood as a form of habitat. The deadwood pool is referred to as wood (or woody) debris (or detritus) (WD) in the majority of the forest ecology literature. Adjectives, which might or might not completely divide the pools, are widely used to define their size and location. For instance, wood debris classified as coarse (CWD, > 10 cm in diameter), fine (FWD, < 10 cm in diameter), downed (DWD, wood in contact with the forest floor), or standing (SWD, wood vertically oriented and attached to roots) is frequently used. Eight different pools are used in a widely used scheme that was first created to distinguish the ecological roles of WD using the structural position of the forest and the size of wood fragments. This ecological scheme was made simpler in an extensive review paper of deadwood C inventory methods. The review combined all aboveground pool types into one compartment and all belowground pool types into one (Figure 4.4).

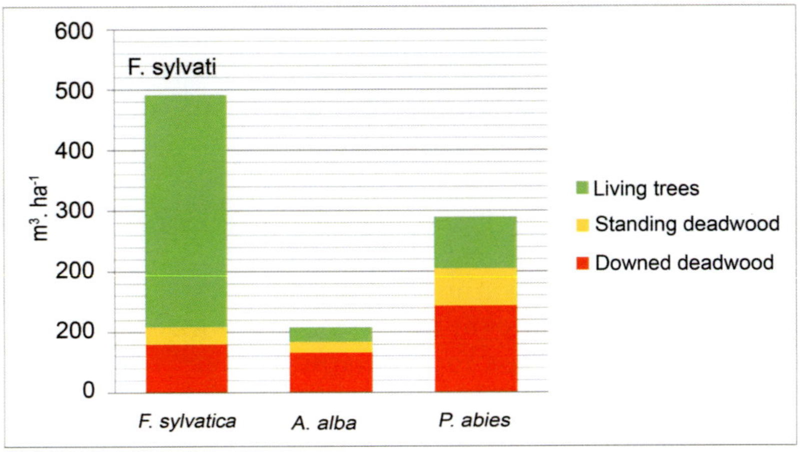

Figure 4.4. Physical characteristics of deadwood (wood density, biomass, and moisture) during the decomposition process in the Žofínský **Primeval Forest**.

Source: P□ív□tivý, T., Šamonil, P. (2021). Variation in Downed Deadwood Density, Biomass, and Moisture during Decomposition in a Natural Temperate Forest. *Forests*. *12*(10). 1352. https://doi.org/10.3390/f12101352.

There are ecological niches connected to deadwood in each of these WD categories that grow into distinctive, if not always unique, communities. Numerous species that both directly or indirectly promote additional decomposition and have the ability to control the transformation of WD into soil C are found in these diverse communities. While research on this topic is expanding, little is known about the high diversity of food webs that are produced when DW is transformed in forests by saprophytic organisms (species that rely on DW at some point in their life cycles). Understanding each pool's decomposition process and the rate at which the different pools change is essential to tracking the destiny of WD or modeling the dynamics of WD over time. Like with the accounting for forest C, these descriptions downplay the contribution of any buried deadwood, wherever it may be found.

4.2.5. Strategies of Forest Management

Forest resource management involves three strategies:

- *Maximum Sustainable Yield (MSY)*: The goal is to extract as much resource as possible without running out of resources from harvests. It implies that felling trees at the quickest developmental stage, the intermediate stage, optimizes the output of timber over time. The drawback of this approach is that it destroys the habitat of species that rely on mature trees to survive, changing the ecology of the forest.

- *Ecosystem-based management*: It tries to control resource extraction in order to reduce damage to the ecosystem and the ecological processes that produce the resources. It's a novel approach to reserve management that benefits both people and biodiversity. It is a management approach that safeguards and restores an ecosystem's structure, function, and species composition while enabling sustainable socioeconomic use. Ecosystems are dynamic and change during succession; they are also vulnerable to unforeseen events or disturbances like earthquakes and fires. These are the fundamental beliefs of ecosystem management. The management must therefore be adaptable and flexible. Thus, adaptive forest management comes into play.
- *Adaptive management*: Any characteristic that increases fitness in those who possess it is considered adaptive in and of itself. In terms of forest management, this entails the methodical testing of various strategies that may be appropriate to enhance forest management over time. Because it involves formulating hypotheses and testing optimal practices for managing forest resources, it blends science and management. It demonstrates how essential humans are to all ecosystems, demonstrating the need for ongoing population control.

Did you get it?

1. **What is the water cycle?**
2. **What is meant by hydrology?**

CLOSER LOOK

HOW FOREST STRUCTURE INFLUENCES THE WATER CYCLE

Forests are a critical component of the global water cycle: Trees draw water from the ground and release it into the atmosphere as vapor through pores in their leaves in a process called transpiration, which can influence temperatures and rainfall worldwide. Forests are also dynamic ecosystems, subject to both natural events, such as pest infestations and droughts, and human activities like logging, which can lead to significant changes in forest structure. Despite the crucial roles forests play, the relationship between forest structure and the global water cycle remains poorly understood.

To address gaps in our understanding of this relationship, *Aron et al.* conducted a comparison of two forest sites in Michigan to investigate how disturbances in forest structure can impact water transport from the land surface to the atmosphere.

The research team chose two neighboring field sites at the University of Michigan Biological Station in Northern Lower Michigan: an undisturbed control site dominated by bigtooth aspen and paper birch, and a site where aspen and birches were intentionally killed by researchers in 2008, resulting in a disturbed site with a more open canopy compared to the control site. The spatial arrangement of trees within a forest affects the amount of light and heat reaching the ground, influencing not only transpiration but also other processes like evaporation and entrainment—the mixing of air above the canopy into the canopy—contributing to the amount of water vapor reaching the atmosphere.

Taking advantage of the fact that each of these processes results in distinct isotopic signals in water vapor, the researchers measured stable water isotopes at six heights in the two forest sites during the spring, summer, and fall of 2016.

The results revealed that the disturbed canopy was both drier and warmer than the undisturbed control site. The control site also exhibited a more stratified isotopic profile, suggesting less vertical mixing of the air in the forests, whereas the more open canopy appeared to encourage more mixing. The differences between the two sites were most prominent in the summer and spring.

The study demonstrates that forest canopy can regulate the rate at which moisture and energy are returned to the atmosphere at a local scale, which can in turn influence water retention and the makeup of forest ecosystems. The results provide important context for researchers interested in modeling how both forest ecology and water cycles will evolve as climate change progresses.

ROLE MODEL

SUZANNE SIMARD

Suzanne Simard is one of the top ecologists in the world, but it took her a long time to get there. Her research that involves tree communication and cooperation took a lot of patience, but it led to sustainable forest practices.

Suzanne Simard (born 1960) is a Canadian scientist who is a professor in the Department of Forest and Conservation Sciences at the University of British Columbia. After growing up in the Monashee Mountains, British Columbia, she received her PhD in Forest Sciences at Oregon State University. Prior to teaching at the University of British Columbia, Simard worked as a research scientist at the British Columbia Ministry of Forests.

Simard is best known for the research she conducted on the underground networks of forests characterized by fungi and roots. She studies how these fungi and roots facilitate communication and interaction between trees and plants of an ecosystem. Within the communication between trees and plants is the exchange of carbon, water, nutrients, and defense signals between trees. Simard is also a leader of TerreWEB, an initiative set to train graduate students and post-doctoral fellows in global change science and its communication.

She used rare carbon isotopes as tracers in both field and greenhouse experiments to measure the flow and sharing of carbon between individual trees and species. She discovered, for instance, that birch and Douglas fir share carbon. Birch trees receive extra carbon from Douglas firs when the birch trees lose their leaves, and birch trees supply carbon to Douglas fir trees that are in the shade.

Mother Trees

Simard identified something called a hub tree, or "mother tree". Mother trees are the largest trees in forests that act as central hubs for vast below-ground mycorrhizal networks. A mother tree supports seedlings by infecting them with fungi and supplying them with the nutrients they need to grow.

She discovered that Douglas firs provide carbon to baby firs. She found that there was more carbon sent to baby firs that came from that specific mother tree than random baby firs not related to that specific fir tree. It was also found that the mother trees changed their root structure to make room for baby trees.

Her book "*Finding the Mother Tree*" asserts that forest ecologies are interdependent with fungal mycelium. She asserts that trees (and other plants) exchange sugars through their respective root systems and through interconnected fungal mycelial structures to share (and at times trade) micronutrients. This is significant in terms of the way existing woods and forests are managed and new plantations established.

Interspecies Cooperation

Simard found that "fir trees were using the fungal web to trade nutrients with paper-bark birch trees over the course of the season." For example, tree species can lend one another sugars as deficits occur within seasonal changes. This is a particularly beneficial exchange between deciduous and coniferous trees as their energy deficits occur during different periods. The benefit of this cooperative underground economy appears to be better overall health, more total photosynthesis, and greater resilience in the face of disturbance.

Science Communication

Suzanne Simard is an advocate of science communication. At the University of British Columbia, she initiated the Communication of Science Program TerreWEB with colleagues Dr. Julia Dordel and Dr. Maja Krzic, which has been training graduate students to become better communicators of their research since 2011. Simard has appeared in videos intended for general audiences, including three TED talks, the short documentary "*Do Trees Communicate*?" and the longer documentary films "*Intelligent Trees*" (where she appears alongside forester and author Peter Wohlleben) and "*Fantastic Fungi.*" New Scientist magazine interviewed Simard in 2021. Suzanne Simard has published a book where she reviews her discoveries about the life of trees and forests along with autobiographical notes.

Simard discussed her work and her book "*Finding the Mother Tree*" on BBC Radio 4's *Woman's Hour* in March 2022.

Popular Culture

Simard's life and work served as the primary inspiration for Patricia Westerford, a central character in Richard Powers' 2018 Pulitzer Prize-winning novel "*The Overstory*," in which Westerford pioneers the controversial idea that trees can communicate with each other and is ridiculed by fellow scientists before eventually being vindicated. Simard's work was referenced in Season 2, Episode 11 of the Apple TV+ series "*Ted Lasso*" when Coach Beard says, "You know, we used to believe that trees competed with each other for light. Suzanne Simard's fieldwork challenged that perception, and we now realize that the forest is a socialist community. Trees work in harmony to share the sunlight." In 2022, Simard appeared as a panelist on *Canada Reads*, advocating for Clayton Thomas-Müller's book "*Life in the City of Dirty Water.*"

CASE STUDY

FOREST ECOSYSTEMS AND GLOBAL CHANGE OF INSUBRIA

Forest ecosystems are challenged worldwide by a series of concomitant processes (e.g., climate shifts; land use change including urbanization; water, air, and soil pollution; habitat fragmentation and degradation; increased long-distance trade and the associated introduction of exotic species). These processes and their interactions can be summarized under the umbrella term of 'global change'. Just as with climate change, global change is and will be affecting not only forests but also other ecosystems and the eco-boundaries between them. Forests are particularly vulnerable to these processes due to the long-term nature of their main structural components (trees). Trees are generally long-lived organisms that live their entire life at a certain location. Tree species, populations, and individuals can thus only slowly catch up with rapid shifts in climate, fire frequencies, and human use of the land. There is evidence that worldwide forests are already suffering due to global changes, as suggested by increasing tree mortality rates associated with drought in various regions and by increased impacts of wind, bark beetles, and wildfires on forests in Europe and bark beetles in North America.

There is a consensus that the various global change drivers and their interactions are posing a severe threat to biodiversity worldwide. Biodiversity conservation, in turn, is fundamental to safeguarding the resilience of ecosystems to environmental changes. This follows from the relationship between biodiversity and the provision of ecosystem services. Without genetic, species, and ecosystem diversity, landscapes lose their capacity to provide clean water and air, safe food, and other primary resources, as well as recreation potential and other services. There is thus a growing recognition that landscape, ecosystem, forest, plant, soil, crop, and animal health are not independent of each other and may well be ultimately related to human well-being. Many studies on the effects of single global change factors on forest ecosystems are appearing. Given the amount of literature produced, there is a tendency for global change scientists to be aware only of the slice of the literature relevant to the factor they are studying, despite the many potential interactions with other global change drivers. Some studies are starting to consider the interactions among global change drivers, although this is often achieved by focusing on a single species rather than on whole ecosystems or on species interactions. Given that global change drivers are not likely to act in isolation, there is a need to consider their potential interrelations. This section provides an overview of the potential effects on forest ecosystems of global change factors and their interactions, focusing on Insubria as a case study.

Global Change and the Biodiversity of Insubric Ecosystems

The generality of the spatial congruency of variation in biodiversity among taxa is a key issue in conservation biogeography because, if species-rich regions tend to be rich not just for one taxonomic group, but for most of them, then bio-indicators that are easier to sample can be used to establish conservation priorities. Given their role as ecosystem engineers, trees are likely to remain important bio-indicators for a variety of other taxa, but we know little about whether trees will be able to track the predicted rapid climate shifts. The issue of finding reliable long-term bio-indicators under global change is further complicated in Insubria by the presence of several endemic species (e.g., *Gentiana insubrica, Ophrys benacensis, Primula longobarda, Ranunculus bilobus, Saxifragatombeanensis*), which need to be preserved in spite of the many threats posed by human beings. Models based on the commendable Swiss Biodiversity Monitoring Program are likely to be more reliable if predicting future occurrences of currently widespread species, but ideally, we would also need to be able to understand how rare species will behave. The longitudinal gradient in insubric mean annual precipitation has not only important consequences for the current biodiversity of the region but also shows how insubric ecosystems may develop depending on whether the future climate will shift towards drier or more humid conditions. On the one hand, drier conditions in future summers would make the insubric region more similar to the Mediterranean. On the other hand, more humid winters would change the insubric climate into a sub-Atlantic one. Evergreen hygrophilous broad-leaved species such as *Ilex aquifolium* and *Prunus laurocerasus* thrive in the western part of Insubria. The invasion of the exotic palm *Trachycarpus fortunei* in Ticino may be rather associated with a decrease in frost days. Other evergreen species more resilient to summer drought (e.g., *Quercus ilex, Viburnum tinus,* and *Olea europaea*) tend to be present towards the drier eastern end of the region. In general, the rarity of water deficits during insubric summers has enabled the cultivation of many evergreen broad-leaved species in parks and historic gardens. Some of these species have gone on to become naturalized in and around urbanized areas, confirming the role of private and historic gardens as hotspots of invasive species introduction.

People, Deadwood and Winter Forest Fires

In addition to a range of ecosystem services, chestnut cultivation can provide insurance against insubric winter forest fires. Cultivated chestnut groves can be less prone to fire than other types of woodland, as they lack a thick understory, which tends to be removed or prevented by grazing in order to facilitate chestnut harvesting. On the other hand, chestnut stands tend to occur close to human settlements, thus increasing the likelihood of forest fire start, but also of fire control. This is often the case throughout the distributional range of chestnut, given its human introduction and cultivation.

Winter forest fires are a peculiarity of insubric ecosystems. In the Mediterranean, winter is normally the rainy season, when fires are thus not common. In Insubria, summer is generally not as dry as in other parts of the Alps, and also winter can often be rich in precipitation (in the form of snow up from a given altitude). However, the arrival of föhn can rapidly turn a humid winter into a fire-risk situation, as this dry and relatively mild wind is able to melt snow in a matter of days. Most fires in Canton Ticino are started by people, either accidentally or intentionally. Although not all fires are started in winter, the majority of the forest fire area of Canton Ticino is burnt during föhn events in the cold season, when atmospheric perturbations arriving from w-nw-n produce dry and occasionally strong winds on the southern side of the Alps, thus increasing the risk that fires will get out of control. A similar situation is found in the Italian part of Insubria, in other regions of the southern Alps, and partly in California.

Winter forest fires affect all components of insubric ecosystems, from fauna to flora, from fungi to soil, from invasive species to native ones. Despite the centuries of fire use (and abuse) by the local population, insubric ecosystems have generally not co-evolved with fire as much as other ecosystems, for example, in western North America. However, it is now recognized by the scientific community that a non-null presence of forest fires could play an ecological role as forest disturbance also in present-day Insubria. For example, forest fires can contribute to creating a mosaic of different forest ages and increasing the amount of deadwood. These structural features are important for many species of a wide range of taxa. Although fire has been shown to facilitate the invasion of many exotic species, for instance, in Mediterranean ecosystems, fires may provide a tool to control the invasion of some exotic evergreen species in Insubria, wherever this is felt as unwelcome by landscape managers, the local population, and other stakeholders. Nonetheless, when fires occur repeatedly, the flora may be impoverished compared to insubric ecosystems where fires do not occur or return after a long time.

Given the generally high human population densities of the Insubric region, an increase in the occurrence of forest fires may create problems at the interface between rural, residential, and urbanized areas. A diverse age-class and size-structure may be obtained also by managing forests rather than burning them. Nonetheless, felling may need subsidies and might not provide a satisfactory amount of deadwood unless measures are taken to ensure this requirement for biodiversity is considered by foresters. Throughout the Alps, deadwood volumes have recently increased thanks to increased awareness of its importance for biodiversity, but also due to a lack of recent forest management.

From 1960 to 2000, the number of forest fires in southern Switzerland has increased. Most of these fires have occurred in low- to middle-altitude Insubric ecosystems, as sub-alpine forests are rarely affected. Forest fires can increase the amount of deadwood in forests if post-fire salvage logging is avoided, but can also decrease it in case of high fire intensity. The question arises as to whether the recent increase in forest fires in southern Switzerland is causally linked to changes in (i) the behavior of the population, (ii) the structure and dynamics of forests and the countryside, (iii) the feedback between fire occurrence and development of fire-prone vegetation, (iv) climate, or any of these factors in combination. A comparative analysis (1951–2010) of the inner alpine area, the northern, western, and southern Alps found for the latter region a recent increase in meteorological fire danger but a decrease in anthropogenic ignitions. In turn, a comparison of fire regimes and socio-economic conditions (1904–2008) for Valais and Ticino found that the reduction since the 1990s in fire ignitions for the insubric part of Ticino can be mainly explained by the banning of burning garden debris in the open and of fireworks and bonfires during the Swiss national day (August 1st; this ban is only effective in the presence of fire hazard).

SUMMARY

- The study of forest ecology focuses on how organisms interact within a forest ecosystem and with their surroundings. This field is vital for those concerned with forest preservation.
- Trees are large, usually long-lived perennial plants with large woody stems and extensive root systems that dominate biological and ecological conditions and processes in local or regional segments of landscapes known as forests.
- Boreal forests, which cover a significant area of the Northern Hemisphere, are characterized by a contiguous green belt of deciduous and coniferous trees.
- The planet's surface layer, known as soil, results from the breakdown of rocks into components. This process is influenced by various factors like wind and water erosion, as well as chemical reactions with salts.
- Sand soil is a type of soil with the largest particles. It usually has big gaps between the particles, which make it easy for water and minerals to move through. Additionally, it contains rock and mineral granules, along with a decent amount of grit.
- Silty soil is known for its rich nutrients and excellent drainage, making it one of the most fertile soil types. It contains more minerals and nutrients compared to sandy soil, but its composition is quite similar otherwise. The particle size is slightly smaller. Loamy soil, on the other hand, is a blend of various soil types, with different proportions of silt, sand, and clay.
- Loamy soil is nutrient-rich, has excellent drainage to prevent water buildup and the rotting of plant roots, and holds water well due to the heavy grittiness provided by the sand.
- The term "water cycle," which is another name for the hydrologic cycle or the H_2O cycle, refers to the ongoing movement of water across the Earth's surface. The bulk of water on Earth doesn't change much over time, but how it is divided up among the main ice, fresh, salty, and atmospheric reservoirs depends on a variety of meteorological factors.
- Hydrology is the scientific branch of physical geography that looks into the water cycle. It mainly deals with the worldwide source, distribution, and features of water.
- The process of photosynthesis is powered by the sun, kicking off the food pyramid. All life processes in trees and other green plants begin with photosynthesis, which generates plant biomass as a byproduct.

MULTIPLE CHOICE QUESTIONS

1. How many parts are there in the forest ecosystem?

a. One

b. Two

c. Three

d. Four

2. On which factor forest type is mainly dependent?

a. Abiotic

b. Size of the forest

c. Shape of trees

d. Products from the trees

3. How is erosion controlled by forests?

a. By reducing the sunlight penetration

b. By reducing the rainfall's force on the soil's surface

c. By reducing the pressure

d. By increasing the rainfall's force on the soil's surface

4. How do forests increase the atmosphere's humidity?

a. By transpiration

b. By inspiration

c. By expiration

d. By oxidation

5. How does the tropical rainforest get its name?

a. Due to less rain

b. Due to heavy rain

c. Due to moderate rain

d. Due to no rain required

REVIEW QUESTIONS

1. Explain the numerous types of forest ecosystems.
2. Identify the insects that are found in tropical rainforests.
3. List the various tropical rainforest birds.
4. Illustrate the forest ecosystem structure.
5. Determine the numerous types of soil.
6. Give an overview of microclimate.
7. What are the cycles of the forest ecosystem?
8. Define the strategies of forest management.

Answers

1. (b) 2. (a) 3. (b) 4. (a) 5. (b)

REFERENCES

1. Alt, D., & Donald, H., (2000). *Roadside Geology of Northern and Central California*. Missoula, MT: Mountain Press.
2. Abebe, M. H., et al., (2006). The role of area enclosures and fallow age in the restoration of plant diversity in northern Ethiopia. *African Journal of Ecology.*
3. Aerts, R., et al., (2007). Restoration of dry Afromontane forest using pioneer shrubs as nurse-plants for *Olea europaea* ssp. cuspidata. *Restoration Ecology*.
4. Yusuf, H., et al., (2011). Assessment of woody species encroachment in the grasslands of Nechisar National Park, Ethiopia. *African Journal of Ecology*.
5. Yimer, F., Ledin, S., & Abdelkadir, A., (2006). Soil property variations in relation to topographic aspect and vegetation community in the south-eastern highlands of Ethiopia. *Forest Ecology and Management*.
6. Tilahun, K., (2006). Analysis of rainfall climate and evapo-transpiration in arid and semi-arid regions of Ethiopia using data over the last half a century. *Journal of Arid Environments.*
7. Tesemma, Z. K., Mohamed, Y. A., & Steenhuis, T. S (2010). *Trends in Rainfall and Runoff in the Blue Nile Basin: 1964–2003*. 24(25):3747 – 3758 24(25):3747 – 3758 DOI:10.1002/hyp.7893
8. Tefera, B., & Stroosnijder, L., (2007). Integrated watershed management: A planning methodology for construction of new dams in Ethiopia. *Lakes & Reservoirs: Research and Management.*
9. Hanjra, M. A., Ferede, T., & Gutta, D. G., (2009). Pathways to breaking the poverty trap in Ethiopia: Investments in agricultural water, education, and markets. *Agricultural Water Management.*
10. Gebregziabher, G., Namara, R. E., & Holden, S., (2009). Poverty reduction with irrigation investment: An empirical case study from Tigray, Ethiopia. *Agricultural Water Management.*
11. Bekele, W., & Drake, L., (2003). Soil and water conservation decision behavior of subsistence farmers in the Eastern Highlands of Ethiopia: A case study of the Hunde-Lafto area. *Ecological Economics.*

CHAPTER 5

The Environmental Factors

LEARNING OBJECTIVES

After studying this chapter, you will be able to:

- Define an environment
- Discuss the physical (nonliving) environment
- Explain the composition of physical environment factors

INTRODUCTORY EXAMPLE

Social structures and cultural practices are greatly influenced by the climate. People can avoid the hottest part of the day in warm climates like the Middle East and Southern Europe by following customs like siestas, which are short afternoon naps. In contrast, people in cold climates preserve food and wear fur clothing. Significant risks also come from natural disasters like hurricanes and floods.

INTRODUCTION

We are a part of the interconnectedness of everything on our planet. The area of biology known as ecology studies the relationships that exist between living things and their surroundings. Documenting and comprehending the effects of human actions that are altering the biosphere on life on Earth is a primary objective of ecology. We need to understand the various kinds of ecological factors to uncover the fascinating truths regarding the interactions between living things and these factors. These elements influence how a population changes dynamically in a given setting. It is critical that people comprehend the various categories of ecological factors.

5.1. DEFINING THE ENVIRONMENT

Learning Objectives

- Define an environment

An organism's environment is everything that surrounds or affects it during its life. In other words, anything around a living thing, such as people, places, and objects, can be either natural or man-made. The French verb "environner," which means to encircle or surround, is the origin of the word "environment." Early man's environment was limited to the physical features of the planet Earth, such as the land (lithosphere), the air (atmosphere), and the water (hydrosphere), as well as biotic communities. However, as time passed and society progressed, man expanded his environment to include social, economic, and political roles. At the organism level, physiological interaction essentially aims to understand how various organisms are adapted to their surroundings for survival, population growth, and reproduction. Every living thing, including viruses and humans, relies on the environment for basic requirements like food, shelter, water, and oxygen. An organism's environment encompasses everything it is exposed to during its lifetime.

In other words, the environment comprises all interactions between the air, water, and land, as well as between these elements and humans, other living things, and material goods. It includes the physical, biological, and interconnected environments. Environmental studies provide an approach to understanding the global environment, the impact of human life on the environment, and vice versa. The environment is a multidisciplinary field that encompasses physics, chemistry, geology, geography, history, economics, physiology, biotechnology, remote sensing, geophysics, soil science, and hydrology, among other subjects. It is truly universal in nature.

The environment is essential to all biotic and abiotic components since it is a part of them. As a result, environmental problems like global warming, ozone layer depletion, forest shrinkage, energy resource depletion, loss of biodiversity, etc., affect everyone. The environment aids in establishing standards for a secure and thriving natural ecosystem.

5.1.1. Definition and Scope

The environment or settings that a person, animal, or plant lives in or uses; it is a collection of circumstances that all living things experience and that include all natural forces that offer opportunities for growth and development as well as risk and harm. The

Keyword

Global warming is the phenomenon of a gradual increase in the average temperature of the earth.

Remember

Environment also deals with the analysis of the processes in the hydrosphere, atmosphere, lithosphere, and organisms that lead to the polluted biosphere.

environment can be defined as the "sum total of living, nonliving components; influences and events, surrounding an organism" in its most comprehensive form. According to some definitions, the environment is "a holistic view of the world as it functions at any point in time, with a multitude of spatial elemental and socio-economic systems distinguished by quality and attributes of space and mode of behavior of abiotic and biotic forms." That definition is more thorough.

- "All external forces, influences, and conditions that affect life, nature behavior, and the growth, development, and maturity of living organisms" are collectively referred to as the environment (Douglas and Holland).
- "The entirety of the circumstances that surround man at any particular point in space and time is referred to as the environment" (C. C. Park).
- The whole spectrum of outside factors influencing an organism—both biological and physical—as well as other organisms, i.e., natural forces that surround a person.
- The entire environmental system, which includes interactions between him and his natural and artificial surroundings as well as the biosphere.

It is argued that the environment is an inseparable whole made up of interacting systems of biological, physical, and cultural elements that are connected both individually and collectively in a variety of ways. The human habitat's varied character, as well as its opportunities and limitations, are determined by physical elements such as space, landforms, water bodies, climate, soils, rocks, and minerals. The biosphere is made up of biological elements such as humans, microorganisms, animals, and plants. Economic, social, and political aspects of culture are essentially man-made characteristics that contribute to the formation of a cultural milieu. Accordingly, an environment is described as either "the social and cultural conditions that influence an individual or community, or (1) the circumstances and conditions (physical conditions) that surround an organism or group of organisms." Since humans live in both the natural and "built," or social, technological, and cultural realms, they are all a part of our environment.

Environmental knowledge is a multidisciplinary body of knowledge, the fundamentals of which directly affect every region of the earth.

Its main characteristics include:

- Conservation and natural resources.

- Maintenance and management of biological diversity.
- Controlling and managing environmental pollution to the permissible limit
- Stabilization of human population and environment.
- Development of alternate sources of renewable energy systems
- Providing a new dimension to the nation's security through conservation, protection, management and maintenance of the environment

As new opportunities for environmental protection and management emerge, it also addresses important issues such as clean and safe drinking water, hygienic living conditions, pollution-free fresh air, fertile land, nutritious food, and the development of sustainable environmental laws. An organism's interaction and relationship with its environment are intricate and multifaceted. No organism can survive on its own without interacting with other living things or biotic and abiotic forms. As a result, the environment of every organism includes other organisms. Our environment is made up of everything we come into contact with or that we require for survival. It is true that different people will see the environment from different perspectives, but overall, it can be said that the physical, biological, and cultural elements that make up the environment interact in a system and are linked both individually and collectively in a variety of ways. The environment is an extremely dynamic entity rather than one that is static. Abiotic and biotic factors alike are constantly changing.

Did you get it?

1. **What does the French word Environner mean?**
2. **What is the importance of biotic and abiotic components in the environment?**

5.2. THE PHYSICAL (NONLIVING) ENVIRONMENT

Learning Objectives

- Describe the categories of physical or abiotic environment

The physical or abiotic environment is separated into the following categories based on its attributes and conditions.

i. Solid i.e., lithosphere (solid earth)

ii. Liquid i.e., hydrosphere (water component)

iii. Gas i.e., atmosphere (gaseous component)

These environments are known as the lithosphere, hydrosphere, and atmosphere. They can also be broken down into smaller units based on different spatial scales, like mountains, plateaus, plains, lakes, rivers, glaciers, and deserts. The climatic conditions that create specific habitats for biological communities, such as tropical, temperate, and polar environments, are also part of the physical environment.

Did you get it?

1. **What is the abiotic environment divided into?**

5.3. COMPOSITION OF THE PHYSICAL ENVIRONMENT FACTORS

Learning Objectives

- Describe climatic factors
- Understand edaphic factors
- Explain topographic factors
- Discuss biotic factors

In an ecosystem, many organisms coexist. Physical, chemical, and biological components are all part of the environment. An external element becomes a factor when it has an impact on an organism's ability to survive. Environmental or ecological factors refer to all of these factors combined. Abiotic (nonliving) and biotic (living) components of these elements can be combined to form an organism's environment.

However, the ecological factors are meaningfully grouped into four classes, which are as follows:

1. Climatic factors
2. Edaphic factors
3. Topographic factors
4. Biotic factors

5.3.1. Climatic Factors

The primary natural variable influencing plant life is the climate. Light, temperature, water, wind, and fire are examples of climatic factors.

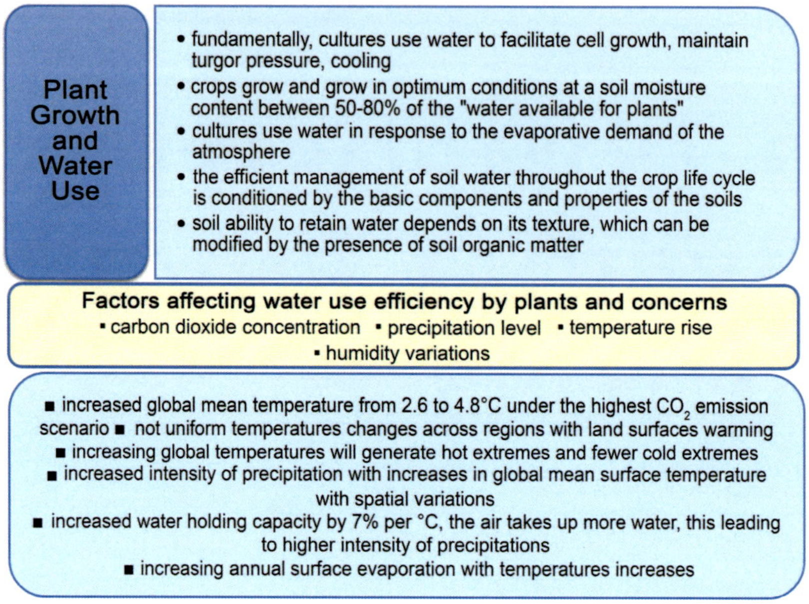

Figure 5.1. Environmental factors affecting a plant.

Source: Gavrilescu, M. (2021). Water, Soil, and Plants Interactions in a Threatened Environment. *Water*. *13*(19), 2746. https://doi.org/10.3390/w13192746.

5.3.1.1. Light

It's widely known that light plays a crucial role in many of the basic physiological functions of plants, such as photosynthesis, transpiration, *seed germination*, and flowering. Visible light is the part of sunlight that our eyes can see. The visible light spectrum ranges from about 400 nm (violet) to 700 nm (red). The wavelengths where photosynthesis is most effective are blue (400–500 nm) and red (600–700 nm). Plants don't absorb light as strongly in the green (500–600 nm) part of the spectrum (Figure 5.2).

Keyword

Seed germination is the process through which a seed develops into a plant. It involves activating a dormant embryo, which then emerges from the seed coat and grows into a seedling.

5.3.1.1.1. Effects of Light on Plants

Based on the tolerance to intensities of light, the plants are divided into two types. They are:

1. **Heliophytes** – Light loving plants. Example: Angiosperms.
2. **Sciophytes** – Shade loving plants. Example: Bryophytes and Pteridophytes.

The environment is dark and the people living there are unaware that the Sun is a celestial energy source located in the deep sea (>500 m). What then is their energy source?

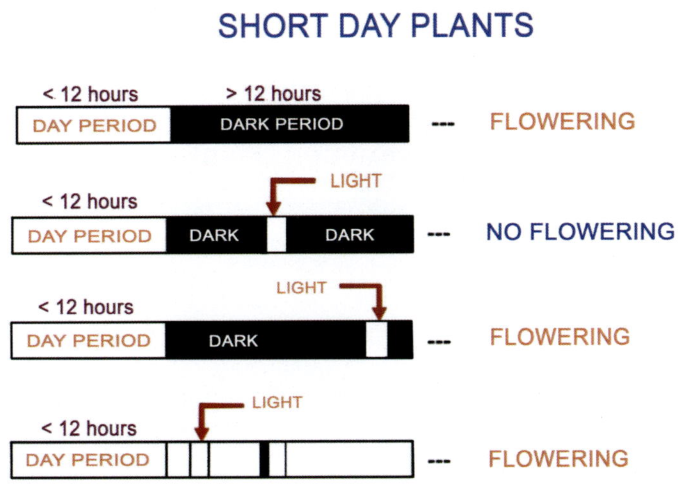

Figure 5.2. Various effects of light upon a green plant.

Source: By EarthDragon – Own work, CC BY-SA 4.0, https://commons.wikimedia.org/w/index.php?curid=117999638.

5.3.1.2. Temperature

One of the key elements that influences practically all of an organism's metabolic processes is temperature. An organism's physiological processes are all dependent on an ideal temperature at which to operate at its maximum metabolic rate. For every living thing, there are three distinct temperature limits.

They are:

1. **Minimum temperature** – Physiological activities are lowest.
2. **Optimum temperature** – Physiological activities are maximum.
3. **Maximum temperature** – Physiological activities will stop.

Raunkiaer divided the world's vegetation into the following four categories according to the average temperature in each area: megatherms, mesotherms, microtherms, and hekistotherms. In deep-sea hydrothermal vents and thermal springs where the average temperature is higher than 100 °C.

Based on the range of thermal tolerance, organisms are divided into two types.

1. Eurythermal: Organisms that can tolerate a wide range of temperature fluctuations.

Example: *Zostera* (A marine Angiosperm) and *Artemisia tridentata*.

2. Stenothermal: Organisms that can tolerate only a small range of temperature variations. Example: Mango and Palm (Terrestrial Angiosperms).

5.3.1.2.1. Thermal Stratification

It typically inhabits aquatic environments. Thermal stratification is the term used to describe how a water body's temperature profile changes as depth increases. Thermal stratifications come in three different varieties (Figure 5.3).

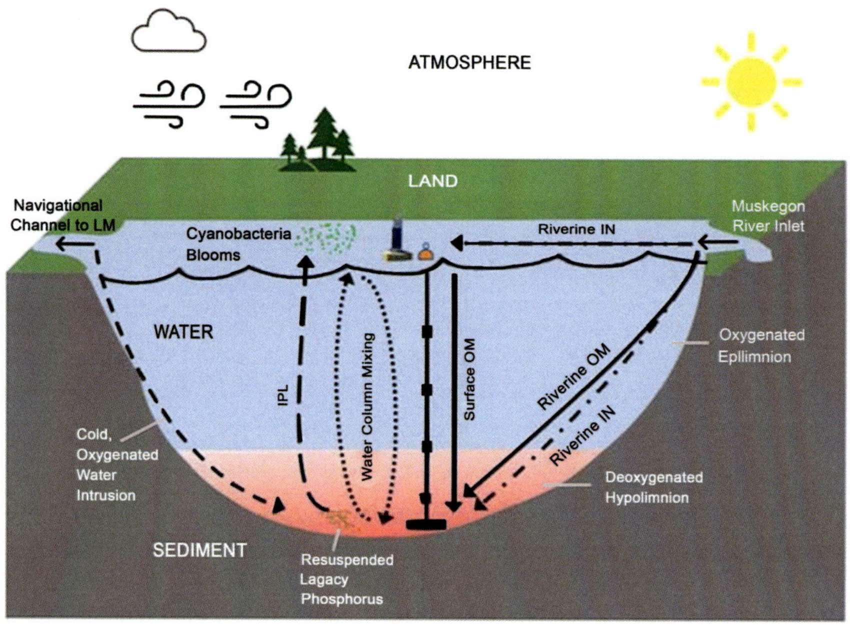

Figure 5.3. Thermal stratification of a pond.

Source: Dugener, N. M., Weinke, A. D., Stone, I. P., Biddanda, B. A. (2023). Recurringly Hypoxic: Bottom Water Oxygen Depletion Is Linked to Temperature and Precipitation in a Great Lakes Estuary. *Hydrobiology*. *2*(2), 410–430. https://doi.org/10.3390/hydrobiology2020027.

1. Epilimnion – The upper layer of warmer water.
2. Metalimnion – The middle layer with a zone of gradual decrease in temperature.
3. Hypolimnion – The bottom layer of colder water.

5.3.1.2.2. Temperature-Based Zonation

The temperature and vegetation on the surface of the earth are influenced by variations in latitude and altitude. Below is an illustration of the latitudinal and altitudinal zonation of vegetation.

- *Latitude:* Latitude is an angle that ranges from 00 at the equator to 900 at the poles (Figure 5.4).
- *Altitude:* How high a place is located above the sea level is called the altitude of the place (Figure 5.5).

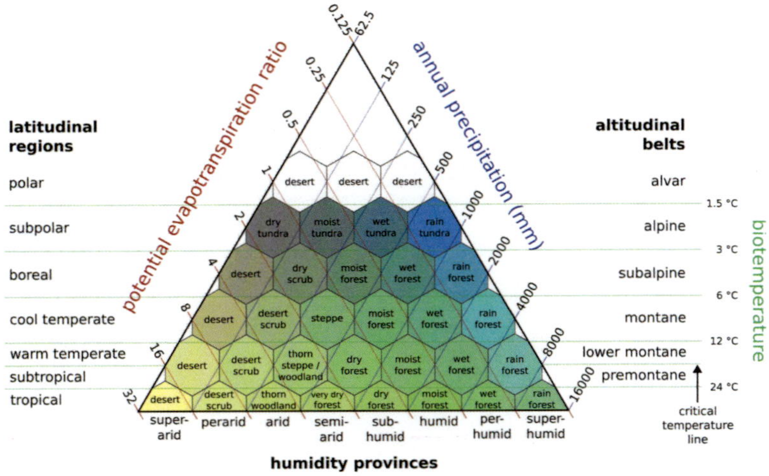

Figure 5.4. Latitudinal zonation of vegetation type.

Source: By Peter Halasz – own work, based on [1–7], and Ecology (textbook) Peter Stiling, 4th ed. P 236, and Textbook of Biodiversity K. V. Krishnamurthy 2003 p 55., CC BY-SA 2.5, https://commons.wikimedia.org/w/index.php?curid=1737503.

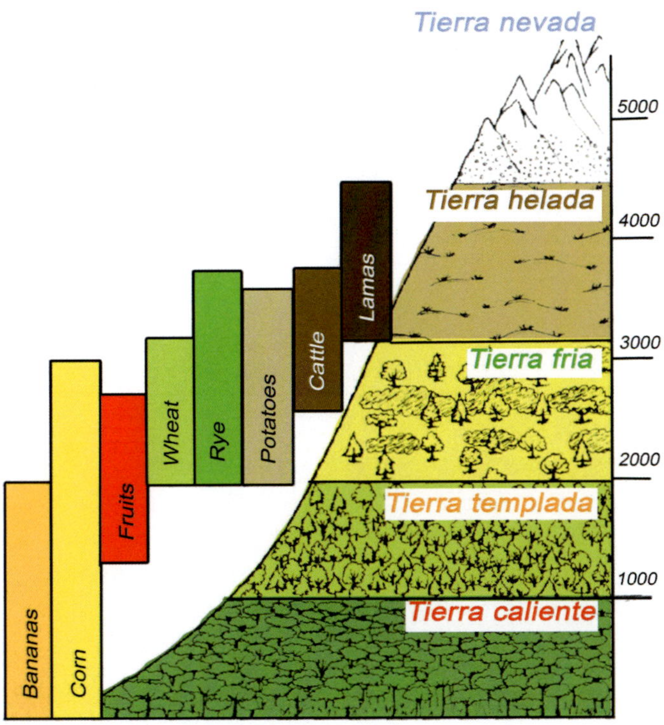

Figure 5.5. Altitudinal zonation of vegetation.

Source: By Chris.urs-o; Maksim; Anita Graser – Wikipedia de, CC BY-SA 3.0, https://commons.wikimedia.org/w/index.php?curid=5905197.

The timber line, also known as the tree line, is an imaginary line that indicates the altitude at which trees cannot grow in a mountain range or other elevated land. The typical tree growth range is between 3000 and 4000 meters above sea level.

5.3.1.2.3. *Effects of Temperature*

The following physiological processes are influenced by temperature:

- Temperature affects the enzymatic action of all the biochemical reactions in a plant body.
- It influences CO_2 and O_2 solubility in the biological systems. Increases respiration and stimulates the growth of seedlings.
- Low temperature with high humidity can spread diseases to plants.
- The varying temperature with moisture determines the distribution of the vegetation types.

5.3.1.3. Water

One of the most significant climatic variables is water. It has an impact on all essential functions of living things. It is thought that during Earth's evolution, only water served as the home for life. Over 70% of the Earth's surface is covered in water. In the natural world, plants have three ways to obtain water: precipitation, soil water, and atmospheric moisture. Water availability affects plant productivity and distribution. Additionally, water quality matters, particularly for aquatic life. In various water bodies, the total salinity is as follows: (i) Inland water (fresh water) contains less than 1%, (ii) 3.5% in seawater, and (iii) over 100% in super-salinized water (Lagoons).

Based on the range of tolerance of salinity, organisms are divided into two types.

1. *Euryhaline:* Organisms that can live in water with a wide range of salinity. Examples: Marine algae and marina angiosperms
2. *Stenohaline:* Organisms that can withstand only a small range of salinity. Example: Plants of estuaries.

5.3.1.3.1. *Global Distribution of Water*

The Earth has water in different parts of it. Around 70% of the Earth is made up of oceans, which reflect blue light, giving the impression that the Earth is blue from space. As a result, Earth is sometimes called the "blue planet." It is estimated that there are 1.386 billion km^3 (333 million cubic miles) of water on Earth. Just 3% of the total water available is freshwater, with the remaining 97% being saltwater. Ninety-seven percent of the water on Earth is salty. Salinity is 35% (or 3± percent) on average. This translates to 35,000 mg of dissolved salt ions per liter of seawater or 34 grams of dissolved salts in 1 kg of seawater. Marginal water is defined as having a salinity of between 0.35% and 1%. On Earth, fresh and saltwater have a ratio of about 40 to 1.

About 69% of all freshwater is found as ice and snow, 30% is found as groundwater, and less than 0.3% is found as liquid. This indicates that 99% of the water is underground and only 1% is usable for human consumption. Lakes hold 87% of the liquid surface freshwater, swamps contain 11%, and rivers contain 2% of the total. There is about 680,000 km^3 of freshwater in the form of glaciers and ice that is always present. These include the ice sheets that contain water resources in the Arctic, Antarctic, and Greenland. You cannot access this water. Groundwater and river water come from naturally occurring glaciers that are melting.

Water that has a salinity of less than 1% of the ocean's is referred to as fresh water, i.e., below roughly 0.35%. Lakes and rivers are great sources of freshwater for human consumption. Lakes (as well as ponds), reservoirs, rivers, streams, and wetlands are examples of surface waters. Rainfall, runoff from melting snow and ice, and base flow from groundwater systems all contribute to the flow of water into these bodies of water. Only 0.3% of the freshwater resources on Earth are found in surface waters. The greatest amount of fresh surface waters (90,000 km^3) is stored in lakes. Compared to water found in rivers or streams, this amounts to forty times more water. An estimated 2.7 million km^2 is covered by lakes. This amounts to about 2% of the land's total area (apart from the polar regions). Although lakes are typically small, over 95% of all lake freshwater is thought to be found in the 145 largest lakes in the world. The largest and deepest lake in the world, Lake Baikal in Russia, holds 27% of all freshwater found in lakes. Lake waters offer a multitude of benefits to the global community, including recreation, transportation, fishing, and water supply.

Lagoons, swamps, bogs, marshes, and mires are common names for these water-saturated environments. The area of wetlands worldwide is approximately four times larger than that of lakes. Merely 10% of the water present in lakes and other surface waters is contained in them.

Remember

Water is used mainly for irrigation (68%), public supply (21%), and industry (11%).

Groundwater is defined as water that is present in the spaces between rocks in the geological strata beneath the Earth's surface. It is found within the cracks and pores in rocks. Approximately 800,042 cubic km, or one-third, of the 13,841,200,000 cubic km of water that exists on Earth is groundwater. Ninety-five percent is groundwater, thirty-five percent is surface water, and fifteen percent is soil moisture. Merely 0.36% of the total freshwater on the planet is accessible for human use. There are approximately 61,234 cubic kilometers of water in the soil as moisture. Subsurface water is made up of both soil moisture and groundwater. The percentage of all natural water resources that are groundwater is 0.58%. It can be found 4 kilometers below the surface of the Earth. The majority of groundwater comes from precipitation, such as rain or snow, and is known as meteoric water. Groundwater is created when precipitation or snowmelt seeps into the Earth and flows downward. Through infiltration through the pores and fissures of permeable rocks, groundwater is stored in various layers of the Earth. It can be found in the pore spaces that exist between the grains of minerals and the underlying layers of rock or sediments. Rainfall is the primary source of groundwater. It creeps in by slowly seeping into the Earth.

(i) *Meteoric Water:* The primary source of groundwater is this. This water is known as "meteoric" or "shooting star water" because it falls in the form of rain and snow. Groundwater is created when water seeps through the pores, joints, and fissures in rocks. By allowing water vapor from circulating air in pore spaces to condense, groundwater can be directly obtained from atmospheric moisture. Another name for this is "condensational water." In arid and semi-arid regions, it serves as the primary water source. There is a pressure differential between the soil and the water vapor in the atmosphere during the summer months due to the temperature differential between the land and the air. As a result, water vapor seeps into the soil. Groundwater is increased by water seeping from lakes, rivers, seas, and artificial channels. This mostly happens in humid areas.

Did you know?

The word "meteoric" (as in the sense of direct atmospheric origin) is used here from the same root as meteorology. The term is from a Greek word that originally referred to astronomical discussions. However, after the publication of Aristotle's book Meteorology, which discussed what we today call earth sciences, the term was eventually used to describe any notable changes appearing in the sky (including meteors, originally thought to be weather phenomena).

(ii) *Connate Water:* Connate water is the liquid found beneath lakes and oceans in the pores and cavities of sedimentary rocks. Another name for it is "sedimentary water." It is the second significant groundwater source. During the process of deposition, this water became trapped in the spaces between sedimentary and volcanic rocks. This water is salty and highly mineralized.

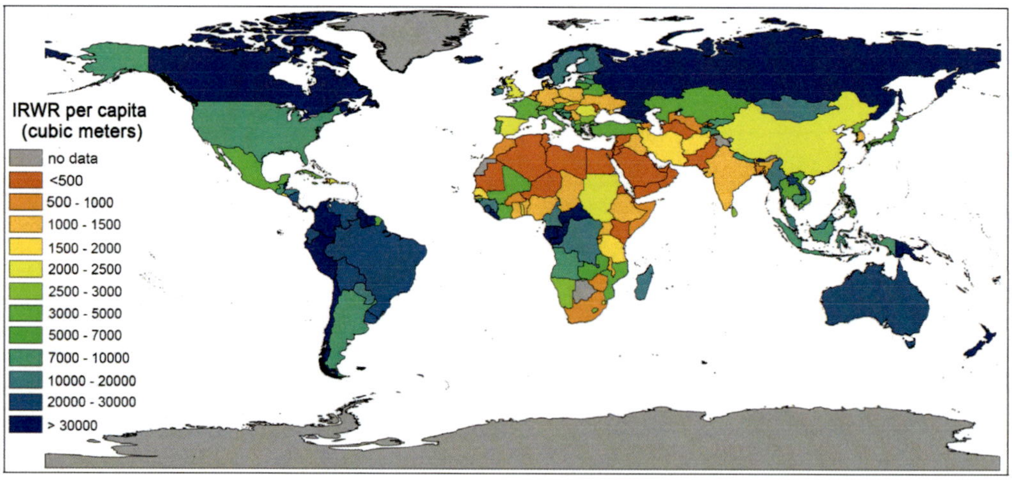

Figure 5.6. Global distribution of water.

Source: Mancosu, N., Snyder, R. L., Kyriakakis, G., Spano, D. (2015). Water Scarcity and Future Challenges for Food Production. *Water*. 7(3), 975–992. https://doi.org/10.3390/w7030975

(iii) *Magmatic Water:* Magmatic water is the water that is transformed into vapors by volcanic action when it enters hot rocks and condenses. Volcanic activity forces hot magma into rocks, where its vapor droplets condense and turn into water. This water is referred to as magmatic water. Sources of this water include geysers, springs, and wells. Although it originates in the earth's interior, it eventually makes its way to the planet's uppermost layers (Figure 5.6).

5.3.1.4. Wind

"Wind" refers to air movement. It is an essential ecological component as well. Various gases, particles, and other elements are present in the atmospheric air. The gases that make up the atmosphere are as follows: argon and other gases make up 0.93%, oxygen makes up 21%, nitrogen makes up 78%, and carbon dioxide makes up 0.04%. Other elements that make up the air include water vapor, gaseous pollutants, dust, smoke particles, microbes, pollen grains, spores, and so forth. An instrument called an anemometer is used to gauge wind speed (Figure 5.7).

5.3.1.4.1. Effects of Wind

- Wind is an important factor in the formation of rain
- Causes wave formation in lakes and oceans, which promotes aeration of water
- Strong wind causes soil erosion and reduces soil fertility
- Increases the rate of transpiration
- Helps in pollination in anemophilous plants
- It also helps in dispersal of many fruits, seeds, spores, etc.
- Strong wind may cause up-rooting of big trees
- Unidirectional wind stimulates the development of flag forms in trees.

Figure 5.7. Flag form in trees.

Source: https://upload.wikimedia.org/wikipedia/commons/8/8a/Windblown_trees%2C_Humphrey_Head_-_geograph.org.uk_-_48659.jpg.

5.3.1.5. Fire

Heat and light are released during the chemical process of combustion, which is what causes fire, an exothermic factor. It develops naturally occasionally because of the friction between the surfaces of the trees, but it is primarily man-made.

Fire is generally divided into

1. **Ground fire** – Which is flameless and subterranean.
2. **Surface fire** – Which consumes the herbs and shrubs.
3. **Crown fire** – Which burns the forest canopy.

5.3.1.5.1. Effects of Fire

- Fire has a direct lethal effect on plants
- Burn scars are suitable places for the entry of parasitic fungi and insects
- It brings out the alteration of light, rainfall, *nutrient cycle*, fertility of soil, pH, soil flora and fauna
- Some fungi that grow in the soil of burnt areas are called pyrophilous.

Example: *Pyronema confluens*.

- Indicators of fire – *Pteris* (fern) and *Pyronema* (fungus) indicate the burnt up and fire-disturbed areas. So they are called indicators of fire.
- Fire break – It is a gap made in the vegetation that acts as a barrier to slow down or stop the progress of fire.
- A natural fire break may occur when there is a lack of vegetation such as a river, lake and canyon found in between vegetation may act as a natural fire break.
- Rhytidome: It is the structural defense of plants against fire. The outer bark of trees which extends to the last formed periderm is called Rhytidome. It is composed of multiple layers of suberized periderm, cortical and phloem tissues. It protects the stem against fire, water loss, and invasion of insects and prevents infections by microorganisms.

Keyword

Nutrient cycle is the movement and exchange of inorganic and organic matter back into the production of matter.

5.3.2. Edaphic Factors

The physical and chemical makeup of the soil that has developed in a specific location is one of the abiotic elements associated with soil, or edaphic factors. Soil science is known as pedology.

Soil

The worn top layer of the Earth that supports plant life is called soil. It is an intricate composite mass made up of various elements such as soil water, soil air, soil organisms, and other components.

One of nature's resources, soil is created when rocks weather and break down. It is described as the sand, silt, and clay matrix that contains some concentration of water and air along with dead organic matter and living biomass. It provides an abundance of both organic and inorganic materials necessary for the support of microscopic communities and plant life. Mineral elements obtained from the weathering and fragmentation of parent rock material are included under inorganic material. Organic wastes, the decomposing remains of plants and animals, and their breakdown products are examples of organic components.

5.3.2.1. Soil Formation

Changes in the parent material (rocks) are the first step in the formation of soil. Raw materials accumulate first, and then organic materials build up on the surface. Soil is formed through a combination of anthropogenic, chemical, biological, and physical processes that modify the parent material. There are two primary processes in the formation of soil.

- *Weathering of rocks* – Weathering is the process by which rocks break down. Physical or chemical processes may be involved. Mechanical forces cause physical weathering. Temperature variations cause rocks to expand or contract, which leads to the formation of fissures and cracks in the rock. This reduces the size of the rocks. Low temperatures cause the water in the cracks in the rock to freeze, leading to the expansion of the rock and its shattering into small pieces. In addition to these, the friction created by hail, rain, and wind also helps break rocks into smaller pieces. Water's action or the activities of soil-dwelling organisms cause chemical weathering. Rocks dissolve in water, leading to chemical changes in the rock. Microbes' organic acids and oxidation-reduction processes aid in the decomposition of rocks and enrich the soil with organic matter.
- *Mineralization and humification* – Through the process of mineralization, the tiny rock fragments or particles created by rock weathering are transformed into a homogeneous matrix. Both plants and animals are crucial to this process. Plant roots encroach on soil particles, further pulverizing them into fine particles. The organic acids and humic acids secreted by the microbes connected to plant roots help further break down rock material into fine soil particles. Actinomycetes, fungi, and bacteria are examples of microorganisms that aid in the decomposition process. Organic compounds are transformed into inorganic nutrients during this process. Minerals, carbon dioxide, and water are among the simple products that are produced when organic compounds break down. We refer to this process as mineralization. The partially decomposed organic matter that remains after mineralization is known as humus; it is a colloidal, dark, amorphous material. Humus

gives the soil a loose texture that aids in aeration and helps in the retention of water and nutrients.

Numerous variables, including the parent material (rock), climate, and topography (relief), affect the process of soil formation (Figure 5.8).

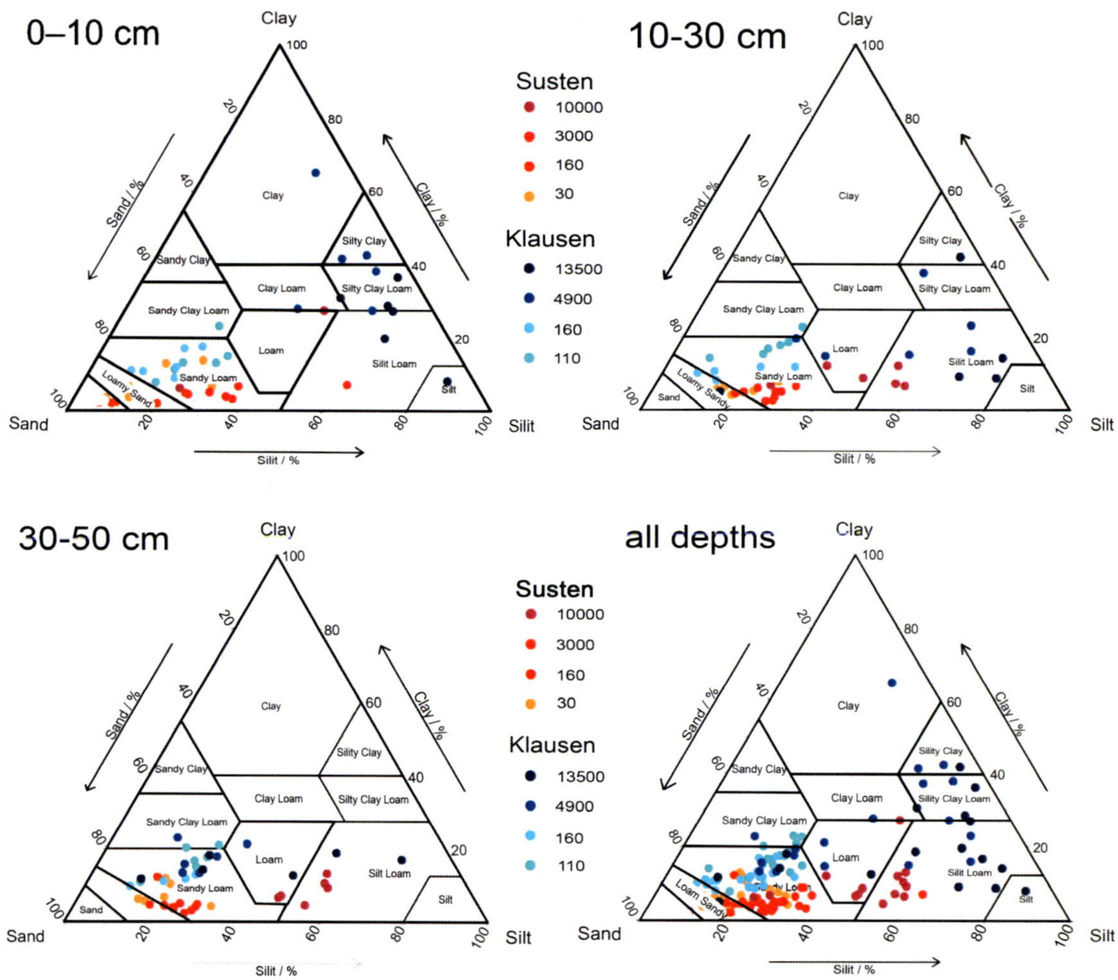

Figure 5.8. Diagrammatic representations of the processes affecting soil formation.

Source: Musso, A., Tikhomirov, D., Plötze, M. L., Greinwald, K., Hartmann, A., Geitner, C., Maier, F., Petibon, F., Egli, M. (2022). Soil Formation and Mass Redistribution during the Holocene Using Meteoric ^{10}Be, Soil Chemistry and Mineralogy. *Geosciences*. *12*(2), 99. https://doi.org/10.3390/geosciences12020099.

5.3.2.2. Factors Affecting Soil Formation Include:

Several factors affect how the soil profile develops. Changes in the parent material, topography, climate change, and organisms are the main examples of these. Each of these elements modifies the properties of soil and imparts a unique profile.

- *Parent Material* – The parent material has a significant influence on the soil's characteristics. The characteristics of the soil are determined by the parent material's chemical makeup. The parent material affects the soil's permeability, color, texture, structure, and mineral content. A coarse-textured soil that was formed from parent material with a coarse grain size is resistant to weathering. The limestone-derived soil has a high concentration of soluble bases and is naturally fertile. Because the parent material had few soluble ions, the soil that was created from it was acidic. This occurs as a result of water passing through the soil, eliminating the bases and replacing them with hydrogen ions. Sandstone-derived soils have little soluble base content.
- *Climate* – Soil development and chemistry are influenced by climate. The pace of chemical reactions during soil formation is regulated by climatic factors. Due to the disruption of dissolved ions' passage through the soil, precipitation also has an impact on the formation of horizons. Due to limited microbial species and less vegetation, the soils in hot, dry desert regions typically lack organic matter. Dry areas experience coarse-textured soil because of a lack of precipitation, which prevents chemical weathering. The removal of organic material from the soil is facilitated by a high annual precipitation rate. In the tundra region, the accumulation of organic matter is impacted by the cold temperatures that limit microbial activity. A high amount of organic matter accumulates in the warm, humid tropics due to the rapid rate of bacterial activity that promotes decomposition. The presence of rising temperatures accelerates biological processes. Each 10° C increase in temperature causes the rate of reaction to double. A maximum temperature of 30 to 35° C is reached by enzyme-catalyzed reactions, which are sensitive to high temperatures. The rate at which soil is eroded and deposited is determined by the water velocity.
- *Topography (elevation)* – As it controls water runoff, it is a factor that has a significant influence on how soil is formed. Runoff produced when raindrops land on the surface is influenced by the slope's length and angle. More effective water erosion occurs on steeper, bare slopes. Soil development is hindered by water flowing across the surface because it pushes away the parent material. Water volume and flow rate both increase in the direction of the slope.
- *Biological Factors* – The soil's biological constituents comprise microorganisms, vegetation, and related fauna. These organisms have a significant impact on the composition of the soil. Microorganisms like bacteria, fungi, actinomycetes, earthworms, tiny arthropods, and burrowing animals (rabbits, moles) are the main members of the soil-forming fauna. The biological activities of soil-dwelling organisms and plant roots contribute to the decomposition of organic matter in the soil. Grass and algae are pioneer species that cause organic matter to accumulate on the parent material's surface. Soils that are less microbially rich, such as desert soils, are also poorly formed and devoid of organic matter.
- *Time* – All pedogenic (soil-forming) processes occur over time. Young soils show only minimal profile development, i.e., possess only horizon A, while mature soils possess all the horizons.

5.3.2.2. Soil Types

Based on soil formation (pedogenesis), the soils are divided into

1. **Residual soils** –These are soils formed by weathering and pedogenesis of the rock.
2. **Transported soils** – These are transported by various agencies.

The important edaphic factors which affect vegetation are as follows:

1. **Soil moisture**: Plants absorbs rain water and moisture directly from the air
2. **Soil water**: Soil water is more important than any other ecological factor affecting the distribution of plants. Rain is the main source of soil water. Capillary water held between pore spaces of soil particles and angles between them is the most important form of water available to plants.
3. **Soil reactions**: Soil may be acidic, alkaline, or neutral in its reaction. The pH value of the soil solution determines the availability of plant nutrients. The optimal pH range of the soil for cultivating crop plants is 5.5 to 6.8.
4. **Soil nutrients**: Soil fertility and productivity are the ability of soil to provide all essential plant nutrients such as minerals and organic nutrients in the form of ions.
5. **Soil temperature**: The soil temperature of an area plays an important role in determining the geographical distribution of plants. Low temperature reduces the use of water and solute absorption by roots.
6. **Soil atmosphere**: The spaces left between soil particles are called pore spaces which contain oxygen and carbon dioxide.
7. **Soil organisms:** Many organisms existing in the soil like bacteria, fungi, algae, protozoans, nematodes, insects, earthworms, etc. are called soil organisms.

Keyword

Soil organisms are a wide range of living organisms that play a vital role in determining soil characteristics.

5.3.2.3. Composition of Soil

The composition of the soil plays a significant role in managing nutrients. Soil water is the primary source of readily available nutrients for plant uptake, while soil minerals and organic matter serve as reservoirs for nutrients. Since many of the microorganisms that reside in the soil require air to carry out the biological processes that release additional nutrients into the soil, soil air also plays a crucial role.

Soil is primarily composed of four elements: air, water, minerals, and organic matter. Approximately 45% minerals, 5% organic matter, 20–30% water, and 20–30% air make up typical soil. At most, these percentages are generalizations. The soil is incredibly dynamic and complex (Figure 5.9).

Remember

The composition of the soil can fluctuate daily, depending on numerous factors such as water supply, cultivation practices, and/or soil type.

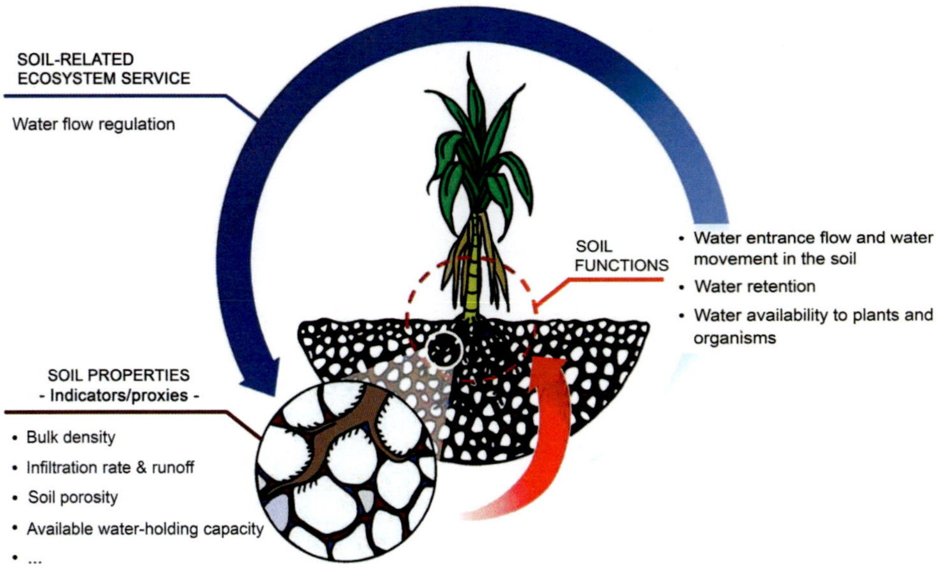

Figure 5.9. Composition of an ideal soil.

Source: Lustosa Carvalho, M., Bonini da Luz, F., de Lima, R. P., Cavalieri-Polizeli, K. M. V., Carvalho, J. L. N., Cherubin, M. R. (2022). Assessment of Soil Physical Quality and Water Flow Regulation under Straw Removal Management in Sugarcane Production Fields. *Sustainability*. *14*(2), 841. https://doi.org/10.3390/su14020841.

Minerals and organic matter that make up the solid phase of soil are typically stable. However, organic matter may be lost from the soil if it is not managed properly. The soil's most dynamic properties are its liquid and gas phases, which are air and water, respectively. As the soil dries or gets wet, the proportions of air and water in it are continually shifting.

5.3.2.4. Soil Texture

You can determine the soil's coarseness or fineness by taking a handful of it. Soil texture refers to the feel that results from the size and relative distribution of mineral particles in the soil. Sand, silt, and clay are the three size categories of the particles that make up soil. The largest particles are made of sand, and the smallest are made of clay. The texture of the soil is determined by the proportions of sand, silt, and clay (Figure 5.10).

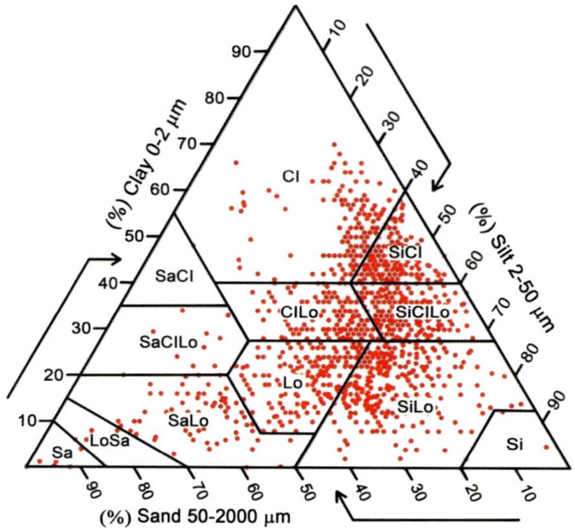

Figure 5.10. Relative sizes of sand, silt, and clay particles.

Key: (Cl: clay; SiCl: silty clay; SiClLo: silty clay loam; SaCl: sandy clay; SaClLo: sandy clay loam; ClLo: clay loam; Si: silt; SiLo: silt loam; Lo: loam; Sa: sand; LoSa: loamy sand; SaLo: sandy loam).

Source: Taghizadeh-Mehrjardi R, Emadi M, Cherati A, Heung B, Mosavi A, Scholten T. Bio-Inspired Hybridization of Artificial Neural Networks: An Application for Mapping the Spatial Distribution of Soil Texture Fractions. Remote Sensing. 2021; 13(5):1025. https://doi.org/10.3390/rs13051025.

It is possible to estimate soil texture in the field or determine it in a laboratory. The phrase "texture by feel" describes the field method. When a moist sample is squeezed in the hand, sandy soils do not ball. When moist, loamy soils will ball up, but when pressed between the thumb and forefinger, they won't form a ribbon longer than five centimeters. When wet, clayey soils also take the shape of a ball. When squeezed between the thumb and forefinger, they also form a ribbon that is longer than five centimeters (Figure 5.11).

The more precise laboratory approach makes use of the relative rates at which various-sized particles fall or settle in a liquid as a result of gravity. This technique can be used to calculate the precise proportions of clay, silt, and sand in soil. To determine the soil texture, these percentages are entered into the soil texture triangle. Next, the name of the soil texture is ascertained using the soil texture triangle. The textural classes of soil are twelve. Because of the usage of this triangle, words like "clay" and "loam" always have the same meaning. Every texture has a corresponding percentage of clay, silt, or sand.

It is uncommon for soil to consist entirely of silt, clay, or sand. Rather, the majority of soils are a blend of the three. For instance, soil with a clay loam texture contains almost equal amounts of sand, silt, and clay, whereas soil with a sandy loam texture contains 15% clay, 20% silt, and 65% sand. When soil is described as loam, it simply means that it contains a specific proportion of sand, silt, and clay; it makes no mention of the presence or absence of organic matter. Understanding the texture aids in soil management (Figure 5.12).

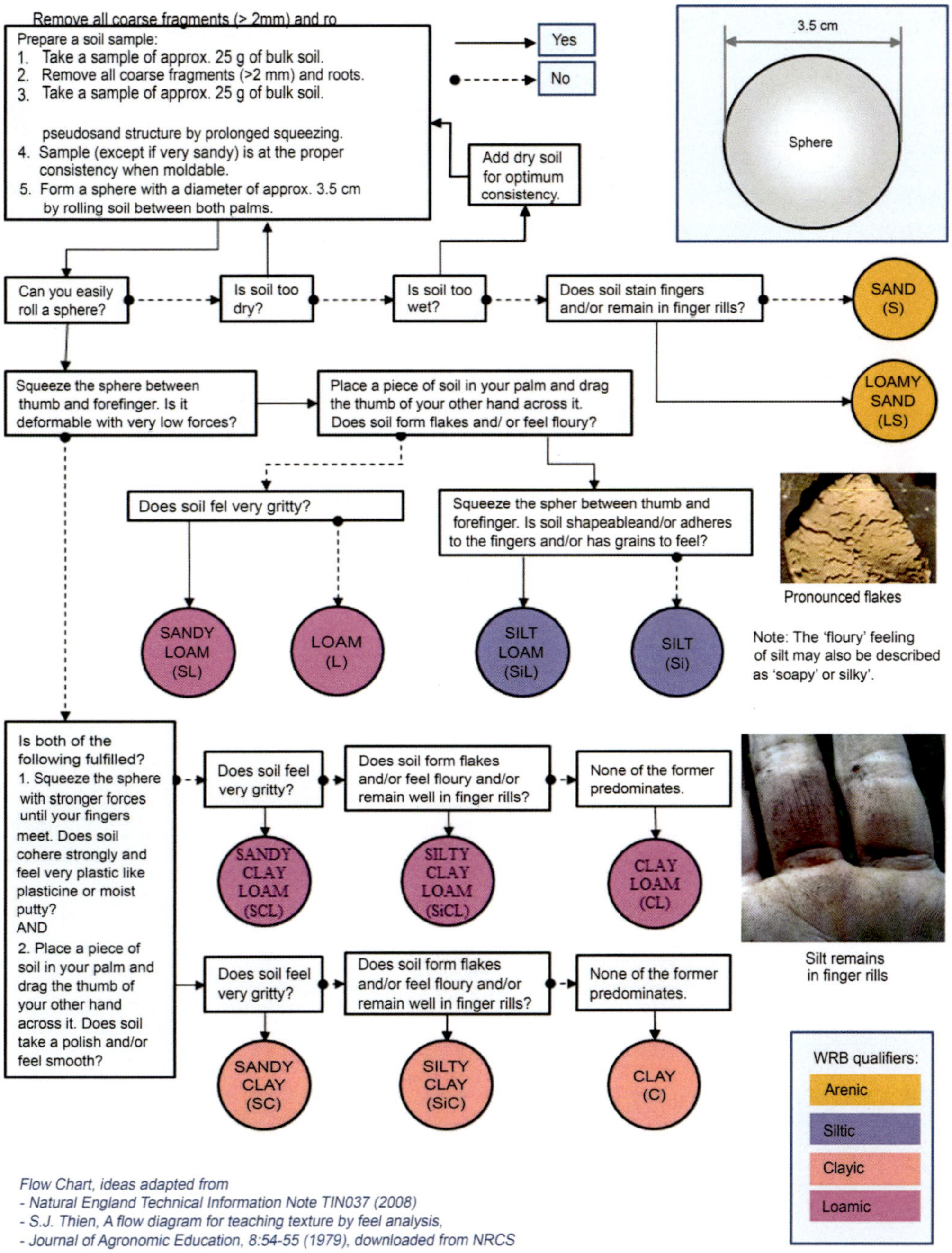

Figure 5.11. Can you make a ribbon of soil?

Source: By Vincent Buness, Margaretha Rau, Peter Schad; photograph of flakes by Peter Schad, photograph of silt in finger rills by Vincent Buness – IUSS Working Group WRB. 2022. World Reference Base for Soil Resources. International soil classification system for naming soils and creating legends for soil maps. 4th edition. International Union of Soil Sciences (IUSS), Vienna, Austria. Page 183, CC BY-SA 4.0, https://commons.wikimedia.org/w/index.php?curid=126727200.

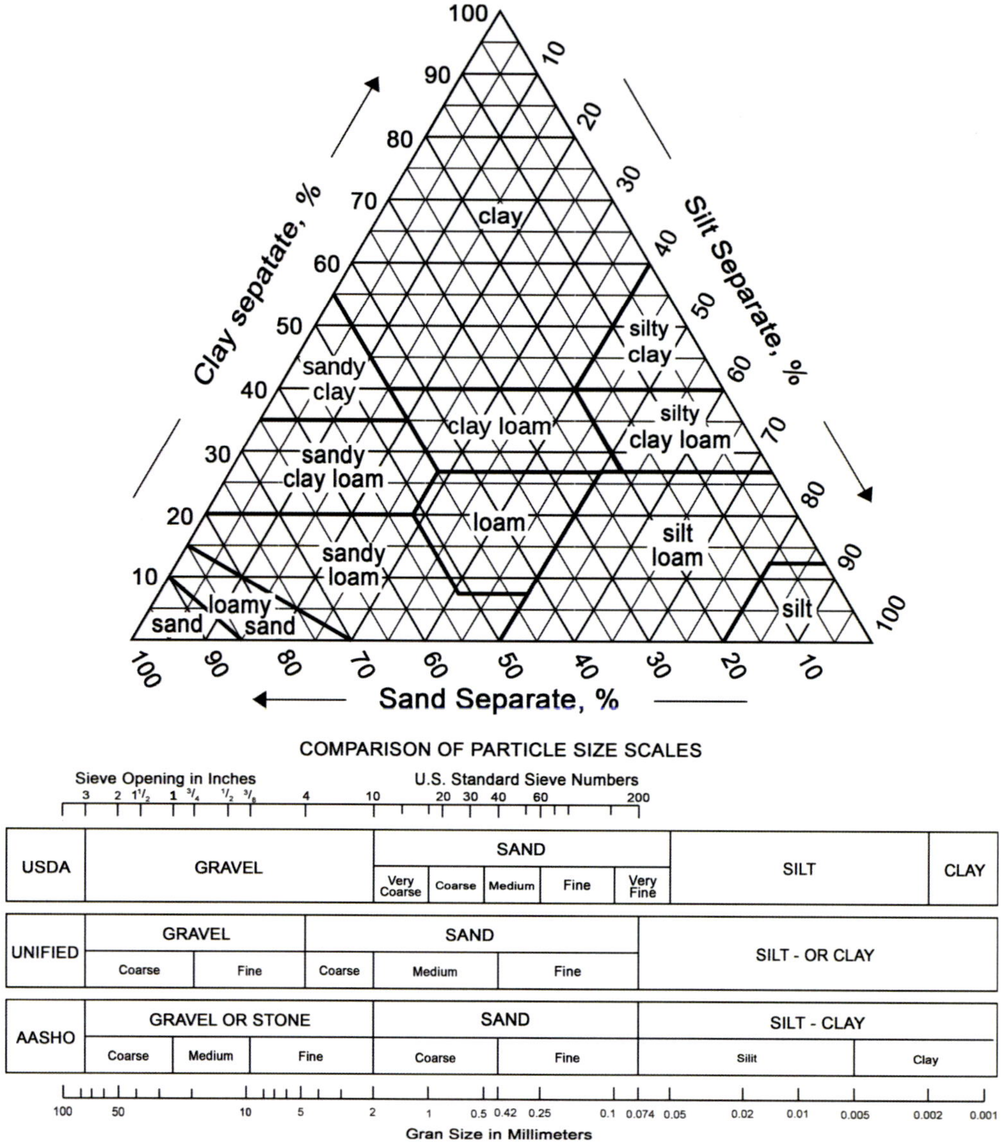

Figure 5.12. Texture triangle, with several examples of how to describe soil texture using the triangle.

Source: By USDA Soil Survey Division Staff – Manually vectorized from File:SoilTextureTriangle.jpgOriginal *Source:* A scanned JPG on the old USDA domain http://soils.usda.gov. According to its full URL this JPG must have been implemented in the online version of the 1993 Soil Survey Manual (SSM)., Public Domain, https://commons.wikimedia.org/w/index.php?curid=109468074.

5.3.2.5. Soil Profile

A soil profile describes the layers (horizons) that comprise the soil from the surface to the depths in terms of both physical and chemical characteristics. The soil has layers that are distinct from the original parent material and have formed over time. The characteristics of a horizon are used to assign labels. Time, climatic variations, the actions of living things, and relief all contribute to the formation of different soil layers. The American Department of Agriculture (USDA) developed the system. Following their location and characteristics, these layers are called O, A, E, B, C, and R. These horizons are commonly found in a mature soil profile (Table 5.1).

Table 5.1. Description of different layers in soil horizon

Layers of soil horizon	Description
O (organic)	The top layer dominated by organic material. It includes a layer of litter and humus. This horizon is not present in cultivated soils, manicured lawns, and severely eroded soils.
A	The mineral soil horizon present at the surface or below an O horizon is referred as topsoil. It contains a mixture of organic and mineral components. It has more organic carbon. This layer supports plants and microbial growth. This layer is formed when organic matter gets transformed into humic materials by soil biota. The humic materials coat the soil particles, coloring them brown and black. The formation of this horizon takes a long time in decades. This horizon increases in thickness and SOM content.
E (alluvial)	Horizon characterized by eluviation i.e., removal of materials such as silicate clay, iron, aluminum, or organic matter. It is commonly not found. Usually pale colored and low in organic matter. The texture is similar to horizon A.
B	Horizon is dominated by the loss of most of the original rock material and shows illuviation i.e., concentration of the silicate clay, iron, aluminum, or humus from higher horizons. These materials get added by leaching from overlying horizons. This zone is formed when dissolved and suspended materials are carried downward to greater depths with percolating rainwater. It has various sub horizons which are denoted as B_{t1}, B_{t2}, B_{tg}.
C	It is the layer of soil excluding hard bedrock. It lacks characteristics of O, A, E or B horizons.
R	The underlying hard bedrock

Source: www.egyankosh.ac.in.

5.3.3. Topographic Factors

Topography refers to the features found on Earth's surface. The climate of any region is influenced by its topography to varying degrees. These factors include solar radiation, temperature, humidity, rainfall, latitude, and altitude. Topography modifies soil conditions and even small-scale microclimates, which impact vegetation. Topographic variables include mountain direction, altitude, latitude, and steepness, among others.

a. Latitudes and Altitudes

Latitudes are the degrees of separation from the equator. The equator experiences the highest temperatures, which progressively decrease towards the poles. From the equator to the poles, various vegetation types are present, as illustrated in Figure 5.13.

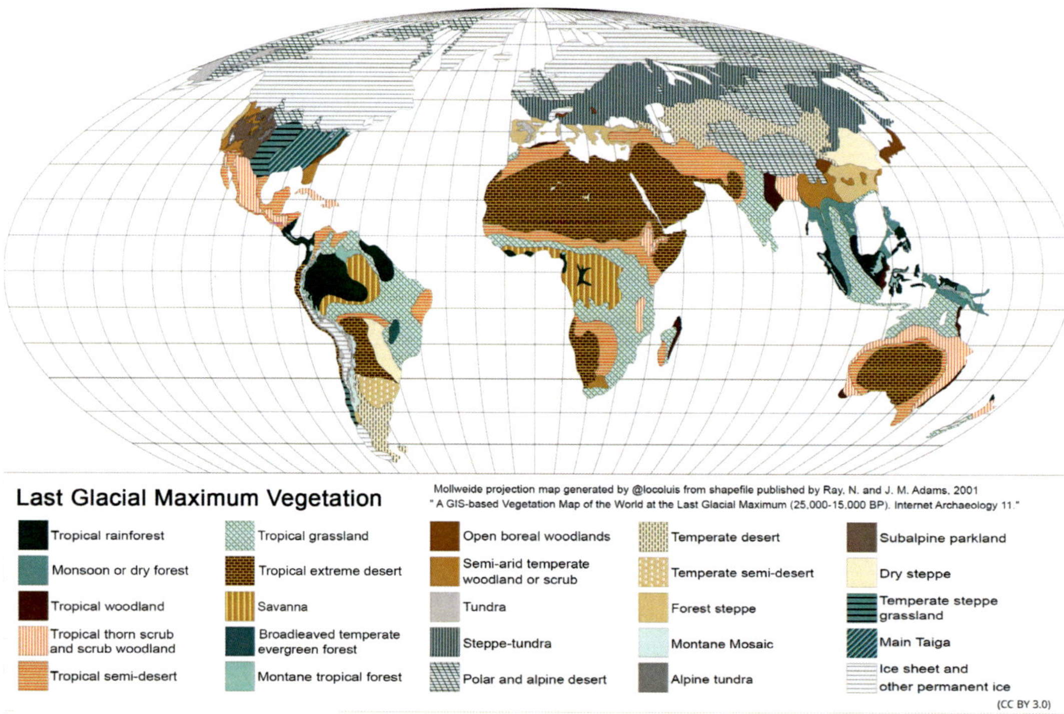

Figure 5.13. Latitudinal and altitudinal vegetation.

Source: By Locoluis – Map generated from shapefile published by Ray, N. and J. M. Adams. 2001, A GIS-based Vegetation Map of the World at the Last Glacial Maximum (25,000–15,000 BP). Internet Archaeology 11. *Source:* http://intarch.ac.uk/journal/issue11/rayadams_toc.html, CC BY 3.0, https://commons.wikimedia.org/w/index.php?curid=42569666.

The height is defined as the height above sea level. The wind speed stays high, air pressure and temperature drop, and light intensity and humidity rise at high altitudes. These elements cause distinct zonation in the vegetation at various altitudes.

b. Direction of the Mountain

The flora and fauna found on the north and south faces of mountains and hills vary due to variations in temperature, *humidity*, rainfall, light intensity, and duration.

Ecotone – The transition zone between two ecosystems. Example: The border between forest and grassland.

Edge effect – Because of the influences of the two habitats' environments, those species can be found in ecotone areas. We refer to this as edge effect. Owls in the ecotone region across a forest and grassland are an instance.

Keyword

Humidity is the concentration of water vapor present in the air.

The amount of rain, wind, and solar radiation that a mountain or hill receives varies depending on its two faces. Due to the heavy rains, the windward part of these two faces has good vegetation, while the leeward part has poor vegetation due to rain shadows (rain deficit). Similarly, the center and edge of aquatic bodies, such as ponds, have different water depths in their soil due to variations in soil slope and wave activity within the body of water. Consequently, different organism species may exist in different areas of the same region.

c. Steepness of the Mountain

Rain can run off the mountain or hill due to its steepness. Water loss consequently results in a water deficit and rapid topsoil erosion, leading to poor vegetation. However, because surface water drains more slowly and water remains in the soil longer in the plains and valleys, there is a greater diversity of vegetation there (Figure 5.14).

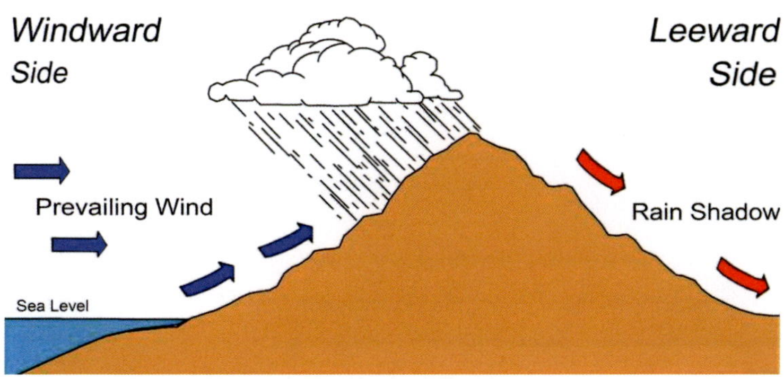

Figure 5.14. Steepness of mountain.

Source: By Meg Stewart – https://www.flickr.com/photos/megstewart/8644087724/, CC BY-SA 2.0, https://commons.wikimedia.org/w/index.php?curid=77241409.

5.3.4. Biotic Factors

Plants and animals interact with one another to form what are known as biotic factors, which can have a significant impact on vegetation. The environment is altered and there may be both direct and indirect impacts. Most plants are sociable organisms that interact with one another and share information. Animals and plants have a similar effect on plant life, either directly or indirectly. Both positive and negative interactions can be used to categorize the various interactions that occur between them (Figure 5.15).

5.3.4.1. Positive Interactions

Positive interaction occurs when one or both of the involved species gain something from the interaction. Commensalism and mutualism are two examples.

a. Mutualism:

It is an interaction between two species of organisms in which both benefit from the obligate association. The following are common examples of mutualism.

Nitrogen fixation: Living in symbiosis, Rhizobium (Bacterium) creates nodules in the roots of leguminous plants. After obtaining nourishment from leguminous plants, the Rhizobium converts atmospheric nitrogen into nitrate, which host plants can use.

Other examples:

- Water fern (*Azolla*) and nitrogen-fixing Cyanobacterium *(Anabaena*).
- Anabaena present in coralloid roots of Cycas (Gymnosperm)
- Cyanobacterium (*Nostoc*) found in the thalloid body of *Anthoceros* (Bryophytes)
- Wasps present in fruits of fig.
- Lichen is a mutual association of an alga and a fungus.
- Roots of terrestrial plants and fungal hyphae: Mycorrhiza

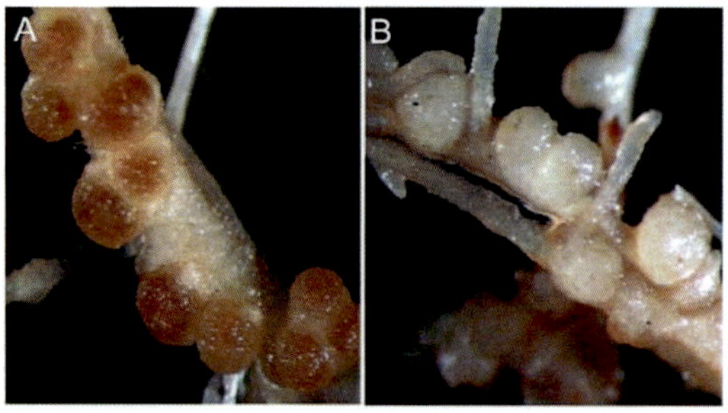

Figure 5.15. A nodulated legume plant root with bacteria.

Source: Lirio-Paredes, J., Ogata-Gutiérrez, K., Zúñiga-Dávila, D. (2022). Effects of Rhizobia Isolated from Coffee Fields in the High Jungle Peruvian Region, Tested on *Phaseolus vulgaris* L. var. Red Kidney. *Microorganisms. 10*(4), 823. https://doi.org/10.3390/microorganisms10040823.

b. *Commensalism:* It is an interaction between two organisms wherein one gains something and the other experiences neither harm nor benefit. The other species is referred to as the host and the species that benefits is known as the commensal.

Epiphytes: Epiphytes are plants that grow on other plants without causing damage to them. Tropical rainforests are typical habitats for them.

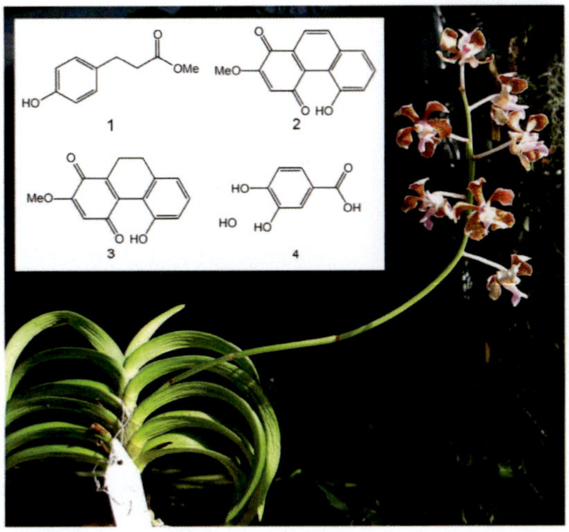

Figure 5.16. An epiphytic plant-Vanda.

Source: Jimoh, T. O., Nuamnaichati, N., Sungthong, R., Chansriniyom, C., Chanvorachote, P., Likhitwitayawuid, K., Chaotham, C., Sritularak, B. (2022). Phytochemicals from *Vanda bensonii* and Their Bioactivities to Inhibit Growth and Metastasis of Non-Small Cell Lung Cancer Cells. *Molecules*. *27*(22), 7902. https://doi.org/10.3390/molecules27227902.

The hygroscopic roots of epiphytic higher plants, or orchids, hold a unique kind of spongy tissue known as velamen, which helps the plant absorb water and nutrients from the atmosphere (Figure 5.16). As a result, it makes food for itself and is independent of the host. They don't do any damage to the host plant—they just use it as support.

- Many orchids, ferns, lianas, hanging mosses, *Peperomia*, money plant, and *Usnea* (Lichen) are some of the examples of epiphytes.
- Spanish Moss – *Tillandsia* grows on the bark of Oak and Pine trees.

5.3.4.2. Negative Interactions

Negative interaction occurs when there is a benefit to one of the interacting species while harm to the other occurs. Examples include competition, parasitism, amensalism, and predation.

a. *Predation:* This involves a dynamic between two animal species, wherein one species hunts, takes down, and consumes the other species. The species that carries out the hunting is identified as the predator, and the species being hunted is referred to as the prey. The predator gains advantages, while the prey suffers negative consequences (Figure 5.17).

Figure 5.17. Pitcher plant--*Nepenthes x ventrata*. Natural hybrid of *N. ventricosa* and *N. alata*.

Source: Dávila-Lara, A., Rodríguez-López, C. E., O'Connor, S. E., Mithöfer, A. (2020). Metabolomics Analysis Reveals Tissue-Specific Metabolite Compositions in Leaf Blade and Traps of Carnivorous *Nepenthes* Plants. *International Journal of Molecular Sciences*. *21*(12), 4376. https://doi.org/10.3390/ijms21124376.

Examples:

- Many plants like *Drosera* (Sundew Plant), *Nepenthes* (Pitcher Plant), *Dionaea* (Venus flytrap), *Utricularia* (Bladderwort), and *Sarracenia* are predators that consume insects and other small animals for their food as a source of nitrogen. They are also known as insectivorous plants.
- Many herbivores are predators. Cattle, camels, goats, etc., frequently browse on the tender shoots of herbs, shrubs, and trees. Generally, annuals suffer more than perennials. Grazing and browsing may cause remarkable changes in vegetation. Nearly 25% of all insects are known as phytophagous (feeding on plant sap and other parts of the plant).
- Many defense mechanisms evolved to avoid their predations by plants. For example, *Calotropis* produces highly poisonous cardiac glycosides, tobacco produces nicotine, coffee plants produce caffeine, and the *Cinchona* plant produces quinine. Thorns of *Bougainvillea,* spines of *Opuntia,* and latex of cacti also protect them from predators (Figure 5.18).

Figure 5.18. Insectivorous plant Utricularia.

Source: Płachno, B. J., Silva, S. R., □wi□tek, P., Dixon, K. W., Lustofin, K., Seber, G. C., Miranda, V. F. O. (2020). Structural Features of Carnivorous Plant (*Genlisea*, *Utricularia*) Tubers as Abiotic Stress Resistance Organs. *International Journal of Molecular Sciences*. *21*(14), 5143. https://doi.org/10.3390/ijms21145143.

b. *Parasitism:* In this type of interaction, the larger partner—the host or plant—provides food for the smaller partner, the parasite. Thus, the host species suffers while the parasitic species gains. The two types of parasitism are holoparasite and hemiparasite, and they are distinguished by the host-parasite relationship.

(i) *Holoparasites:* Holoparasites are organisms whose entire nutritional needs are met by the host plants. Another name for them is total parasites (Figure 5.19).

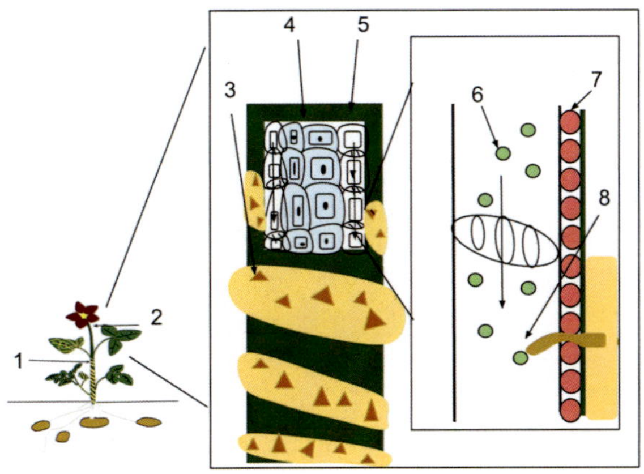

Figure 5.19. Cuscuta uses haustoria to penetrate the vascular system of its host plant and remove sugars and nutrients from the host's phloem. [Note: twining direction is reversed from that shown in this illustration.] (1) Cuscuta plant; (2) Host plant; (3) Cuscuta leaves; (4) Ground tissue; (5) Phloem; (6) Sugars and nutrients; (7) Epidermal tissue; (8) A Cuscuta haustorium growing into the phloem of the host plant.

Source: By Maliyanbinette – Own work, CC BY-SA 4.0, https://commons.wikimedia.org/w/index.php?curid=40500038.

Examples:

1. *Cuscuta* is a total stem parasite of the host plant *Acacia*, *Duranta* and many other plants. *Cuscuta* even gets flower-inducing hormone from its host plant.
2. *Balanophora, orobanche* and *Rafflesia* are the total root parasites found on higher plants.

(ii) *Hemiparasites:* Hemiparasites are those organisms that produce their own food through photosynthesis and only use water and minerals from their host plant. Another name for them is partial parasites.

Examples:

- *Viscum* and *Loranthus* are partial stem parasites.
- *Santalum* (Sandal Wood) is a partial root parasite

The parasitic plants develop their own internal haustorial roots to absorb nutrients from the host plant's vascular tissues.

c. *Competition:* It is an interaction between two species or organisms that causes harm to both. In a population with an irregular distribution, competition is at its most severe. Competition is divided into two categories: intraspecific and interspecific.

- *Intraspecific competition:* It is an interaction between members of the same species. Because every member of the species has the same needs for food, habitat, pollination, etc., this competition is quite intense. To meet their needs, they have developed comparable adaptations.
- *Interspecific competition:* It involves interactions between members of different species. Many types of grasses thrive in grasslands because there is little competition for resources like water and nutrients. During a drought, when

there is a water shortage, a struggle for survival ensues between the various grassland species. The amount of nutrients, the availability of water, and the ability to migrate to new areas all influence survival in these two competitions. Various herbivorous species, their larvae, and grasshoppers compete for forage or fodder plants. In a forest, trees, shrubs, and herbs compete with each other for light, water, nutrients, pollination, and the dispersal of fruits and seeds. Tiny fish and the Utricularia (Bladderwort) compete for small crustaceans and insects.

d. *Amensalism:* It is an instance of interspecific interaction wherein one species is hindered without causing any harm or benefit to the other. The secretion of specific substances known as allelopathic substances is what causes the inhibition. Antibiosis is another name for amensalism.

- *Penicillium notatum* produces penicillin to inhibit the growth of a variety of bacteria, especially *Staphylococcus*.
- *Trichoderma* inhibits the growth of the fungus *Aspergillus*.
- Roots and hulls of Black Walnut *Juglans nigra* secrete an alkaloid Juglone which inhibits the growth of seedlings of Apple, Tomato, and *Alfalfa* around it.

5.3.4.3. Interspecific interactions/ Co-evolutionary dynamics

i. **Mimicry:** This phenomenon occurs when a living organism alters its shape, appearance, behavior, or structure to resemble another living organism to defend itself and improve its chances of survival. While animal mimicry frequently serves as protection, floral mimicry typically attracts pollinators. Because of its shape, sudden heritable mutation, and ability to preserve natural selection, mimicry is an evolutionary phenomenon of great significance (Figure 5.20).

Figure 5.20. Mimicry by a leaf insect.

Source: By Sandilya Theuerkauf – Photo by Sandilya Theuerkauf, Wynaad, CC BY-SA 2.5, https://commons.wikimedia.org/w/index.php?curid=720559.

Example:

- The plant, *Ophrys* an orchid, the flower looks like a female insect to attract the male insect to get pollinated by the male insect and it is otherwise called 'floral mimicry '.
- *Carausium morosus* – stick insect or walking stick. It is a protective mimicry.
- *Phyllium frondosum* – leaf insect, another example of protective mimicry.

ii. Myrmecophily: Ants occasionally find refuge in certain trees, including acacia, mango, litchi, and jamun. These ants protect the plants from any intruders, and the plants give these ants food and shelter in return. We call this phenomenon myrmecophily. Acacia and acacia ants are examples (Figure 5.212).

Figure 5.21. Myrmecophily.

Source: By Geert Orye – https://www.flickr.com/photos/11741671@N03/3125149283/sizes/z/in/photostream/, CC BY 2.0, https://commons.wikimedia.org/w/index.php?curid=17541019.

iii. Co-evolution: The interaction between organisms, when it continues for generations, involves reciprocal changes in the genetic and morphological characteristics of both organisms. This type of evolution is called co-evolution. It is a kind of co-adaptation and mutual change among interactive species (Figure 5.22).

Figure 5.22. Co-evolution.

Source: By Charles J. Sharp – Own work, from Sharp Photography, sharpphotography, CC BY 3.0, https://commons.wikimedia.org/w/index.php?curid=12374160.

Examples:

- Corolla length and proboscis length of butterflies and moths (*Habenaria* and Moth).
- Bird's beak shape and flower shape and size.

Did you get it?

1. **Define the term 'ecological factor'.**
2. **What are the 4 ecological factors?**

CLOSER LOOK

FIVE ENVIRONMENTAL FACTORS THAT AFFECT HEALTH

In simple terms, the environment can be defined as everything surrounding us. Our environment consists of both living and nonliving entities, known as the biotic and abiotic components of the environment. These environmental factors are identifiable elements within the environment that influence an organism's survival, functions, and development. Environmental factors encompass all aspects that alter the natural environment. Some elements are visible, while others are not. Air, water, climate, soil, natural vegetation, and landforms are all environmental factors. By definition, environmental factors influence daily life and play a crucial role in creating health disparities among different geographic regions. The interaction between human and ecological factors significantly affects human health, either directly or indirectly.

The primary five environmental factors which affect the health and well-being of human life are:

Pollution

Pollution is defined as introducing pollutants, organic molecules, or other unsafe materials into the Earth's atmosphere. This can either be caused by natural events (like forest fires) or by man-made activities (like emissions from automobiles, factories, nuclear wastes, etc.)

The different types of pollution include:

- Air pollution
- Soil pollution
- Noise pollution
- Water pollution
- Radioactive pollution

Deforestation

Deforestation refers to the cutting down of trees and the destruction of forests. These can either be for human habitation and use or natural causes.

Deforestation has a tremendous negative impact on the environment. The most dramatic effect is a loss of trees and forests, which drives climate change, and several other problems have been raised for the ecosystem.

Solid Waste Pollution

Solid waste pollution is also referred to as a type of pollution, along with air, water, noise, and soil pollution.

Solid waste pollution refers to the improper disposal of solid waste, including garbage, construction waste, metal pieces, glass and plastic products, dead and decaying matter, solid agricultural waste, medical waste, radioactive waste, etc.

Global Warming

The phenomenon of a gradual increase in the average temperature of the Earth's atmosphere and its water bodies is termed global warming. The leading cause of this environmental issue is the increased volumes of greenhouse gases, such as carbon dioxide and methane released by burning fossil fuels, emissions from vehicles, industries, and other human activities.

Change in Climatic Conditions

There are several causes of climate change. The most significant effect on the climate is the increasing trend in the concentration of greenhouse gases and global warming in the atmosphere.

Climate change is having an adverse impact on the environment. These include the physical, chemical, and biological characteristics of ecosystems.

Other than natural disasters, humans are also responsible for climate change. The burning of fossil fuels and emissions from vehicles and industries are a few of them.

A change in the global climate includes:

- Typhoons
- Rising temperature
- An average to more rainfall
- Melting of global ocean currents and lots more.

ROLE MODEL

JEAN-BAPTISTE LAMARCK

Jean-Baptiste Lamarck was born in Northern France on August 1, 1744. He was the youngest of eleven children born to Philippe Jacques de Monet de La Marck and Marie-Françoise de Fontaines de Chuignolles, a noble but not rich family. Most men in Lamarck's family went into the military, including his father and older brothers. However, Jean's father pushed him toward a career in the Church, so Lamarck went to a Jesuit college in the late 1750s. When his father died in 1760, Lamarck rode off to a battle in Germany and joined the French army.

He quickly rose through the military ranks and became a commanding lieutenant over troops stationed in Monaco. Unfortunately, Lamarck was injured during a game he was playing with his troops, and after surgery made the injury worse, he was decommissioned. He then went off to study medicine with his brother but decided along the way that the natural world, and particularly botany, was a better choice for him.

Biography

In 1778, he published *Flore française*, a book that contained the first dichotomous key that helped identify different species based on contrasting characteristics. His work earned him the title of "Botanist to the King," which was given to him by Comte de Buffon in 1781. He was able to then travel around Europe and collect plant samples and data for his work.

Turning his attention to the animal kingdom, Lamarck was the first to use the term "invertebrate" to describe animals without backbones. He began collecting fossils and studying all sorts of simple species. Unfortunately, he became completely blind before he finished his writings on the subject, but he was assisted by his daughter so he could publish his works on zoology.

His most well-known contributions to zoology were rooted in the Theory of Evolution. Lamarck was the first to claim that humans had evolved from a lower species. In fact, his hypothesis stated that all living things built up from the most simple all the way up to humans. He believed that new species spontaneously generated and body parts or organs that were not used would just shrivel up and go away. His contemporary, Georges Cuvier, quickly denounced this idea and worked hard to promote his own nearly opposite ideas.

Jean-Baptiste Lamarck was one of the first scientists to publish the idea that adaptation occurred in species to help them better survive in the environment. He went on to assert that these physical changes were then passed down to the next generation. While this is now known to be incorrect, Charles Darwin used these ideas when forming his theory of Natural Selection.

Personal Life

Jean-Baptiste Lamarck had a total of eight children with three different wives. His first wife, Marie Rosalie Delaporte, gave him six children before she died in 1792. However, they did not marry until she was on her deathbed. His second wife, Charlotte Victoire Reverdy, gave birth to two children but died two years after they were married. His final wife, Julie Mallet, did not have any children before she died in 1819.

It is rumored that Lamarck may have had a fourth wife, but it has not been confirmed. However, it is clear that he had one deaf son and another son who was declared clinically insane. His two living daughters took care of him on his deathbed and were left poor. Only one living son was making a good living as an engineer and had children at the time of Lamarck's death.

SUMMARY

- An organism's environment encompasses everything that surrounds or impacts it throughout its life. This includes people, places, and objects, which can be natural or man-made.
- The environment is vital to both living (biotic) and nonliving (abiotic) components. Environmental issues like global warming, ozone layer depletion, deforestation, energy resource depletion, and loss of biodiversity affect everyone. The environment sets standards for a safe and flourishing natural ecosystem.
- In an ecosystem, various organisms coexist with physical, chemical, and biological components. External elements impact survival. Environmental factors encompass all these elements. Abiotic and biotic components combine to form an organism's environment.
- Water is one of the most significant climatic variables. Water impacts essential functions of all living things. During Earth's evolution, only water was believed to support life. More than 70% of the Earth's surface is water. In nature, plants obtain water through precipitation, soil water, and atmospheric moisture.
- Connate water is the liquid found beneath lakes and oceans in the pores and cavities of sedimentary rocks. Another name for it is "sediment water." It is the second significant groundwater source.
- Magmatic water is the water that is transformed into vapors by volcanic action when it enters hot rocks and condenses. Volcanic activity forces hot magma into rocks, where its vapor droplets condense and turn into water.
- "Wind" refers to air movement. It is an essential ecological component as well. Various gases, particles, and other elements are present in atmospheric air.

MULTIPLE CHOICE QUESTIONS

Tick the correct answer.

1. The major source of water for the soil is

a. groundwater

b. capillary water

c. precipitation

d. combined water

2. Soil particles arranged in order of increasing size are

a. sand-silt-clay

b. clay-sand-silt

c. silt-clay-sand

d. clay-silt-sand

3. **Soil that is made up of particles that have been carried in water and then deposited, is called**
   a. subsoil
   b. clay
   c. silt
   d. loam
4. **Which of these factors is biotic?**
   a. photoperiod
   b. CO_2 content of the soil
   c. texture and porosity of soil
   d. rainfall
5. **The factor governing the structure of the earth's surface is**
   a. topographic
   b. edaphic
   c. biotic
   d. temperature
6. **Plants that can withstand a wide range of temperatures are called**
   a. stenothermic
   b. monothermic
   c. mesothermic
   d. eurythermic
7. **Environment includes**
   a. Abiotic component
   b. Biotic component
   c. Oxygen and Nitrogen
   d. Abiotic and Biotic component
8. **Among the following climatic factors, which one has the least effect on a terrestrial ecosystem?**
   a. Temperature variation
   b. Wind
   c. Conditions of sunlight
   d. Availability of water
9. **The method of soil conservation in which rocks are piled up to slow down the flow of water which prevents gullies and further soil loss is called**
   a. Mulching
   b. Contour barriers

c. Rock dam

d. Terrace Farming

10. A man made ecological factor is

a. plowing

b. farming

c. grazing

d. fire

Answers

1. (c)	2. (d)	3. (c)	4. (b)	5. (a)
6. (d)	7. (d)	8. (b)	9. (c)	10. (d)

REVIEW QUESTIONS

1. What is the definition of the environment?
2. What are the categories in which plants are divided?
3. What are the three different varieties of thermal stratifications
4. What is connate water?
5. Which of the soil properties helps in the exchange of nutrients in plants?
6. Describe the effects of light on plants.
7. What makes up the organic matter component of soil?
8. What is the proportion of fresh and saline water on earth?
9. Describe how wind and wind current are produced.

REFERENCES

1. Arnaud, C. H., (2010). Exposing the exposome. Chemical & Engineering News (Vol. 88, No. 33, pp. 42–44). American Chemical Society.
2. Asthma and Its Environmental Triggers, (2006). National Institute of Environmental Health Sciences.
3. Buck, L. G. M., Yeung, E., Sundaram, R., et al., (2013). The exposome – exciting opportunities for discoveries in reproductive and perinatal epidemiology. Paediatric and Perinatal Epidemiology, 27(3), 229–236.
4. Gallagher, J., (2015). Cancer is Not Just 'Bad Luck' But Down to Environment, Study Suggests. BBC.
5. Lioy, P. J., & Rappaport, S. M., (2011). Exposure science and the exposome: An opportunity for coherence in the environmental health sciences. Environmental Health Perspectives, 119(11), A466, 467.

6. Miller, G., W., & Jones, D. P., (2014). The nature of nurture: Refining the definition of the exposome. Toxicological Sciences, 137(1), 1, 2.
7. Patel, C. J., Bhattacharya, J., & Butte, A. J., (2010). An environment-wide association study (EWAS) on type 2 diabetes mellitus. PLOS One, 5(5), e10746.
8. Rappaport, S. M., & Smith, M. T., (2010). Epidemiology. Environment and disease risks. Science, 330(6003), 460, 461.
9. Sharma, P. D., (2005). Ecology and Environment. Rastogi Publication.
10. Wild, C. P., (2005). Complementing the genome with an "exposome": The outstanding challenge of environmental exposure measurement in molecular epidemiology. Cancer Epidemiology, Biomarkers & Prevention, 14(8), 1847–1850.

CHAPTER 6

Agricultural Ecology

LEARNING OBJECTIVES

After studying this chapter, you will be able to:

- Discuss agroecology
- Understand ecological agriculture
- Identify human impacts on the global nitrogen cycle
- Describe the ecological impacts of increased global food production

INTRODUCTORY EXAMPLE

Agroecology shares much in common with other approaches to sustainable farming. Agroecology is farming that "centers on food production that makes the best use of nature's goods and services while not damaging these resources." Farming thrives when it works with local ecosystems, for example, improving soil and plant quality through available biomass and biodiversity, rather than battling nature with chemical inputs. Agroecological farmers seek to improve food yields for balanced nutrition, strengthen fair markets for their produce, enhance healthy ecosystems, and build on ancestral knowledge and customs.

Promoters of agroecology strive to nurture a healthy landscape in which to grow the world's food and fiber. They are guided by an ethos of bio and cultural diversity featuring small farmer-centered applied research and policies that protect their livelihoods. Worldwide, scientists, grassroots organizations, NGOs, consumers, universities, and public agencies are working with farmers to construct sustainable and nutritious food systems based on agroecology.

There are now unprecedented opportunities to advance agroecology globally. Too frequently, the corporate food system has negative impacts on people's health, the environment, and the well-being of family farmers. Agroecology is recognized as both a mitigation and adaptation strategy for climate change. Consumers are increasingly demanding healthier food and a closer connection to food producers. Social movements around the globe – many with significant leadership by women's and indigenous organizations – are coalescing in campaigns for a healthy food system built on an environmental and human rights ethos. The demand for agroecology is rising.

INTRODUCTION

Agriculture helps to meet the basic needs of humans and their civilization by providing food, clothing, shelter, medicine, and recreation. Hence, agriculture is the most important enterprise in the world. It is a productive unit where the free gifts of nature, namely land, light, air, temperature, and rainwater, etc., are integrated into a single primary unit indispensable for human beings. Secondary productive units, namely animals including livestock, birds, and insects, feed on these primary units and provide concentrated products such as meat, milk, wool, eggs, honey, silk, and lac. Agriculture provides food, feed, fiber, fuel, furniture, raw materials, and materials for and from factories; provides a free fare and fresh environment, abundant food for driving out famine; favors friendship by eliminating fights. Satisfactory agricultural production brings peace, prosperity, harmony, health, and wealth to individuals of a nation by driving away distrust, discord, and anarchy. It helps to elevate the community consisting of different castes and clauses; thus it leads to better social, cultural, political, and economic lives.

Agricultural development is multidirectional with galloping speed and rapid spread concerning time and space. After the Green Revolution, farmers started using improved cultural practices and agricultural inputs in intensive cropping systems with labor-intensive programs to enhance the production potential per unit land, time, and input. It provided the suitable environment for all these improved genotypes to foster and manifest their yield potential in newer areas and seasons. Agriculture consists of growing plants and rearing animals in order to yield, produce, and thus it helps to maintain a biological equilibrium in nature.

6.1. AGROECOLOGY

Learning Objectives

- Explain the principles of agroecology
- Describe the biodiversification of agroecosystems
- Discuss agroecology and the design of sustainable agroecosystems

The concept of sustainable agriculture is a relatively recent response to the decline in the quality of the natural resource base associated with modern agriculture. Today, the question of agricultural production has evolved from a purely technical one to a more complex one characterized by social, cultural, political, and economic dimensions. The concept of sustainability, although controversial and diffuse due to existing conflicting definitions and interpretations of its meaning, is useful because it captures a set of concerns about agriculture which is conceived as the result of the co-evolution of socioeconomic and natural systems. A wider understanding of the agricultural context requires the study of agriculture, the global environment, and social systems given that agricultural development results from the complex interaction of a multitude of factors. It is through this deeper understanding of the ecology of agricultural systems that doors will open to new management options more in tune with the objectives of truly sustainable agriculture.

The sustainability concept has prompted much discussion and has promoted the need to propose major adjustments in conventional agriculture to make it more environmentally, socially, and economically viable and compatible. Several possible solutions to the environmental problems created by capital and technology-intensive farming systems have been proposed, and research is currently in progress to evaluate alternative systems. The main focus lies on the reduction or elimination of agrochemical inputs through management changes to assure adequate plant nutrition and plant protection through organic nutrient sources and *integrated pest management*, respectively.

Although hundreds of more environmentally prone research projects and technological development attempts have taken place, and many lessons have been learned, the thrust is still highly technological, emphasizing the suppression of limiting factors or the symptoms that mask an ill-producing agroecosystem. The prevalent philosophy is that pests, nutrient deficiencies, or other factors are the cause of low productivity, as opposed to the view that pests or nutrients only become limiting if conditions in the agroecosystem are not in equilibrium. For this reason, there still prevails a narrow view that specific causes affect productivity, and overcoming the limiting factor via new technologies continues to be the main goal. This view has diverted agriculturists from realizing that limiting factors only represent symptoms of a more systemic disease inherent to imbalances within the agroecosystem and from an appreciation of the context and complexity of agroecological processes, thus underestimating the root causes of agricultural limitations.

On the other hand, the science of agroecology, which is defined as the application of ecological concepts and principles to the design and management of sustainable agroecosystems, provides a framework to assess the complexity of agroecosystems. The idea of agroecology is to go beyond the use of alternative practices and to develop agroecosystems with minimal dependence on high agrochemical and energy inputs, emphasizing complex agricultural systems in which ecological interactions and synergisms between biological components provide the mechanisms for the systems to sponsor their soil fertility, productivity, and crop protection.

Keyword

Integrated pest management is a science-based process that uses a combination of biological, cultural, and chemical practices to control insect pests in agriculture.

6.1.1. Principles of Agroecology

In the search to reinstate more ecological rationale into agricultural production, scientists and developers have disregarded a key point in the development of a more self-sufficient and sustaining agriculture: a deep understanding of the nature of agroecosystems and the principles by which they function. Given this limitation, agroecology has emerged as the discipline that provides the basic ecological principles for how to study, design, and manage agroecosystems that are both productive and natural resource conserving, and that are also culturally sensitive, socially just, and economically viable.

Did you know?

The term agroecology first appeared in 1928 under the pen of American agronomist Basil Bensin. His understanding of agroecology then referred exclusively to the application of methods of ecology to the processes of agronomic research.

Agroecology goes beyond a one-dimensional view of agroecosystems—their genetics, agronomy, edaphology, and so on—to embrace an understanding of ecological and social levels of coevolution, structure, and function. Instead of focusing on one particular component of the agroecosystem, agroecology emphasizes the interrelatedness of all agroecosystem components and the complex dynamics of ecological processes.

Agroecosystems are communities of plants and animals interacting with their physical and chemical environments that have been modified by people to produce food, fiber, fuel, and other products for human consumption and processing. Agroecology is the holistic study of agroecosystems, including all environmental and human elements. It focuses on the form, dynamics, and functions of their interrelationships and the processes in which they are involved. An area used for agricultural production, e.g., a field, is seen as a complex system in which ecological processes found under natural conditions also occur, e.g., nutrient cycling, predator/prey interactions, competition, symbiosis, and successional changes. Implicit in agroecological research is the idea that, by understanding these ecological relationships and processes, agroecosystems can be manipulated to improve production and to produce more sustainably, with fewer negative environmental or social impacts and fewer external inputs. The design of such systems is based on the application of the following ecological principles:

1. Enhance recycling of biomass and optimize nutrient availability and balance nutrient flow.
2. Securing favorable soil conditions for plant growth, particularly by managing organic matter and enhancing soil biotic activity.
3. Minimizing losses due to flows of solar radiation, air, and water by way of microclimate management, water harvesting, and soil management through increased soil cover.
4. Species and genetic diversification of the agroecosystem in time and space.
5. Enhancing beneficial biological interactions and synergies among agrobiodiversity components, thus resulting in the promotion of key ecological processes and services.

These principles can be applied through various techniques and strategies. Each of these will have different effects on productivity, stability, and resilience within the farm system, depending on the local opportunities, resource constraints, and, in most cases, on the market. The ultimate goal of agroecological design is to integrate components so that overall biological efficiency is improved, biodiversity is preserved, and the agroecosystem's productivity and self-sustaining capacity are maintained. The goal is to design a quilt of agroecosystems within a landscape unit, each mimicking the structure and function of natural ecosystems.

6.1.2. Biodiversification of Agroecosystems

From a management perspective, the agroecological objective is to provide balanced environments, sustained yields, biologically mediated soil fertility, and natural pest regulation through the design of diversified agroecosystems and the use of low-input technologies. Agroecologists are now recognizing that intercropping, agroforestry, and other diversification methods mimic natural ecological processes and that the sustainability of complex agroecosystems lies in the ecological models they follow. By designing farming systems that mimic nature, optimal use can be made of sunlight, soil nutrients, and rainfall.

Agroecological management must lead management to optimal recycling of nutrients and organic matter turnover, closed energy flows, water and soil conservation, and balance pest-natural enemy populations. The strategy exploits the complementarities and synergisms that result from the various combinations of crops, trees, and animals in spatial and temporal arrangements.

In essence, the optimal behavior of agroecosystems depends on the level of interactions between the various biotic and abiotic components. By assembling a functional biodiversity, it is possible to initiate synergisms that subsidize agroecosystem processes by providing ecological services such as the activation of soil biology, the recycling of nutrients, the enhancement of beneficial arthropods and antagonists, and so on. Today there is a diverse selection of practices and technologies available, which vary in effectiveness as well as in strategic value. Key practices are those of a preventative nature and act by reinforcing the "immunity" of the agroecosystem through a series of mechanisms.

Various strategies to restore agricultural diversity in time and space include crop rotations, cover crops, intercropping, crop/livestock mixtures, and so on, which exhibit the following ecological features:

1. *Crop Rotations*: Temporal diversity incorporated into cropping systems, providing crop nutrients and breaking the life cycles of several insect pests, diseases, and weed life cycles.
2. *Polycultures*: Complex cropping systems in which two or more crop species are planted within sufficient spatial proximity to result in competition or complementation, thus enhancing yields.
3. *Agroforestry Systems*: An agricultural system where trees are grown together with annual crops and/or animals, resulting in enhanced complementary relations between components increasing multiple uses of the agroecosystem.
4. *Cover Crops*: The use of pure or mixed stands of legumes or other annual plant species under fruit trees to improve soil fertility, enhancing biological control of pests, and modifying the orchard microclimate.
5. Animal integration in agroecosystems aids in achieving high biomass output and optimal recycling.

All of the above diversified forms of agroecosystems share in common the following features:

a. Maintain vegetative cover as an effective soil and water conserving measure, met through the use of no-till practices, mulch farming, and use of cover crops and other appropriate methods.

b. Provide a regular supply of organic matter through the addition of organic matter (manure, compost, and promotion of soil biotic activity).

c. Enhance nutrient recycling mechanisms through the use of livestock systems based on legumes, etc.

d. Promote pest regulation through enhanced activity of biological control agents achieved by introducing and/or conserving natural enemies and antagonists.

Research on diversified cropping systems underscores the great importance of diversity in an agricultural setting. Diversity is of value in agroecosystems for a variety of reasons:

- As diversity increases, so do opportunities for coexistence and beneficial interactions between species that can enhance agroecosystem sustainability.
- Greater diversity often allows better resource-use efficiency in an agroecosystem. There is better system-level adaptation to habitat heterogeneity, leading to complementarity in crop species needs, diversification of niches, overlap of species niches, and partitioning of resources.
- Ecosystems in which plant species are intermingled possess an associated resistance to herbivores, as in diverse systems there is a greater abundance and diversity of natural enemies of pest insects keeping in check the populations of individual herbivore species.

- A diverse crop assemblage can create a diversity of microclimates within the cropping system that can be occupied by a range of non-crop organisms —including beneficial predators, parasites, pollinators, soil fauna, and antagonists— that are of importance for the entire system.
- Diversity in the agricultural landscape can contribute to the conservation of biodiversity in surrounding natural ecosystems.
- Diversity in the soil performs a variety of ecological services such as nutrient recycling, detoxification of noxious chemicals, and regulation of plant growth.
- Diversity reduces risk for farmers, especially in marginal areas with more unpredictable environmental conditions. If one crop does not do well, income from others can compensate.

6.1.3. Agroecology and the Design of Sustainable Agroecosystems

Most people involved in the promotion of sustainable agriculture aim to create a form of agriculture that maintains productivity in the long term by:

- optimizing the use of locally available resources by combining the different components of the farm system, i.e., plants, animals, soil, water, climate, and people, so that they complement each other and have the greatest possible synergetic effects.
- reducing the use of off-farm, external, and non-renewable inputs with the greatest potential to damage the environment or harm the health of farmers and consumers, and a more targeted use of the remaining inputs used with a view to minimizing variable costs.
- relying mainly on resources within the agroecosystem by replacing external inputs with nutrient cycling, better conservation, and an expanded use of local resources.
- improving the match between cropping patterns and the productive potential and environmental constraints of climate and landscape to ensure the long-term sustainability of current production levels.
- working to value and conserve biological diversity, both in the wild and in domesticated landscapes, and making optimal use of the biological and genetic potential of plant and animal species; and
- taking full advantage of local knowledge and practices, including innovative approaches not yet fully understood by scientists although widely adopted by farmers.

Agroecology provides the knowledge and methodology necessary for developing an agricultural approach that is, on the one hand, environmentally sound and, on the other hand, highly productive, socially equitable, and economically viable. Through the application of agroecological principles, the basic challenge for sustainable agriculture to make better use of internal resources can be easily achieved by minimizing the external inputs used, and preferably by regenerating internal resources more effectively through diversification strategies that enhance synergisms among key components of the agroecosystem.

The ultimate goal of agroecological design is to integrate components so that overall biological efficiency is improved, biodiversity is preserved, and the agroecosystem productivity and its self-regulating capacity are maintained. The goal is to design an agroecosystem that mimics the structure and function of local *natural ecosystems*; that is, a system with high species diversity and a biologically active soil, one that promotes natural pest control, nutrient recycling, and high soil cover to prevent resource losses.

KEYWORD

The natural ecosystem is a community of living and nonliving entities that occurs freely in nature. Agroecology provides guidelines to develop diversified agroecosystems that take advantage of the effects of the integration of plant and animal biodiversity. Such integration enhances complex interactions and synergisms, optimizing ecosystem functions and processes, such as biotic regulation of harmful organisms, nutrient recycling, and biomass production and accumulation. This allows agroecosystems to support their own functioning. The end result of agroecological design is improved economic and ecological sustainability of the agroecosystem, with the proposed management systems specifically tailored to the local resource base and operational framework of existing environmental and socioeconomic conditions. In an agroecological strategy, management components are directed towards highlighting the conservation and enhancement of local agricultural resources (germplasm, soil, beneficial fauna, plant biodiversity, etc.) by emphasizing a development methodology that encourages farmer participation, use of traditional knowledge, and adaptation of farm enterprises that fit local needs and socioeconomic and biophysical conditions.

Did you get it?

1. What is agroecology?
2. What are the principles of agroecology?

6.2. ECOLOGICAL AGRICULTURE

Learning Objectives

- Explain the principles of ecological agriculture
- Discuss livestock and nutrient cycling
- Learn about methane emissions from paddy farming
- State the methane fluxes in wetlands

Ecological agriculture involves growing food in diverse systems that are 100% organic (chemical free) and avoid monocultures. It focuses on maintaining soil fertility by having a mix of plant species, including nitrogen fixers, throughout the year. The use of multifunctional trees and perennial plants is also emphasized. Fertilizers and sprays are derived from plant materials like compost teas. This farming approach is based on natural systems.

Ecological agriculture is truly sustainable as it can regenerate and refertilize degraded agricultural soils worldwide, ensuring continuous food production on the land. In contrast, conventional agriculture degrades and depletes the soil, posing challenges to food security. The environmental and social costs of the so-called "progress" in agriculture over the past decades have been significant.

Furthermore, ecological agriculture surpasses industrial agriculture in terms of yield, providing a continuous output from the land instead of relying on one or two monoculture harvests per year. It is more energy and cost-efficient when compared to conventional farming practices that heavily rely on chemical inputs. Ultimately, ecological agriculture offers a better output for the resources invested.

6.2.1. Principles of Ecological Agriculture

The term ecological agriculture is similar to agroecology as it applies ecological principles and approaches to agricultural ecosystems. While not excluding organic agriculture, it is a broader concept that may be closer to 'sustainable' agriculture. Ecology is the 'missing science' in traditional agricultural education and research. In going about the work of agriculture, we purposefully disturb natural ecosystems (or formerly natural ecosystems) to produce food and fiber crops and animal products of various kinds. These agroecosystems can be approached as any other ecosystem.

6.2.1.1. Strengths of Natural Ecosystems

Natural ecosystems exhibit certain strengths or characteristics. These include the following:

- *Efficiency:* Efficient energy flows are characteristic of natural systems. The sun's energy captured by green plants is then used by many organisms, as fungi and bacteria decompose organic residues and are then fed upon by other organisms, which are themselves fed upon by others higher up the food web. Natural ecosystems also tend to be efficient in capturing and using rainfall and in mobilizing and cycling nutrients. This helps to keep the ecosystem from 'running down' through the excessive loss of nutrients and at the same time helps maintain the quality of the groundwater and surface waters. Precipitation tends to enter the porous soil, rather than runoff, providing water to plants as well as recharge to groundwater, slowly releasing water to streams and rivers.

Remember

Manure handling and soil management practices should be made effective in capturing and recycling nutrients.

- *Diversity:* A great biological diversity, both above ground and in the soil, characterizes many natural ecosystems in temperate and tropical regions. This provides checks and balances, nutrient availability to plants, checks on disease outbreaks, etc. For example, competition for resources and specific antagonisms (such as antibiotic production) from the multitude of soil organisms usually keep soil-borne plant diseases from severely damaging a natural grassland or forest.
- *Self-sufficiency:* A consequence of efficiency and diversity is that natural terrestrial ecosystems are self-sufficient—requiring only inputs of sunlight and rainfall.
- *Self-regulation:* Because of the great diversity of organisms, outbreaks (or huge population increases) of diseases or insects that severely damage plants or animals are uncommon. In addition, plants have a number of defense mechanisms that help protect them from attack.
- *Resiliency:* Disturbances occur in all ecosystems— natural or not. The stronger ones are more resistant to disturbances and can bounce back quickly.

6.2.2. Livestock and Nutrient Cycling

In some crop-livestock systems, greater amounts of nutrients are taken out than go in. There is not enough animal manure for all cropland, and access to inorganic fertilizer is limited. Nutrient cycling patterns vary greatly across ecological zones, among different animal species and crop combinations, among different groups of farmers, and also from year to year. In the context of nutrient intake and use by ruminants, the extent of "nutrient mining" is not yet well understood. Some people feel that transferring nutrients from grazed pasture to manured fields has long-term negative effects on mixed farming systems. For example, the cut-and-carry system in Indonesia, based on forages from field boundaries and roadsides, has been operating for two centuries.

Nutrients must be added to sustain crop yields on inherently infertile soils. Since animals are part of many farming systems, their manure is a low-cost source of nutrients. The fertilizing value of manure depends on factors such as its total nutrient content (which depends on fodder quality), manure storage and application practices, and the rate of mineralization of nutrients in the manure.

In many areas, manure supply is influenced by factors such as herd size, grazing rights to communal range, and access to manure exchange contracts with mobile herders. Better manure management depends on available labor to collect, process, transport, and spread manure on fields. Cash is needed to build animal enclosures and to buy animals and carts for transport. Two possible scenarios may be considered. One is to make nutrient cycling more efficient by mobilizing on-farm resources, e.g., legume fallows, soil conservation, and better manure management. The second involves greater reliance on external inputs such as fertilizers, mechanization, and small-scale irrigation. Both scenarios may cover when population density increases land values, making fertility maintenance profitable. Better access to markets and greater inflows of non-farm income can also help the intensification process.

Keyword

Manure is organic matter that is used as organic fertilizer in agriculture.

Inadequate feed supplies, low returns to fertilizer use, inappropriate policies, and poor communication among researchers, policymakers, and farmers are some of the constraints to improve agricultural productivity. There is a need to quantify nutrient balances and develop models to predict trends in existing and improved farming systems. Ways to permit wider use of farm inputs and to develop markets to assure farm profitability need to be found. Technologies are needed to improve the management of natural vegetation and the nutritive quality of feeds and to reduce nutrient losses from manure and urine.

6.2.3. Methane Emission from Paddy Farming

Methane has been recognized as a powerful greenhouse gas, second only to CO_2, accounting for almost 20% of total global warming. Methane reacts with hydroxyl radicals in the troposphere.

While the global source strength of CH_4 is known within acceptable ranges, the contribution of individual sources is still uncertain. Methane is released into the atmosphere from a variety of natural and anthropogenic sources. Rice production accounts for 10–40% of the global methane source strength. Annual rice production is expected to increase greatly over the next three decades to meet the demand of the growing world population. As a result, the CH_4 source strength of wetland rice will continue to increase unless new technologies or mitigation options are developed that reduce methane emissions while allowing the necessary increases in rice yield.

The methane concentration in the atmosphere has more than doubled since the Industrial Revolution, from 800 parts per billion by volume (ppbv) to 1720 ppbv. The methane increase slowed down during the eighties. The world average increase was 12–14 ppbv per year in 1983 and 9–10 ppbv per year in 1990. The reason for the slowing down of the methane increase may be that the sources have not diminished during the eighties but there has been an increase, at least locally, of the major sink for methane oxidation in the troposphere (by OH radicals).

Methane is a product of thermogenic processes in fossil organic material in the Earth's sedimentary basins. This fossil methane leaks into the atmosphere during the production, transportation, and distribution of coal, oil, and gas. Methane is also emitted as a result of the incomplete combustion of fossil fuels and the burning of charcoal or wood. Methane is produced at the surface of the Earth in anaerobic environments. This is the case, for example, in shallow water and marshes, in wet rice fields, and in the rumen of cattle or in manure storage. This is called biogenic methane. From shallow water and marshes or rice fields, it is emitted through bubbling or the stems of aquatic plants. Part of the methane is oxidized in the water. It escapes from ruminants through belching. Landfills are another source of methane. It escapes through holes and cracks in the soil. Part of this methane is oxidized in the soil cover. Till now, the estimates of the contribution of rice agriculture to atmospheric methane have largely been based on field measurements in the 1980s, mainly in subtropical regions (California, Spain, and Italy), although most wetland rice is grown in the tropics. Field measurements for global estimates have usually been made during the growing phase of rice only.

Most of the atmospheric methane is of biogenic origin. Flooded rice fields are a significant source of atmospheric methane. Flooding a soil cuts off the soil's oxygen supply that stimulates anaerobic decomposition of organic matter and concomitant methane formation. Methane produced in the rice soil is released to the atmosphere through ebullition and rice plant-mediated diffusion. Expanding irrigation and increased production of wetland rice has enhanced methane fluxes to the atmosphere from rice fields. Annual methane emissions from rice fields to the atmosphere have been estimated to range from less than 20 to 100 million tons. If rice production increases up to 1 billion tons in the next century, it will further increase concomitant methane emissions.

The total global CH_4 emission source is about 600 Tg/year [Tg, teragrams (10^12g)] and the emission of 1 kg of CH_4 to the present-day atmosphere is 21 times more effective in perturbing the radiation balance than is an emission of 1 kg of CO_2. Wetland rice fields are among the important sources of atmospheric CH_4, with an estimated source strength of 60±40

Tg.year1 (Houghton et al., 1995). Microbial production in anoxic wetland rice soils is a chief source of atmospheric CH_4. As much higher CH_4 emissions from well-managed irrigated rice fields in the wet than in the dry season could not be explained by seasonal differences in temperature, it is hypothesized that high CH_4 emissions in the wet season are caused by low grain-to-biomass ratios. In a greenhouse experiment, removing spikelets to reduce the plants' capacity to store photosynthetically fixed C in grains led to increased CH_4 emissions, presumably via extra C inputs to the soil. Unfavorable conditions for spikelet formation in the wet season may likewise explain high methane emissions. This relationship between reduced grain filling and CH_4 emission provides opportunities to mitigate CH_4 emissions by optimizing rice productivity.

Figure 6.1 shows how productivity levels may vary among different areas depending on local environmental factors. As grain yield increases from areas 1 to 4, methane formation and emission increase according to the broken arrow; this happens due to increasing inputs of plant material into the soil, and by faster soil organic matter turnover. Within a given area, the numbered line shows a hypothetical yield-methane emission relation. Along this line, CH_4 emissions can be minimized by minimizing leakage of photosynthetic products from the plant and by optimizing storage of C in grain. It is known that reducing anthropogenic CH_4 emissions can minimize global warming over the next several decades. This suggests that rice agriculture may facilitate such an approach, perhaps even if production of this major staple crop of the world continues to increase.

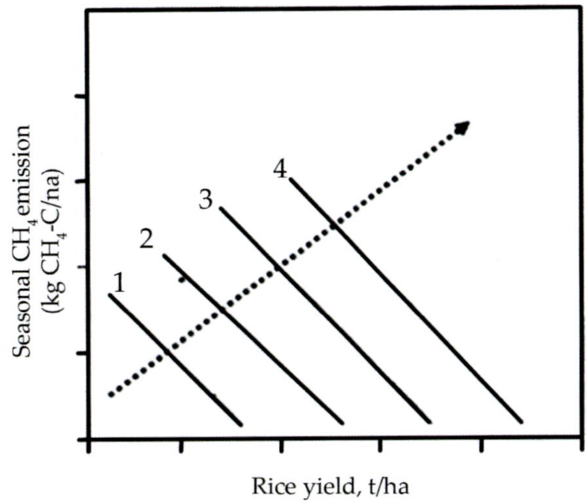

Figure 6.1. Hypothetical relationships between grain yield and seasonal methane emission for areas with specific sets of environmental conditions causing differences in overall agronomic productivity (lines 1–4) and across such areas, fur "average production" (bold broken arrow).

Source: Optimizing grain yields reduces CH4 emissions from rice paddy fields, H. A. C. Denier van der Gon, M. J. Kropff, N. van Breemen, +4, and H. H. van Laar Authors Info and Affiliations, August 20, 2002, 99 (19) 12021–12024, https://doi.org/10.1073/pnas.192276599.

Several studies of methane emissions from rice fields have been conducted in China, India, Indonesia, Japan, the Philippines, and Thailand, coordinated by the International Rice Research Institute. The studies aim to determine the methane source strength of major rice ecologies, its controlling factors and processes, and mitigation options. Factors controlling methane fluxes include (1) abiotic conditions such as temperature and pH; (2) organic matter content and quality; (3) the type of rice cultivation; and (4) cultural and agronomic practices.

Some promising mitigation options include better water management, organic amendments, fertilization, improved cultural practices, and rice cultivars.

6.2.4. Methane Fluxes in Wetlands

The composition of the atmosphere is strongly affected by microbial activities in soils and sediments as well as those of the terrestrial plant cover. Natural wetlands form an important component of the terrestrial ecosystem in the Northern Hemisphere. They constitute only about 3% of the total land surface area but have an enormous deposit of soil organic carbon, estimated at about 20% of the global total. In a large-scale integrated study, researchers looked at the biological, physical, and chemical controls on CH_4 and CO_2 emissions from process level to landscape and large scales. This study revealed that seasonal soil temperature is the best predictor of CH_4 emissions at the large scale, that the influence of vascular plant productivity is highly species-specific, and that the region is a net source of GHG emission throughout the year, despite the strong mid-summer uptake of CO_2. Of the eight important quantified sources of methane (wetlands, rice, ruminant animals, termites, fossil fuel, biomass burning, landfill), northern peat lands lead with a global estimate of methane emission of about 40 Tg yr^1.

Keyword

Soil organic carbon is a major contributor to overall soil health, agriculture, climate change, and food solutions.

Some important factors involved in gas emissions from wetlands include hydrology (levels of the water table), nutritional sources, microbial activities, temperature, and illumination. The surface vegetation plays a critical role in the temporal organization of gas release, and circadian control of plant physiology is significant. There is an interdependence and interplay among the separate biological, physical, and chemical controls on the carbon budget of the plant-soil-microbe system. Soil temperature is the most influential factor across large temporal and spatial scales. The water table level is usually an important control on seasonal variations in CH_4 flux at individual sites but is not a good predictor of fluxes at the large scale; at this scale, the water table probably acts more like an on-off switch. In general, the effects of vascular plants on CH_4 fluxes seem to depend on the species composition. The global warming effects of wetlands, traditionally regarded as sinks for CO_2 (as evidenced by the slow accumulation of peat) and as sources of CH_4 as a consequence of anaerobic processes of decay and decomposition, are highly complex. CH_4 seems to have a greater radiative forcing effect on the atmosphere than does CO_2.

Did you get it?

1. What are the principles of ecological agriculture?
2. What are the factors controlling methane flux?

6.3. HUMAN IMPACT ON THE GLOBAL NITROGEN CYCLE

Learning Objectives

- Describe the nitrogen cycle
- Explain human-driven nitrogen fixation
- Learn about impacts on the atmosphere
- Describe nitrogen saturation and ecosystem functioning
- Learn about the effects on biodiversity and the species mix

Human activities are greatly increasing the amount of nitrogen cycling between the living world and the soil, water, and atmosphere. In fact, humans have already doubled the rate of nitrogen entering the land-based nitrogen cycle, and that rate is continuing to climb. This human-driven global change is having serious impacts on ecosystems around the world because nitrogen is essential to living organisms and its availability plays a crucial role in the organization and functioning of the world's ecosystems. In many ecosystems on land and sea, the supply of nitrogen is a key factor controlling the nature and diversity of plant life, the population dynamics of both grazing animals and their predators, and vital ecological processes such as plant productivity and the cycling of carbon and soil minerals. This is true not only in wild or unmanaged systems but in most croplands and forestry plantations as well. Excessive nitrogen additions can pollute ecosystems and alter both their ecological functioning and the living communities they support.

Most of the human activities responsible for the increase in global nitrogen are local in scale, from the production and use of nitrogen fertilizers to the burning of fossil fuels in automobiles, power generation plants, and industries. However, human activities have not only increased the supply but enhanced the global movement of various forms of nitrogen through air and water. Because of this increased mobility, excess nitrogen from human activities has serious and long-term environmental consequences for large regions of the Earth.

The impacts of human domination of the nitrogen cycle that we have identified with certainty include:

- Increased global concentrations of nitrous oxide (N_2O), a potent greenhouse gas, in the atmosphere as well as increased regional concentrations of other oxides of nitrogen (including nitric oxide, NO) that drive the formation of photochemical smog.
- Losses of soil nutrients such as calcium and potassium that are essential for long-term soil fertility.
- Substantial acidification of soils and of the waters of streams and lakes in several regions.
- Greatly increased transport of nitrogen by rivers into estuaries and coastal waters where it is a major pollutant.

We are also confident that human alterations of the nitrogen cycle have:

- Accelerated losses of biological diversity, especially among plants adapted to low-nitrogen soils, and subsequently, the animals and microbes that depend on these plants.
- Caused changes in the plant and animal life and ecological processes of estuarine and nearshore ecosystems, and contributed to long-term declines in coastal marine fisheries.

6.3.1. The Nitrogen Cycle

Nitrogen is an essential component of proteins, genetic material, chlorophyll, and other key organic molecules. All organisms require nitrogen in order to live. It ranks fourth behind oxygen, carbon, and hydrogen as the most common chemical element in living tissues. Until human activities began to alter the natural cycle (Figure 6.2), however, nitrogen was only scantily available to much of the biological world. As a result, nitrogen served as one of the major limiting factors that controlled the dynamics, biodiversity, and functioning of many ecosystems.

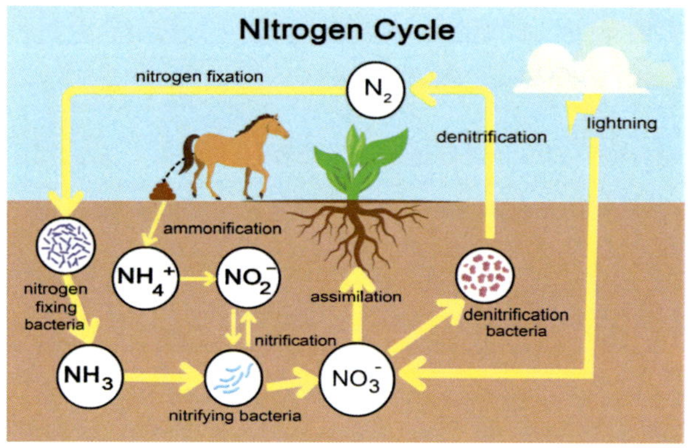

Figure 6.2. Simplified diagram of the nitrogen cycle.

Source: https://sciencenotes.org/wp-content/uploads/2023/08/Nitrogen-Cycle-1024x683.png.

The Earth's atmosphere is 78% nitrogen gas, but most plants and animals cannot use nitrogen gas directly from the air as they do carbon dioxide and oxygen. Instead, plants and all organisms from the grazing animals to the predators to the decomposers that ultimately secure their nourishment from the organic materials synthesized by plants must wait for nitrogen to be fixed, that is, pulled from the air and bonded to hydrogen or oxygen to form inorganic compounds, mainly ammonium (NH_4) and nitrate (NO_3), that they can use.

The amount of gaseous nitrogen being fixed at any given time by natural processes represents only a small addition to the pool of previously fixed nitrogen that cycles among the living and nonliving components of the Earth's ecosystems. Most of that nitrogen, too, is unavailable, locked up in soil organic matter partially rotted plant and animal remains that must be decomposed by soil microbes. These microbes release nitrogen as ammonium or nitrate, allowing it to be recycled through the food web. The two major natural sources of new nitrogen entering this cycle are nitrogen-fixing organisms and lightning.

Nitrogen-fixing organisms include a relatively small number of algae and bacteria. Nitrogen-fixing organisms include a relatively small number of algae and bacteria. Many of them live freely in the soil, but the most important ones are bacteria that form close symbiotic relationships with higher plants. Symbiotic nitrogen-fixing bacteria, such as the Rhizobia, for example, reside and operate in nodules on the roots of peas, beans, alfalfa, and other legumes. These bacteria produce an enzyme that allows them to convert gaseous nitrogen directly into forms usable by plants. Lightning may also indirectly convert atmospheric nitrogen into nitrates, which then fall onto the soil. Quantifying the rate of natural nitrogen fixation before human alterations of the cycle is challenging but necessary for assessing the impacts of human-induced changes to the global nitrogen cycle. The standard unit of measurement for analyzing the global nitrogen cycle is the teragram (abbreviated Tg), which is equivalent to a million metric tons of nitrogen. Worldwide, lightning, for instance, fixes less than 10 Tg of nitrogen per year, perhaps even less than 5 Tg. Microbes are the primary natural sources of new biologically available nitrogen. Prior to the widespread cultivation of legume crops, terrestrial organisms likely fixed between 90 and 140 Tg of nitrogen per year. Therefore, a reasonable upper limit for the rate of natural nitrogen fixation on land is approximately 140 Tg of N per year.

6.3.2. Human-Driven Nitrogen Fixation

During the past century, human activities clearly have accelerated the rate of nitrogen fixation on land, effectively doubling the annual transfer of nitrogen from the vast but unavailable atmospheric pool to the biologically available forms. The major sources of this enhanced supply include industrial processes that produce nitrogen fertilizers, the combustion of fossil fuels, and the cultivation of soybeans, peas, and other crops that host symbiotic nitrogen-fixing bacteria. Furthermore, human activity is also speeding up the release of nitrogen from long-term storage in soils and organic matter (Figure 6.3).

Figure 6.3. Nitrogen is the major factor limiting many terrestrial ecosystems, including most of those in the temperate zone, such as this oak savannah.

Source: By Professor X. – Own work. Uploaded by Professor X, Public Domain, https://commons.wikimedia.org/w/index.php?curid=282227.

The number and identities of the plant and animal species that live in such terrestrial ecosystems, and the functioning of the ecosystem, depend on the rate of nitrogen supply to the ecosystem.

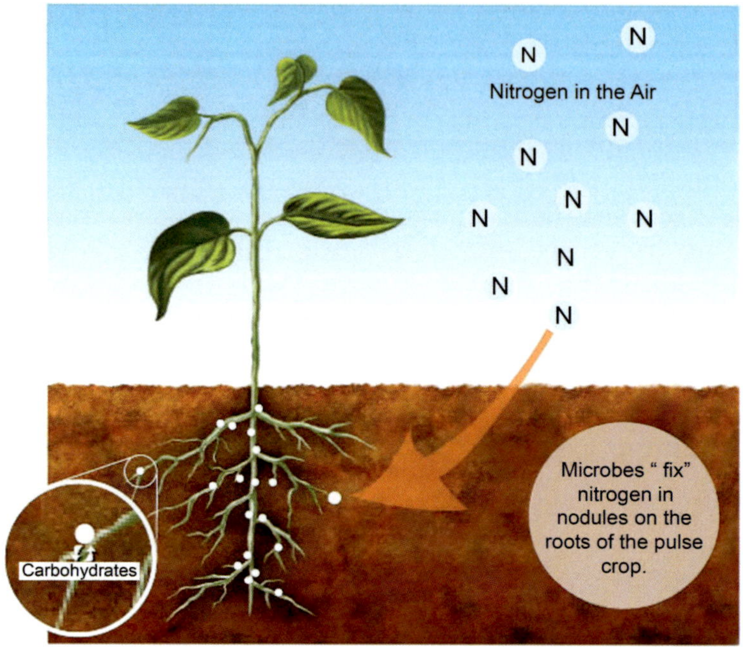

Figure 6.4. Nitrogen fixation and root nodule reduction and herbicide misuse.

Source: https://blogger.googleusercontent.com/img/b/R29vZ2xl/AVvXsEhaV_AbotsWw_R92DaEfzbwX7l_CkF8_SqaaGBxGeJF41seI7xBIWylzLFnjlWrZGtKlCgOQhknwMq9g7XL4HTYvK2no8TBflCdLJMPZs97DBktBtUaR1QdaUKByWHBFiplGUJFboJw-epp/s1600/Nitrogen-fixation.jpg.

6.3.2.1. Nitrogen Fertilizer

Industrial fixation of nitrogen for use as fertilizer currently totals approximately 80 Tg per year and represents by far the largest human contribution of new nitrogen to the global cycle (Figure 6.4). That figure does not include manures and other organic nitrogen fertilizers, which represent a transfer of already-fixed nitrogen from one place to another rather than new fixation.

The process of manufacturing fertilizer by industrial nitrogen fixation was first developed in Germany during World War I, and fertilizer production has grown exponentially since the 1940s. In recent years, the increasing pace of production and use has been truly phenomenal. The amount of industrially fixed nitrogen applied to crops during the decade from 1980 to 1990 more than equaled all industrial fertilizers applied previously in human history.

Until the late 1970s, most industrially produced fertilizer was applied in developed countries. Use in these regions has now stabilized while fertilizer applications in developing countries have risen dramatically. The momentum of human population growth and increasing urbanization ensures that industrial fertilizer production will continue at high and likely accelerating rates for decades in order to meet the escalating demand for food.

6.3.2.2. Nitrogen-Fixing Crops

Nearly one-third of the Earth's land surface is devoted to agricultural and pastoral uses, and humans have replaced large areas of diverse natural vegetation with monocultures of soybeans, peas, alfalfa, and other leguminous crops and forages. Because these plants support symbiotic nitrogen-fixers, they derive much of their nitrogen directly from the atmosphere and greatly increase the rate of nitrogen fixation previously occurring on those lands. Substantial levels of nitrogen fixation also occur during the cultivation of some non-legumes, notably rice. All of this represents new, human-generated stocks of biologically available nitrogen. The quantity of nitrogen fixed by crops is more difficult to analyze than industrial nitrogen production. Estimates range from 32 to 53 Tg per year. As an average, 40 Tg will be used here.

6.3.2.3. Fossil Fuel Burning

The burning of fossil fuels such as coal and oil releases previously fixed nitrogen from long-term storage in geological formations back to the atmosphere in the form of nitrogen-based trace gases such as nitric oxide. High-temperature combustion also fixes a small amount of atmospheric nitrogen directly. Altogether, the operations of automobiles, factories, power plants, and other combustion processes emit more than 20 Tg per year of fixed nitrogen to the atmosphere. All of it is treated here as newly fixed nitrogen because it has been locked up for millions of years and would remain locked up indefinitely if not released by human action.

6.3.2.4. Mobilization of Stored Nitrogen

Besides enhancing fixation and releasing nitrogen from geological reservoirs, human activities also liberate nitrogen from long-term biological storage pools such as soil organic matter and tree trunks, contributing further to the proliferation of biologically available nitrogen. These activities include the burning of forests, wood fuels, and grasslands, which emit more than 40 Tg per year of nitrogen; the draining of wetlands, which sets the stage for oxidation of soil organic matter that could mobilize 10 Tg per year or more of nitrogen; and land clearing for crops, which could mobilize 20 Tg per year from soils.

There are substantial scientific uncertainties about both the quantity and the fate of nitrogen mobilized by such activities. Taken together, however, they could contribute significantly to changes in the global nitrogen cycle.

6.3.2.5. Human versus Natural Nitrogen Fixation

Overall, fertilizer production, legume crops, and fossil fuel burning deposit approximately 140 Tg of new nitrogen into land-based ecosystems each year, a figure that equals the upper estimates for nitrogen fixed naturally by organisms in these ecosystems. Other human activities liberate and make available half again that much nitrogen. From this evidence, it is fair to conclude that human activities have at least doubled the transfer of nitrogen from the atmosphere into the land-based biological nitrogen cycle. This extra nitrogen is spread unevenly across the Earth's surface: Some areas such as northern Europe are being altered profoundly while others such as remote regions in the Southern Hemisphere receive little direct input of human-generated nitrogen. Yet no region remains unaffected. The increase in fixed nitrogen circulating around the globe and falling to the ground as wet or dry deposition is readily detectable, even in cores drilled from the glacial ice of Greenland.

6.3.3. Impacts on the Atmosphere

One major consequence of human-driven alterations in the nitrogen cycle has been regional and global change in the chemistry of the atmosphere (Figure 6.5). Specifically, increased emissions of nitrogen-based trace gases such as nitrous oxide, nitric oxide, and ammonia (NH_3) have occurred.

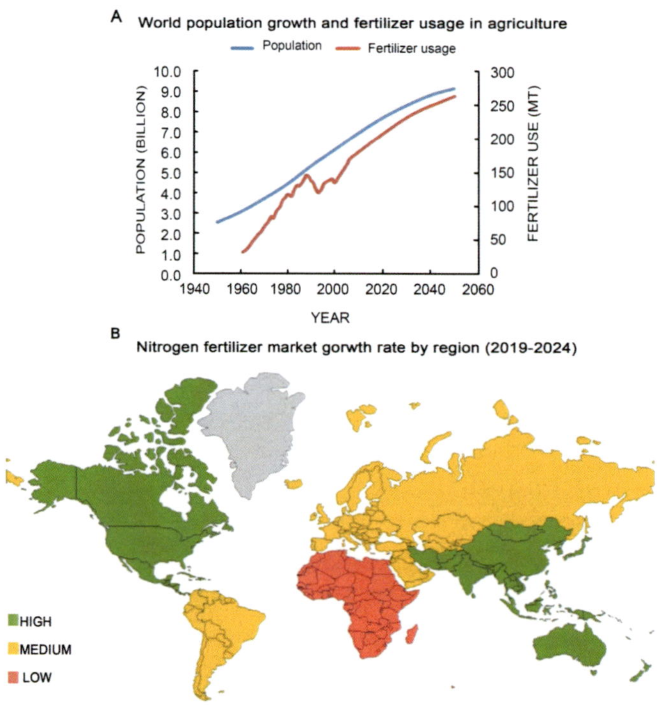

Figure 6.5. Human activities are responsible for a large proportion of the global emissions of nitrogen-containing trace gases, including 40% of the nitrous oxide, 80% or more of nitric oxide, and 70% of ammonia releases. The result is increasing atmospheric concentrations of the greenhouse gas nitrous oxide, of the nitrogen precursors of smog, and of biologically available nitrogen that falls from the atmosphere to fertilize ecosystems.

Source: Imran, A., Hakim, S., Tariq, M., Nawaz, M. S., Laraib, I., Gulzar, U., Hanif, M. K., Siddique, M. J., Hayat, M., Fraz, A., & Ahmad, M. (2021). Diazotrophs for Lowering Nitrogen Pollution Crises: Looking Deep Into the Roots. Front. *Microbiol. 12*, 637815. doi: 10.3389/fmicb.2021.637815.

Although such releases have received less attention than increased emissions of carbon dioxide and various sulfur compounds, the trace nitrogen gases cause environmental effects both while airborne and after they are deposited on the ground. For instance, nitrous oxide is long-lived in the atmosphere and contributes to the human-driven enhancement of the greenhouse effect that likely warms the Earth's climate. Nitric oxide is an important precursor of acid rain and photochemical smog.

Some of the human activities discussed above affect the atmosphere directly. For instance, essentially all of the more than 20 Tg per year of fixed nitrogen released in automobile exhausts and in other emissions from fossil fuel burning is emitted to the atmosphere as nitric oxide. Other activities indirectly enhance emissions to the atmosphere. Intensive fertilization of agricultural soils can increase the rates at which nitrogen in the form of ammonia is volatilized and lost to the air. It can also speed up the microbial breakdown of ammonium and nitrates in the soil, enhancing the release of nitrous oxide. Even in wild or unmanaged lands downwind of agricultural or industrial areas, rain or windborne deposition of human-generated nitrogen can spur increased emissions of nitrogen gases from the soils.

Keyword

Greenhouse gas is a gas that absorbs infrared radiation from the Earth's surface and reradiates it back to the Earth's surface.

6.3.3.1. Nitrous Oxide

Nitrous oxide is a very effective heat-trapping gas in the atmosphere, in part because it absorbs outgoing radiant heat from the Earth in infrared wavelengths that are not captured by the other major greenhouse gases, water vapor, and carbon dioxide. By absorbing and reradiating this heat back toward the Earth, nitrous oxide contributes a few percent to overall greenhouse warming. Although nitrous oxide is unreactive and long-lived in the lower atmosphere, when it rises into the stratosphere, it can trigger reactions that deplete and thin the stratospheric ozone layer that shields the Earth from damaging ultraviolet radiation.

The concentration of nitrous oxide in the atmosphere is currently increasing at the rate of two to three-tenths of a percent per year. While that rise is clearly documented, the sources of the increase remain unresolved. Both fossil fuel burning and the direct impacts of agricultural fertilization have been considered and rejected as the major source. Rather, there is a developing consensus that a wide array of human-driven sources contribute systematically to enrich the terrestrial nitrogen cycle. These dispersed sources include fertilizers, nitrogen-enriched groundwater, nitrogen-saturated forests, forest burning, land clearing, and even the manufacture of nylon, nitric acid, and other industrial products. The net effect is increased global concentrations of a potent greenhouse gas that also contributes to the thinning of the stratospheric ozone layer.

6.3.3.2. Nitric Oxide and Ammonia

Unlike nitrous oxide, which is unreactive in the lower atmosphere, both nitric oxide and ammonia are highly reactive and therefore much shorter-lived. Thus changes in their atmospheric concentrations can be detected only at local or regional scales.

Nitric oxide plays several critical roles in atmospheric chemistry, including catalyzing the formation of photochemical (or brown) smog. In the presence of sunlight, nitric oxide and oxygen react with hydrocarbons emitted by automobile exhausts to form ozone, the most dangerous component of smog. Ground-level ozone has serious detrimental effects on human health as well as the health and productivity of crops and forests.

Nitric oxide, along with other oxides of nitrogen and sulfur, can be transformed in the atmosphere into nitric acid and sulfuric acid, which are the major components of acid rain. Although a number of sources contribute to nitric oxide emissions, combustion is the dominant one. Fossil fuel burning emits more than 20 Tg per year of nitric oxide. Human burning of forests and other plant material may add about 10 Tg, and global emissions of nitric oxide from soils, a substantial fraction of which are human-caused, total 5 to 20 Tg per year. Overall, 80% or more of nitric oxide emissions worldwide are generated by human activities, and in many regions, the result is increased smog and acid rain. In contrast to nitric oxide, ammonia acts as the primary acid-neutralizing agent in the atmosphere, having an opposite influence on the acidity of aerosols, cloud water, and rainfall. Nearly 70% of global ammonia emissions are human-caused. Ammonia volatilized from fertilized fields contributes an estimated 10 Tg per year; ammonia released from domestic animal wastes about 32 Tg; and forest burning some 5 Tg.

6.3.4. Nitrogen Saturation and Ecosystem Functioning

There are limits to how much plant growth can be increased by nitrogen fertilization. At some point, when the natural nitrogen deficiencies in an ecosystem are fully relieved, plant growth becomes limited by the scarcity of other resources such as phosphorus, calcium, or water. When the vegetation can no longer respond to further additions of nitrogen, the ecosystem reaches a state described as nitrogen saturation. When an ecosystem is fully nitrogen-saturated and its soils, plants, and microbes cannot use or retain any more, all new nitrogen deposits will be dispersed to streams, groundwater, and the atmosphere.

Nitrogen saturation has a number of damaging consequences for the health and functioning of ecosystems. These impacts first became apparent in Europe almost two decades ago when scientists observed significant increases in nitrate concentrations in some lakes and streams and also extensive yellowing and loss of needles in spruce and other conifer forests subjected to heavy nitrogen deposition. These observations led to several field experiments in the U.S. and Europe that have revealed a complex cascade of effects set in motion by excess nitrogen in forest soils.

As ammonium builds up in the soil, it is increasingly converted to nitrate by bacterial action, a process that releases hydrogen ions and helps acidify the soil. The buildup of nitrate enhances emissions of nitrous oxides from the soil and also encourages leaching of highly water-soluble nitrate into streams or groundwater. As these negatively charged nitrates seep away, they carry with them positively charged alkaline minerals such as calcium, magnesium, and potassium. Thus human modifications to the nitrogen cycle decrease soil fertility by greatly accelerating the loss of calcium and other nutrients that are vital for plant growth. As calcium is depleted and the soil acidified, aluminum ions are mobilized, eventually reaching toxic concentrations that can damage tree roots or kill fish if the aluminum washes into streams. Trees growing in soils replete with nitrogen but starved of calcium, magnesium, and potassium can develop nutrient imbalances in their roots and leaves. This may reduce their photosynthetic rate and efficiency, stunt their growth, and even increase tree deaths.

Nitrogen saturation is much further advanced over extensive areas of northern Europe than in North America because human-generated nitrogen deposition is several times greater there than in even the most extremely affected areas of North America. In the nitrogen-saturated ecosystems of Europe, a substantial fraction of atmospheric nitrate deposits move from the land into streams without ever being taken up by organisms or playing a role in the biological cycle.

In contrast, in the northeastern U.S., increased leaching of nitrates from the soil and large shifts in the nutrient ratios in tree leaves generally have been observed only in certain types of forests. These include high-elevation sites that receive large nitrogen deposits and sites with shallow soils containing few alkaline minerals to buffer acidification. Elsewhere in the U.S., the early stages of nitrogen saturation have been seen in response to elevated nitrogen deposition in the forests surrounding the Los Angeles Basin and in the Front Range of the Colorado Rockies. Some forests have a very high capacity to retain added nitrogen, particularly regrowing forests that have been subjected to intense or repeated harvesting, an activity that usually causes severe nitrogen losses. Overall, the ability of a forest to retain nitrogen depends on its potential for further growth and the extent of its current nitrogen stocks. Thus, the impacts of nitrogen deposition are tightly linked to other rapidly changing human-driven variables such as shifts in land use, climate, and atmospheric carbon dioxide and ozone levels.

6.3.5. Effects on Biodiversity and the Species Mix

Limited supplies of biologically available nitrogen are a common challenge in most natural ecosystems, and many native plant species are well-suited to thrive under this condition. Introducing new nitrogen supplies to these ecosystems can lead to significant changes in the dominant species and a decrease in overall species diversity as certain plant species, adapted to high nitrogen levels, outcompete others. For instance, in England, nitrogen fertilizers applied to experimental grasslands have resulted in the increased dominance of a few nitrogen-responsive grasses, leading to the decline of many other plant species. The plant species count decreased by more than fivefold at the highest fertilization rate. Similarly, in North America, biodiversity reductions have been observed in grasslands in Minnesota and California due to fertilization. In Western Europe, human-induced nitrogen deposition has been linked to substantial biodiversity losses in formerly species-rich heathlands over the past few decades. In the Netherlands, high human population density, intensive livestock operations, and industries have combined to generate the highest rates of nitrogen deposition in the world. One well-documented consequence has been the conversion of species-rich heathlands to species-poor grasslands and forests. Not only has the species richness of the heath been reduced, but also the biological diversity of the landscape has been diminished because the modified plant communities now resemble the composition of communities occupying more fertile soils. The unique species assemblage adapted to sandy, nitrogen-poor soils is being lost from the region. Losses of biodiversity driven by nitrogen deposition can, in turn, affect other ecological processes.

Did you get it?

1. **What is the impact of human domination on the nitrogen cycle?**
2. **What role does nitric oxide play in atmospheric chemistry?**

6.4. ECOLOGICAL IMPACTS OF INCREASED GLOBAL FOOD PRODUCTION

Learning Objectives

- Learn about the effects of agricultural practices on the environment
- Describe the sustainable agricultural growth

We can survive without many things, but we cannot survive without food, the production of which has become a serious environmental concern. The rapidly growing world's population requires increased food production, which is one of the greatest causes of environmental degradation throughout the world. Of particular concern is meat production, as modern practices of animal raising directly contribute to water and air pollution and increase carbon dioxide emissions, while crop production for animal feed and the use of land for grazing threaten biodiversity and wildlife species.

Factory farming has reduced the amount of land needed for meat production; however, these farms are serious air and water pollutants. The waste from these animals ends up in nature and poses a constant risk of drinking water contamination, seriously affecting the air quality of nearby areas. One solution to the problem of animal waste lies in its use for the production of biofuel, which can then be used for electricity production, but this practice is the exception rather than the rule.

Mass meat production has also been shown to be one of the main contributors to carbon dioxide emissions which in turn are the main cause of climate change. The meat industry is estimated to be responsible for about 9% of total carbon dioxide emissions which are a result of emissions of various gases from the farms as well as from the microbial activities after application of animal waste as fertilizers.

Lastly, animal husbandry poses a serious threat to the local ecosystems and biodiversity due to the use of the land for grazing and animal feed production. As much as one-quarter of the Earth's surface is used for grazing and about one-third of arable land is used to produce animal feed. As a result, the wildlife species struggle with a lack of habitat, while some are even threatened with extinction. Unfortunately, animal husbandry is not the only environmental problem of food production. Crop production negatively affects the environment too. It uses even more land, while the use of pesticides, artificial fertilizers and other chemicals seriously pollutes the environment and threatens wildlife species. Avoidance or reduction of meat consumption therefore does not solve the problem of the harmful effects of food production on the environment.

The key solution to the problem of food production and its effects on the environment also lies in choosing locally produced food as less fossil fuel is used for its transportation as well as choosing organic over non-organic food. Organic food supplements are produced without the use of chemicals which means that they dramatically reduce water, soil, and air pollution. Furthermore, organic food production reduces the pressure on the ecosystems by avoiding the use of toxic agricultural chemicals as well as farming in harmony with nature.

6.4.1. Effects of Agricultural Practices on Environment

Humans have benefited from nature since their existence. It cannot be said that the relationship between humans and nature was initially negative. Humans organized local areas while dealing with agricultural activities, and for a long time, they applied agricultural processes without damaging the balance of nature. The results of science and technology, like various activities and methods, acquired attributes that disrupted nature. For a certain period, human beings lived unaware of the damage they were causing to nature. Initially, the detriment of humans on the environment was not considered due to the environment's renewability feature, even though it was believed that the environment would eradicate impurities. Over time, the increase in harm to the environment exceeded the renewability capacity of the environment, leading to rapid deterioration.

On the other hand, pollution and corruption resulting from common human activities affected the quantity and quality of agricultural production, sparking important discussions on agriculture-environment relationships. There are many plants used for medical purposes. *Matricaria recutita L.* is a well-known medicinal plant species used in several medicinal preparations and its essential oil. It has been reported that agricultural practices impact sesquiterpenoid compounds of *M. recutita* L. Animals also benefit from nature and, in turn, affect humans directly or indirectly. Among all the parameters used to assess the quality of hay, fungal diversity and the production of breathable dust were the most sensitive parameters to agricultural practices and climatic factors.

As environmental conditions affect agricultural practices, agricultural practices also have effects on the environment. Namely, agriculture affects the global flow of greenhouse gases. The main reason for the destruction of forest land is to obtain agricultural land. As a result of obtaining agricultural land, greenhouse gases are created. These greenhouse gases represent the second major negative impact after the negative effects of greenhouse gases created by the use of fossil fuels. Forests store carbon 20 to 40 times more effectively than agricultural lands, and most of this carbon is released into the atmosphere when forests are destroyed to make way for agricultural land.

6.4.1.1. Negative Effects of Agricultural Applications

Environment is described as the external medium where humans, animals, and plants live together. Environment consists of two parts: human and natural. Environmental pollution occurs due to irregular urbanization and unconscious industry and applications. Also, the balance between humans and the natural environment in which humans live is disrupted. The main reasons for pollution are irregular and rapid industry, urbanization, organic and inorganic wastes left in the environment, unintended usage of agricultural lands, and incorrect agricultural applications. Misuse of pesticides and chemical fertilizers, irrigation, tillage, and plant hormone applications are some of the incorrect practices. Additionally, stubble burning, planting without rotation, and inappropriate animal waste disposal are considered mistakes. Evaluating the effects of best management practices (BMPs) in agricultural watersheds is often complicated by significant temporal variability in weather and hydrologic conditions. BMPs are increasingly being utilized by decision-makers to reduce agricultural non-point source pollution while enhancing productivity for farmers. There is a need to consider local-level policies and practices, informed by a combination of participatory approaches and sound science at an appropriate scale.

Remember

Drinking water should not contain more than 20 ppm nitrate.

6.4.1.2. Pesticide Usage

Pesticides are used for the elimination of harmful insects, microorganisms, and other pests. When they mix with soil, water, air, and food, they can affect both human health and the natural balance, ultimately becoming an environmental problem. Pesticide runoff is a significant contributor to surface-water contamination. Pesticides not only kill their target but also many harmless organisms. Modeling stream water pollution by herbicides in agricultural areas is a critical issue due to numerous and incompletely known processes involved. It has been reported that alternative implementation designs combining the use of herbaceous riparian buffers with other practices capable of altering nutrient and pesticide loads, riparian hydrology, and in-stream habitat are needed.

Furthermore, fields, streams, lakes, groundwater, and seas can become storage areas for various poisons over time. There are hundreds of pesticides used worldwide. According to the World Health Organization (WHO), out of 700 mostly used pesticides, 33 are very dangerous, 48 are quite dangerous, 118 are moderately dangerous, and 239 are less dangerous. Developed countries have a 75% rate of pesticide usage.

6.4.1.3. Chemical Fertilizer Usage

The fertilizers that are used to improve plant growth can cause environmental pollution in case of excessive or incorrect usage. Using high amounts of nitrogen fertilizer can result in soil leaching, contaminating groundwater, drinking water, streams, and the sea. This can also impact water organisms and disrupt the natural balance of the environment.

Furthermore, lettuce and spinach grown in soils with high nitrogen levels can accumulate NO_2 and NO_3, as well as certain carcinogenic substances like nitrosamine.

For this purpose, many European countries have limitations on nitrogen fertilizer usage in groundwater conservation regions. Excessive use of phosphorus fertilizers also disrupts the natural balance by increasing phosphate levels in water. Excess phosphorus is harmful to domestic plants.

6.4.1.4. Irrigation

Irrigation is crucial for achieving high agricultural yields and quality in arid and semi-arid regions. Improper irrigation practices can lead to environmental problems such as the rise of groundwater salinity, the mixing of fertilizers and chemical residues with irrigation water, the accumulation of trace elements in water sources, and soil erosion. Water contaminated in this way can cause diseases and harm living organisms, posing a significant environmental issue.

Furthermore, excessive irrigation for agricultural production purposes can result in soil salinity and desertification. It can be argued that agricultural policies influencing land use also impact the extent of soil erosion in agricultural regions by altering the economic conditions of agricultural production.

6.4.1.5. Soil Tillage

Wrong soil tillage with regards to field location, soil structure, and climate conditions causes erosion. This situation not only causes inefficient soils, but it also pollutes streams, fills up dams with soil, and creates serious environmental problems. Cultivation of natural ecosystems has led to a marked decline in soil C storage, such that conservation agricultural practices are widely recommended as options to increase soil C storage, thereby mitigating climate change.

6.4.1.6. Rotation

Bioenergy crops play an ecologically and economically fundamental role as an alternative to agri-food production and as renewable energy sources. Little attention has been focused on soil quality following the conversion of agricultural lands to biomass crops. Agricultural applications that lack rotation due to a lack of knowledge or economic reasons cause a decrease in soil fertility, degradation, increased disease, and harm to the soil, as well as erosion.

6.4.1.7. Plant Hormone Usage

The term plant hormone refers to organic substances created by plants that can have a significant impact, even at very low intensities, on growth, development, and yield. Using plant hormones is harmless when the appropriate dosage and timing are observed, but the same hormones could have toxic effects if used carelessly. One of the most commonly used hormones is 2,4-D. The permitted amount of this hormone varies from one country to another. For example, Sweden does not permit any residue of 2,4-D, while Germany allows 2.0 ppm in citrus species and 0.1 ppm for other products..

6.4.1.8. Stubble Burning

For the elimination of stems, hay, and especially secondary products, stubble is burned to prepare for seed sowing. However, it is evident that stubble burning causes very significant environmental problems. It leads to wind and water erosion, damages natural vegetation, and renders the soil infertile by depleting vitality in the topsoil. Due to these environmental harms, stubble burning is prohibited by laws in many countries.

6.4.1.9. Animal Wastes

Animal production has brought about numerous changes in various industry sectors. These changes have raised a wide range of concerns regarding the impacts of animal wastes on the environment. In animal husbandry, especially poultry farming, negative effects on the environment arise from manure, urine, and the processing wastes of animals and animal products. These organic wastes contaminate the soil and water bodies, in addition to the dust, gas, and odor effects on the environment. Animal wastes play a significant role in environmental pollution.

6.4.1.10. Positive Effects of Agricultural Practices

While agriculture has negative effects on the environment, it also has positive impacts. For example, regions with common agricultural practices experience various favorable environmental effects such as supporting natural life, oxygen production, and climate regulation depending on the region and ecology. In fertilized fields, oxygen is produced through photosynthesis, increasing the atmospheric oxygen levels. Therefore, cereal production areas generate 12 tons of oxygen per hectare. Oxygen production in agricultural areas surpasses that of forests and vacant lands. In these areas, air pollution decreases due to the absorption of CO_2.

Modern agricultural practices use many kinds of chemicals such as fertilizers, pesticides, cleaners, and crop preservatives to produce and maintain large amounts of high-quality food. However, each of these chemicals has dangerous and unforeseen side effects, such as toxicity to non-target organisms, leading to ecological imbalances. As mentioned above, improper agricultural practices can result in environmental pollution on a significant scale. In other words, agricultural techniques, especially modern ones, could lead to environmental pollution if humans are not vigilant. Consequently, humanity has adopted a new perspective to mitigate the negative impacts of agriculture.

Sustainable agriculture, a novel agricultural approach, appears to be environmentally friendly and is endorsed by developed countries. Environmentally friendly agriculture encompasses three main practices: good agricultural practices, organic agriculture, and precision agriculture. Additionally, practices like crop rotation, planting legumes for nitrogen fixation, and fallowing help mitigate the adverse effects of agriculture on climate change. It is imperative to encourage research on agricultural practices using sustainable and ecologically friendly methods.

Water and air are essential resources for agriculture and all vital activities. A clean, beautiful, green, and healthy environment, free from pollution in the air, water, and soil, as well as noise and other contaminants, is the primary requirement of modern society and a guarantee for the future.

6.4.2. Sustainable Agricultural Growth

Sustainable agriculture is the production of food, fiber, or other plant or animal products using farming techniques that protect the environment, public health, human communities, and animal welfare. This form of agriculture enables us to produce healthful food without compromising future generations' ability to do the same.

The primary benefits of sustainable agriculture are discussed below.

6.4.2.1. Environmental Preservation

Sustainable farms produce crops and raise animals without relying on toxic chemical pesticides, synthetic fertilizers, genetically modified seeds, or practices that degrade soil, water, or other natural resources. By growing a variety of plants and using techniques such as crop rotation, conservation tillage, and pasture-based livestock husbandry, sustainable farms protect biodiversity and foster the development and maintenance of healthy ecosystems.

6.4.2.2. Protection of Public Health

Food production should never come at the expense of human health. Since sustainable crop farms avoid hazardous pesticides, they can grow fruits and vegetables that are safer for consumers, workers, and surrounding communities. Likewise, sustainable livestock farmers and ranchers raise animals without dangerous practices like the use of nontherapeutic antibiotics or arsenic-based growth promoters. Through careful, responsible management of livestock waste, sustainable farmers also protect humans from exposure to pathogens, toxins, and other hazardous pollutants.

6.4.2.3. Sustaining Vibrant Communities

A critical component of sustainable agriculture is its ability to remain economically viable, providing farmers, farmworkers, food processors, and others employed in the food system with a livable wage and safe, fair working conditions. Sustainable farms also bolster local and regional economies, creating good jobs and building strong communities.

6.4.2.4. Upholding Animal Welfare

Sustainable farmers and ranchers treat animals with care and respect, implementing livestock husbandry practices that protect animals' health and well-being. By raising livestock on pasture, these farmers enable their animals to move freely, engage in instinctive behaviors, consume a natural diet, and avoid the stress and illness associated with confinement.

6.4.2.5. Industrial Agriculture

Unfortunately, most food produced in the US is no longer grown or raised on sustainable farms. During the mid-1900s, US agriculture began to industrialize, becoming increasingly mechanized and reliant upon resource-intensive inputs like synthetic fertilizers and chemical pesticides. Over time, farms became larger, more specialized, and centralized, creating a process of extreme consolidation that drove many small farms out of business, and ultimately resulted in market control by a handful of powerful corporations. Although industrial agriculture now produces great quantities of food at low prices, it is able to do so only by implementing practices that threaten the environment, human health, rural communities, and animal welfare.

6.4.2.6. Industrial Crop Production

Today, industrial crops are produced on huge monocrop farms, which rely extensively on chemical pesticides, synthetic fertilizers, and genetically modified crop varieties. These practices deplete and degrade soil, reduce biodiversity, and generate air and water pollutants that degrade the environment and threaten the health of farmworkers, neighbors, and consumers.

6.4.2.7. Industrial Livestock Production

The majority of meat, eggs, and dairy products are now produced in enormous industrial livestock facilities. Also known as factory farms or CAFOs (concentrated animal feeding operations), these facilities confine thousands (and, in some cases, hundreds of thousands) of animals in cramped conditions without access to the outdoors. In addition to compromising animal welfare, factory farms generate a huge amount of waste, which pollutes air, water, and soil, degrading the natural environment and threatening public health.

6.4.2.8. Sustainable Food Future

Although industrial agriculture currently dominates the US food system, public awareness of the problems caused by this model has grown rapidly, building extensive support for sustainable agriculture, creating a robust market for sustainable foods, and inspiring formidable demand for agricultural policy and regulatory reform.

Did you get it?

1. **What are the negative effects of agricultural applications?**
2. **What are the positive effects of agricultural applications?**

CLOSER LOOK

CONSTRAINTS TO ECOLOGICAL AGRICULTURE IN THE US

The basic principles and many practices of ecological agriculture are well known. So why aren't they being practiced on most farms in the US and abroad? For more complex rotations to be used on a majority of conventional farms, there need to be large and easily accessible markets for the additional products—for example, something other than corn and soybeans in the US corn belt. And while it makes ecological sense to reunite animals with the land that raises their feeds, decisions made by large corporations about where to place processing facilities and how they want 'their' animals raised make it difficult for farmers to sell large numbers of animals outside that industrial system that has developed. (It should be noted that the buy-local movement has led to the development of new markets for some farmers through the food service industry. How large and long-lasting an effect this will have is yet to be seen.) Another constraint in the US is that certain commodities receive subsidy payments while others do not. Why leave a relatively dependable system for one with many economic unknowns?

A major constraint that we may face in the future has recently emerged—the use of crops to produce energy. The price of corn increased by 60% in the fall of 2006, mainly due to the increased use of the crop to produce ethanol. The USDA has projected that ethanol production will consume 20% of the US corn crop in 2006. With numerous ethanol plants currently under construction, ethanol will claim a growing share of the corn market and will compete with corn's use as animal feed and feedstock for various byproducts such as high fructose syrup. Given a more profitable corn crop, it will be challenging to transition corn belt farmers to more intricate rotations. With the accelerated drive to produce ethanol (and other products) from crop residues, less organic matter will be reintroduced to the land, resulting in less fertile soils.

While it is true that many farmers are adopting ecological approaches on their farms, few of these are large-scale growers aiming to sell their products in the undifferentiated wholesale commodity markets. Currently, there are government subsidies available for such farmers to adopt more environmentally friendly practices like using cover crops or establishing riparian buffer zones. However, even when environmentally friendly practices or systems are employed, a fully developed ecologically based system is rarely achieved. Therefore, farmers who implement a truly ecologically based system are typically small-scale growers who also aim to secure a premium price above wholesale commodity prices—such as operators of Community Supported Agriculture farms (CSAs) and organic farmers who sell at farmers' markets, farm stands on-site, and to local restaurants.

Over the past decade, there has been a huge increase in interest among farmers in developing value-added products—from bottling milk, to making yogurt, to farmstead cheeses, to processing meat and poultry, and so on. These farmers still represent only a small niche in US agriculture, although very important in some areas and definitely still growing. But the increasing desire for organic foods has been noticed by the 'big guys,' and we now have what should be a non sequitur—organic factory farms. Very large-scale organic dairies as well as vegetable and fruit farms are now common. And of all the outlets now getting interested in organic food, Walmart is jumping in as part of its effort to attract upscale customers. Thus, the small-scale farmer raising crops and animals ecologically needs to cultivate a local constituency—regardless of whether value is added to the products sold—and/or enter into values-based marketing networks with other farmers using their own brand(s).

ROLE MODEL

DAVID LOBELL

David Lobell is an agricultural ecologist and a 2013 MacArthur Fellow. He uses advanced satellite, remote sensing, and geographic information systems (GIS) technologies to research ways to sustainably feed the world's growing population. His latest work helps measure the direct impact of climate change on global food systems.

Early Work

David had an innate curiosity about food from a young age, despite his urban upbringing.

"Everybody has some sort of interest in food at some level, even if it's just eating. I grew up very removed from any sort of agriculture, and that probably had some sort of perverse impact on me becoming really fascinated by it because I just was not exposed to it as a kid."

David's first experiences doing field work cemented his desire to continue to work with food and agriculture. As a graduate student, he began working with the International Maize and Wheat Improvement Center in Mexico (Centro Internacional de Mejoramiento de Maiz y Trigo, or CIMMYT). CIMMYT works to reduce worldwide poverty and hunger by increasing the productivity of maize (corn) and wheat, both important food staples.

Most Exciting Part Of Your Work

Much of David's time is spent trying to make sense of raw data on agricultural productivity.

"I like to do puzzles and work with numbers. For me, the most rewarding thing day in [and] day out is figuring out what the data is saying and what patterns there are in these large, noisy data sets that we work with. Sometimes it feels like we're completing a jigsaw puzzle."

More importantly, David and his team are excited to see when their analysis influences the decisions of farmers, plant breeders, and agronomists. These everyday decisions change the way our food is grown.

"The most rewarding for me is when somebody who's working in agriculture tells me that my work helped them change what they were doing or think about something differently. That's when I get to feeling like I'm making an impact."

David's recent findings, for example, helped farmers adapt their techniques to cope with higher overall temperatures due to climate change. This allows them to maintain productivity and continue feeding large populations despite the changing environmental context.

Most Demanding Part of Your Work

Over the course of his career, the challenges David faces have shifted considerably.

"When I was getting into this work about ten years ago, the biggest challenge was that nobody really cared about agriculture, but that didn't really bother me. Still, it was very hard to find an audience and raise money in academia . . . Over time, agriculture and food have both become much more popular [fields of study] as we've seen food prices go up. So that's less of a challenge now."

David attributes the increased interest in agriculture and food security to the overall rise in attention to climate change in recent years.

How Do You Define Geography?

"I've always heard varying definitions. One time I interviewed in a geography department and they told me that their definition of geography is whatever a geographer is doing! To me, geography is anything that has place or space as its central element.

"A lot of what we do in agriculture is getting information about spatial variation. Anytime you have spatial information associated with your analysis that is inherently place-based."

Geo-Connection

The geographic notion of variation across space is central to David and his team's research strategy.

"For us, examining different locations is a valuable method for testing theories or calibrating models. Comparing various places provides a wealth of information. In our studies, we analyze thousands of different locations, each with slightly different attributes. We leverage this variation to gain insights."

Moreover, "the significance of place and local context" plays a significant role in seeking solutions to the diverse problems David and his team address.

Geographic tools such as GIS, remote sensing, and satellite imagery are crucial in the data-collection process, enabling David and his team to delve deeper into spatial patterns. David estimates that approximately half of his work involves satellite data.

Therefore, You Want to Pursue a Career as an Agricultural Ecologist. David encourages individuals interested in agriculture to utilize opportunities to explore agricultural regions using online tools like Google Earth, a practice he regularly incorporates in his classes at Stanford University.

"Explore the agricultural areas of the world on Google Earth. There is much to learn by observing the landscapes there."

Especially for students, David stresses the importance of thinking creatively, particularly in the sciences.

"There's a notion that all science is done by certain robots out there who know exactly what they're doing and it's a very linear path. But there's a real creative element of science, as well. Once you become an expert at something, you can really think creatively and [about] what else you can do. It's hard to see the light at the end of that tunnel when you're in middle school or high school, to think that you could one day be the person creating things no one thought was possible, but it's really a lot of fun in the end."

Get Involved

Organizations like the United States Department of Agriculture (USDA) and the Food and Agriculture Organization (FAO) release periodic updates on the state of food security or agricultural production in different regions of the world, making it easy to follow the latest developments in the field.

CASE STUDY

AGROFORESTRY TO IMPROVE FARM PRODUCTIVITY IN MALI

Agroforestry is a collective name for land-use systems and technologies where woody perennials (trees, shrubs, palms, bamboos, etc.) are deliberately used on the same land-management units as agricultural crops and/or animals, in some form of spatial arrangement or temporal sequence. In agroforestry systems, there are both ecological and economic interactions between the different components. Agroforestry can also be defined as a dynamic, ecologically based, natural resource management system that, through the integration of trees on farms and in the agricultural landscape, diversifies and sustains production for increased social, economic, and environmental benefits for land users at all levels. In particular, agroforestry is crucial to smallholder farmers and other rural people because it can enhance their food supply, income, and health. Agroforestry systems are multifunctional systems that can provide a wide range of economic, sociocultural, and environmental benefits.

Challenge

In Mali, approximately 65% of the total population lives in rural areas, and some 80% of the nation's labor force is engaged in agriculture. Rain-fed agriculture encounters limitations such as the availability of arable land, extreme and unpredictable weather, pest damage, and poor soil quality. Many districts experience ongoing deficits of the key subsistence crops—maize, sorghum, and millet.

In the last 30 years, rapidly increasing population density has placed intense pressure on soil resources, rendering traditional methods of rotational agriculture—long bush fallow periods, where cultivations are shifted to another field to allow regeneration of the land—obsolete. Continual exploitation of agricultural land has severely degraded soil quality. Alleviating hunger and sustaining healthy human and animal populations mandate new and improved agricultural methods.

Response

Agroforestry has shown great results to combat these pressures. It integrates trees and/or herbaceous or woody perennial shrubs into agricultural systems to enhance specific ecological processes. Trees and shrubs are harvested or fulfill a variety of purposes: providing fencing and building materials, preventing soil erosion, modifying micro-climatic conditions, enhancing the habitat for beneficial arthropods, providing animal fodder, food, shade, fuel, medicine, or fertilizer for other crops. Science-based agroforestry interventions have a 30-year history of improving sub-Saharan African farmers' lives.

Improved fallows involve planting fast-growing woody and herbaceous plant species for short time periods—6 months to 2 years—to replenish soil fertility prior to additional grazing or crop cultivation. Plants used in improved fallows are selected for their rapid growth (biomass production) and soil fertility enhancement (nitrogen fixing), among other characteristics. Short-duration improved fallows replace the traditional rotational farming practice of five to seven years long bush fallows for maintaining soil fertility and yields in sub-Saharan Africa.

Southern Mali's Koutiala region is known for its extremely poor growing conditions, characterized by highly eroded soils, hard iron-oxide rich pans, and low levels of plant nutrients and organic matter. Research conducted between 1996 and 1998 in the region combined on-farm experimentation in N'Goukan village and research plots at N'Tarla Agricultural Research Station. The research evaluated the impact on soil fertility and maize yields of two tree species (*Gliricidia sepium* and *Pteocarpus erinaceus)* and a nitrogen-fixing leguminous shrub (*Stylosanthes hamata*). For both on-farm and experiment station research, treatments were compared individually and in combination with a natural grass fallow and other commonly used soil fertility management practices.

Two separate soil improvement strategies were tested: improved fallow, which involved treatments grown on site, and biomass transfer, with treatments grown off-site and brought to research plots for application. At the on-farm research site, treatments were planted and allowed to grow for two growing seasons in a short fallow cycle. After two growing seasons, the agroforestry treatments were cut at a uniform height (30–40 cm), and biomass was applied to the soil surface and incorporated into the soil before planting a local maize variety. For the experiment station research, the same agroforestry treatments were applied using two separate biomass application rates: "Rate One" was the equivalent of 50 kilograms (kg) of nitrogen per hectare (ha) per year; "Rate Two" was a one-time application of biomass equivalent to the agroforestry treatments' biomass potential at three to four years of age.

The impact of the three agroforestry treatments (*Gliricidia sepium, Pterocarpus erinaceus*, and *Stylosanthes hamata*) was measured at the application rates and compared with the following commonly used treatments: the locally recommended rates of manure application (10 tons/ha); the conventional annual application rate for mineral fertilizer (100 kg/ha); and a natural grass fallow. The same local maize variety was planted in the plots at the experiment station and grown to maturity. Stover biomass, the non-edible part of the plant usually used as cattle fodder, and grain yields were measured after harvest in 1997 and 1998, and a range of soil quality assessments were conducted before and after harvest to evaluate the experimental treatments' effect on soil fertility.

Results

- In the on-farm experimentation, maize produced the greatest amount of grain and stover when followed by a two-year improved fallow period using *Gliricidia* alone or in combination with *Stylosanthes,* a plant that also provides high-quality cattle fodder.

- The agroforestry treatments had a significant effect on crop yields: maize grain yields from the *Gliricidia* treatment plots were eight percent higher than regional averages, which are 2 tons/ha.
- *Gliricidia* plus *Stylosanthes* treatments hold the greatest promise as short-rotation improved fallow species for enhancing maize productivity and providing livestock fodder without the use of synthetic fertilizer inputs. These treatments yield the second-highest amount of maize stover (7 tons/ha), and the highest maize grain yield of any of the treatments (2.9 tons/ha).
- In contrast, maize yields from the use of synthetic fertilizer, using the conventional annual application rate (100 kg urea/ha), were 1.7 tons/ha in 1997 and 2.6 tons/ha in 1998, not outperforming the *Gliricidia* plus *Stylosanthes* treatment.
- On average, yields from all on-farm plots with the agroforestry treatments were higher than the traditional grass fallow control, which produced only 4 tons/ha of corn stover and 0.7 ton/ha of maize grain (lower than the current regional mean).
- The data suggest that annual applications of smaller quantities of biomass from agroforestry treatments (Rate One) produce better and more consistent yields than one-time application of large biomass quantities (Rate Two).
- The three species' decomposition rates occurred with *Gliricidia* exhibiting the fastest rate of decomposition followed by *Stylosanthes* and then *Pterocarpus*. Data on decomposition rates further suggest that the combination of *Gliricidia* and *Stylosanthes* was the best candidate for providing essential plant nutrients and improving yields over the short term.
- Due to the study's limited two-year duration, it was not possible to document treatments' lasting effects on the soil fertility parameters, including soil carbon, nitrogen, and phosphorus.
- Soil analysis showed that soil carbon and nitrogen levels were significantly, though temporarily, enhanced in the *Gliricidia* treatment and that soil carbon concentration was reduced for the fertilizer-applied treatment.
- Data indicate that the agroforestry treatments of *Gliricidia* plus *Stylosanthes* have the best potential for alleviating food insecurity in the region because they enhance maize yield and short-term soil fertility and produce the preferred fodder.

In Mali and across Africa, agroforestry has shown real and lasting results when implemented in a thoughtful and sustained manner. From improved soil fertility to increased crop yields and food security, agroforestry shows that simple measures can have profound effects on communities.

SUMMARY

- The concept of sustainable agriculture is a recent response to the decline in natural resource quality linked to modern agriculture.
- In the quest to reintroduce more ecological reasoning into agricultural production, scientists and developers have overlooked a crucial aspect in the advancement of a more self-sustaining agriculture: a profound comprehension of the nature of agroecosystems and the principles governing their operation.
- Agroecosystems are communities of plants and animals interacting with their physical and chemical environments that have been modified by people to produce food, fiber, fuel, and other products for human consumption and processing.
- Ecological agriculture involves cultivating food in diverse, 100% organic systems without chemicals or monocultures. It enhances soil fertility by maintaining a diverse mix of plant species, including nitrogen fixers, year-round, and utilizes multifunctional trees and perennial plants.
- The term ecological agriculture is akin to agroecology, applying ecological principles to agricultural ecosystems. Efficiency and diversity lead to self-sufficient natural terrestrial ecosystems, needing only sunlight and rainfall as inputs.
- Human activities are significantly increasing the nitrogen cycling between the living world and the soil, water, and atmosphere.
- Nitrogen is crucial for proteins, genetic material, chlorophyll, and other essential organic molecules. All organisms need nitrogen to survive, ranking fourth in living tissues after oxygen, carbon, and hydrogen.
- Bioenergy crops are vital ecologically and economically, serving as alternatives to agri-food production and renewable energy sources.

MULTIPLE CHOICE QUESTIONS

1. **In which year did the term agroecology first appear?**
   a. 1938.
   b. 1928.
   c. 1929.
   d. 1939.
2. **Who coined the word agroecology?**
   a. Basil Bensin.
   b. William Albrecht.
   c. Sanford Soverhill Atwood.
   d. James Nathan Calloway.

3. **Which one is the principle of AE?**
   a. The improvement of agro-food process.
   b. Constant improvement of Environmental performance
   c. Both a and b.
   d. None of the above.
4. **To which principle does the complementarity amongst the elements of agroecosystems belong?**
   a. Synergy.
   b. Biodiversity.
   c. (Recycling.
   d. Economic diversification.
5. **What is recycling in reference to AE?**
   a. Recycling of biomass.
   b. Recycling of nutrients.
   c. Both a and b.
   d. None of the above.

REVIEW QUESTIONS

1. What is ecological agriculture? Write down its principles.
2. How does animal agriculture cause global warming? Explain.
3. Define agroecology.
4. Discuss the biodiversification of agroecosystems.
5. Write about human impacts on the global nitrogen cycle.
6. Describe the ecological impacts of increased global food production.

Answers

1. (b) 2. (a) 3. (c) 4. (a) 5. (c)

REFERENCES

1. Stephen, R. G., (2000). *Agroecosystem Sustainability: Developing Practical Strategies (Advances in Agroecology).* Hardcover.
2. Francis, et al., (2003). Agroecology: The ecology of food systems. *Journal of Sustainable Agriculture, 22*(3), 99–118.
3. Garí, J. A., (2004). *Plant Diversity, Sustainable Rural Livelihoods, and the HIV/AIDS Crisis.* Bangkok: UNDP & FAO, 2004. Published in English and Chinese.

4. Suárez-Torres, J., Suárez-López, J. R., López-Paredes, D., Morocho, H., Cachiguango-Cachiguango, L. E., & Dellai, W., (2017). *Agroecology and Health: Lessons from Indigenous Populations*. Current Environmental Health Reports.
5. Wezel, A., Bellon, S., Doré, T., Francis, C., Vallod, D., & David, C., (2009). Agroecology as a science, a movement and a practice: A review. *Agronomy for Sustainable Development, 29*(4), 503–515.
6. Wibblemann, et al., (2013). *Mainstreaming Agroecology: Implications for Global Food and Farming Systems*. "Archived copy" (PDF).

CHAPTER 7

Population Ecology

LEARNING OBJECTIVES

After studying this chapter, you will be able to:

- Understand the overview of population ecology
- Discuss population growth
- Define population interaction
- Describe intra-specific competition
- Identify inter-specific competition
- Explain population dynamics

INTRODUCTORY EXAMPLE

A population is a group of interacting organisms of the same species and includes individuals of all ages or stages: pre-reproductive juveniles and reproductive adults. Most populations have a mix of young and old individuals. Quantifying the numbers of individuals of each age or stage gives the demographic structure of the population. In addition to demographic structure, populations vary in total number of individuals, called population size, and how densely packed together those individuals are, called population density. A population's geographic range has limits, or bounds, established by the physical limits that the species can tolerate, such as temperature or aridity, and by the encroachment of other species. Population ecologists often first consider the dynamics of population size change over time, of whether the population is growing in size, shrinking, or remaining static over time.

INTRODUCTION

A population is a collection of individuals of the same species that live together in a region. Population ecology is the study of populations (especially population abundance) and how they change over time. Crucial to this study are the various interactions between a population and its resources. A population can decline because it lacks resources or it can decline because it is prey to another species that is increasing in numbers. Populations are limited by their resources in their capacity to grow; the maximum population abundance (for a given species) an environment can sustain is called the carrying capacity. As a population approaches its carrying capacity, overcrowding means that there are fewer resources for the individuals in the population and this results in a reduction in the birth rate. A population with these features is said to be density-dependent. Of course, most populations are density-dependent to some extent, but some grow (almost) exponentially and these are, in effect, density-independent. Ecological models that focus on a single species and the relevant carrying capacity are single species models. Alternatively, multi-species or community models focus on the interactions of specific species.

The discipline of population ecology holds a great deal of philosophical interest. To begin with, we encounter all the usual problems in the philosophy of science, often with new and interesting twists, as well as other problems that seem peculiar to ecology. Some of the familiar problems from the philosophy of science include the nature of explanation and its relationship to laws, and whether higher-level sciences (like ecology) are reducible to lower-level sciences (like biochemistry). Some of the philosophical problems that arise from within population ecology include whether there is a balance of nature and how the uneasy relationship between the mathematical and empirical sides of the discipline might be understood.

7.1. OVERVIEW OF POPULATION ECOLOGY

Learning Objectives

- Describe the properties of a population
- Illustrate the genetic variation within local populations
- Learn about the effects of the modes of reproduction: sexual and asexual

Population ecology is the study of the processes that affect the distribution and abundance of animal and plant populations. A population is a subset of individuals of one species that occupies a particular geographic area and, in sexually reproducing species, interbreeds. The geographic boundaries of a population are easy to establish for some species but more difficult for others. For example, plants or animals occupying islands have a geographic range defined by the perimeter of the island. In contrast, some species are dispersed across vast expanses, and the boundaries of local populations are more difficult to determine. A continuum exists from closed populations that are geographically isolated from, and lack exchange with, other populations of the same species to open populations that show varying degrees of connectedness.

7.1.1. Properties of a Population

The three main properties of a population are Population size (the number of individuals in an area), density (the number of individuals per unit or volume), and dispersion (the arrangement of its individuals).

- *Population size* is represented by the letter N, and it equals the total number of individuals in a population. The larger a population is, the greater its genetic variation and therefore its potential for long-term survival. Increased population size can, however, lead to other issues, such as overuse of resources leading to a population crash.
- *Population density* refers to the number of individuals in a particular area. A low-density area would have more organisms spread out. High-density areas would have more individuals living closer together, leading to greater resource competition.

Population dispersion yields helpful information about how species interact with each other. Researchers can learn more about populations by studying the way they are distributed or dispersed.

Keyword

Natural selection is the process through which populations of living organisms adapt and change.

Population distribution describes how individuals of a species are spread out, whether they live in close proximity to each other or far apart, or clustered into groups.

- Uniform dispersion refers to organisms that live in a specific territory. One example would be penguins. Penguins live in territories, and within those territories, the birds space themselves out relatively uniformly.
- Random dispersion refers to the spread of individuals such as wind-dispersed seeds, which fall randomly after traveling.
- Clustered or clumped dispersion refers to a straight drop of seeds to the ground, rather than being carried, or to groups of animals living together, such as herds or schools. Schools of fish exhibit this manner of dispersion.

7.1.2. Genetic Variation within Local Populations

Did you know?

In the 1940s, ecology was divided into autecology—the study of individual species in relation to the environment—and synecology—the study of groups of species in relation to the environment.

In sexually reproducing species, each local population contains a distinct combination of genes. As a result, a species is a collection of populations that differ genetically from one another to a greater or lesser degree. These genetic differences manifest themselves as differences among populations in morphology, physiology, behavior, and life histories; in other words, genetic characteristics (genotype) affect expressed, or observed, characteristics (phenotype). *Natural selection* initially operates on an individual organismal phenotypic level, favoring or discriminating against individuals based on their expressed characteristics. The gene pool (total aggregate of genes in a population at a certain time) is affected as organisms with phenotypes that are compatible with the environment are more likely to survive for longer periods, during which time they can reproduce more often and pass on more of their genes.

The amount of genetic variation within local populations varies tremendously, and much of the discipline of conservation biology is concerned with maintaining genetic diversity within and among populations of plants and animals. Some small isolated populations of asexual species often have little genetic variation among individuals, whereas large sexual populations often have great variation. Two major factors are responsible for this variety: mode of reproduction and population size.

7.1.3. Effects of Modes of Reproduction: Sexual and Asexual

In sexual populations, genes are recombined in each generation, and new genotypes may result. Offspring in most sexual species inherit half of their genes from their mother and half from their father, and their genetic makeup is therefore different from either parent or any other individual in the population. In both sexually and asexually reproducing species, mutations are the single most important source of genetic variation. New favorable mutations that initially appear in separate individuals can be recombined in many ways over time within a sexual population.

In contrast, the offspring of an asexual individual are genetically identical to their parent. The only source of new gene combinations in asexual populations is mutation. Asexual populations accumulate genetic variation only at the rate at which their genes mutate. Favorable mutations arising in different asexual individuals have no way of recombining and eventually appearing together in any one individual, as they do in sexual populations.

Remember

Population ecology is a subfield of ecology that deals with the dynamics of species populations and how these populations interact with the environment, such as birth and death rates, and by immigration and emigration.

Did you get it?

1. **What is population ecology?**
2. **What are the three main properties of a population?**

7.2. POPULATION GROWTH

Learning Objectives

- Discuss the life histories and the structure of populations
- Discuss life tables and the rate of population growth
- Evaluate exponential and geometric population growth
- Describe logistic population growth

The size of a population for any species is not a static parameter. It keeps changing with time, depending on various factors including food availability, predation pressure, and adverse weather. In fact, it is these changes in population density that give us some idea of what is happening to the population – whether it is flourishing or declining. Whatever might be the ultimate reasons, the density of a population in a given habitat during a given period fluctuates due to changes in four basic processes, two of which (natality and immigration) contribute to an increase in *population density* and two (mortality and emigration) to a decrease.

Natality refers to the number of births during a given period in the population that are added to the initial density.

Mortality is the number of deaths in the population during a given period. Immigration is the number of individuals of the same species that have come into the habitat from elsewhere during the period under consideration.

Emigration is the number of individuals in the population who left the habitat and went elsewhere during the time period under consideration.

So, if N is the population density at time t, then its density at time t +1 is:

$$N_{t+1} = N_t + [(B + I) - (D + E)]$$

Population density tends to increase when the sum of births and immigrants (B + I) exceeds the sum of deaths and emigrants (D + E). Typically, births and deaths play the most crucial roles in influencing population density, with the other factors becoming significant only in specific situations. For example, during the colonization of a new habitat, immigration might have a more substantial impact on population growth than birth rates.

7.2.1. Life Histories and the Structure of Populations

An organism's life history is the sequence of events related to survival and reproduction that occur from birth through death. Populations from different parts of the geographic range that a species inhabits may exhibit marked variations in their life histories. The patterns of demographic variation seen within and among populations are referred to as the structure of populations. These variations include breeding frequency, the age at which reproduction begins, the number of times an individual reproduces during its lifetime, the number of offspring produced at each reproductive episode (clutch or litter size), the ratio of male to female offspring produced, and whether reproduction is sexual or asexual. These differences in life history characteristics can have profound effects on the reproductive success of individuals and the dynamics, ecology, and evolution of populations. Of the many differences in life history that occur among populations, age at the time of first reproduction is one of the most important for understanding the dynamics and evolution of a population. All else being equal, natural selection will favor individuals that reproduce earlier than other individuals in the population within species. This is because by reproducing earlier, an individual's genes enter the gene pool (the sum of a population's genetic material at a given time) sooner than those of other individuals that were born at the same time but have not reproduced. Nonetheless, the "all else being equal" qualification is an important one because delayed reproductive strategies that ensure larger and more robust offspring may be selected for in some species of long-lived organisms. Precocial development (unusually early maturation) to reproduction may be favored; however, if the genes of early reproducers begin to spread throughout the population. Individuals whose genetic makeup allows them to reproduce earlier in life will come to dominate a population if there is no counterbalancing advantage to those individuals that delay reproduction until later in life.

Not all populations, however, are made up of individuals that reproduce very early in life. In the course of a lifetime, an individual must devote energy and resources to physiological demands other than reproduction. This is referred to as the cost of reproduction. To reproduce successfully, a plant may first have to grow to a certain height and outcompete its neighbors, and an animal may have to devote energy to growth so that it can reach a size at which it can fend off predators and successfully compete for mates. In many populations, individuals that delay reproduction have a better chance of surviving and leaving offspring than those that attempt to reproduce early. The opposing demands of growth, defense, and

Keyword

Population density is a geographical term that measures the number of people per unit of land area.

Remember

Population growth reflects the change in the number of individuals over a period of time. The population growth rate is affected by birth and death rates, which in turn are related to resources in their environment or outside factors such as climate and disasters. Decreased resources will lead to decreased population growth.

reproduction are balanced within the constraints of different environments to produce populations that have a diverse range of life history strategies.

Populations can often be divided into one of two extreme types based on their life history strategy. Some populations, known as r-selected, are considered opportunistic because their reproductive behavior involves a high intrinsic rate of growth (r)—individuals give birth once at an early age to many offspring. Populations that follow this strategy are often shaped by an extremely variable and uncertain environment. Because mortality occurs randomly in this setting, the quantity of progeny rather than the quality of care serves the species better. In another strategy, known as K-selected, populations tend to remain near the carrying capacity (K), the maximum number of individuals that the environment can sustain. Individuals in a K-selected population give birth at a later age to fewer offspring. This equilibrial life history is exhibited in more stable environments where reproductive success depends more on the fitness of the offspring than on their numbers.

7.2.2. Life Tables and the Rate of Population Growth

Differences in life history strategies, which include an organism's allocation of its time and resources to reproduction and care of offspring, greatly affect population dynamics. As mentioned earlier, populations in which individuals reproduce at an early age have the potential to grow much faster than populations in which individuals reproduce later. The impact of the age of first reproduction on population growth can be observed in the life tables for a specific species. Life tables were originally developed by insurance companies to provide a method for determining how long a person of a specific age could be expected to live. They are utilized not only by demographers of human populations but also by plant, animal, and microbial ecologists to forecast the life expectancies of nonhuman populations, as well as the effects of variation on demography and population growth. The size of individuals in a closed population (a population in which neither immigration nor emigration occurs) is controlled by the rates of birth (natality), growth, reproduction, and death (mortality). Life tables are constructed to assess how these rates impact the overall growth rate of a population.

7.2.2.1. Survivorship Curves

Life tables follow the fate of a group of individuals all born within the same population in the same year. Of this group, or cohort, only a certain number of individuals will reach each age, and there is an age above which no individuals ever survive. Plotting the number of those members of the group that are still alive at each age results in a survivorship curve for the population. Survivorship curves are usually displayed on a semilogarithmic rather than an arithmetic scale (Figure 7.1).

There are three general types of survivorship curves. Species such as humans and other large mammals, which have fewer numbers of offspring but invest much time and energy in caring for their young (K-selected species), usually have a Type I survivorship curve. This relatively flat curve reflects low juvenile mortality, with most individuals living to old age. A constant probability of dying at any age, shown by the Type II survivorship curve, is evident as a straight line with a constant slope that decreases over time toward zero. Certain lizards, perching birds, and rodents exhibit this type of survivorship curve. In some species that produce many offspring but provide little care for them (r-selected species), mortality is greatest among the youngest individuals. The Type III survivorship curve indicative of this life history is initially very steep, which is reflective of very high mortality among the young, but flattens out as those individuals who reach maturity survive for a relatively longer time; it is exhibited by animals, such as many insects or shellfish. Many populations have survivorship patterns that are more complex than or fall in between these three idealized curves. For example, passerine birds (perching birds such as finches) commonly suffer high mortality during the first year of life and a lower, more constant rate of death in subsequent years.

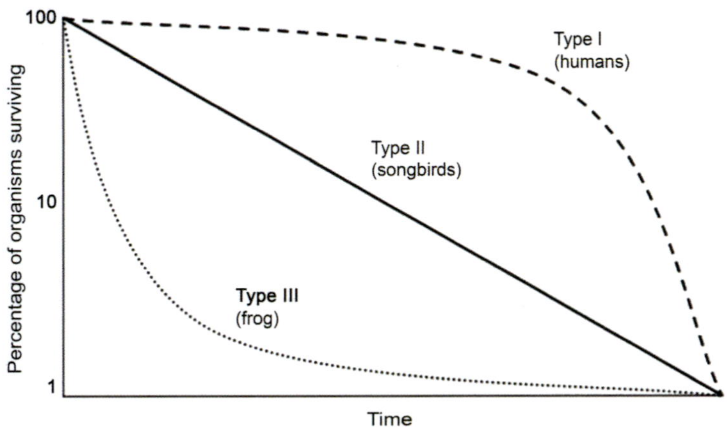

Figure 7.1. Survivorship curve Type I, II, and III survivorship curves. A survivorship curve is the graphic representation of the number of individuals in a population that can be expected to survive to any specific age.

Source: By Rayhusthwaite at English Wikipedia, CC BY 3.0, https://commons.wikimedia.org/w/index.php?curid=6623210.

7.2.2.2. Calculating Population Growth

Life tables are also used to study population growth. The average number of offspring left by a female at each age, together with the proportion of individuals surviving to each age, can be used to evaluate the rate at which the size of the population changes over time. These rates are used by demographers and population ecologists to estimate population growth and to evaluate the effects of conservation efforts on endangered species.

The average number of offspring that a female produces during her lifetime is called the net reproductive rate (R_0). If all females survived to the oldest possible age for that population, the net reproductive rate would simply be the sum of the average number of offspring produced by females at each age. In real populations, however, some females die at every age. The net reproductive rate for a set cohort is obtained by multiplying the proportion of females surviving to each age (l_x) by the average number of offspring produced at each age (m_x) and then adding the products from all the age groups: $R_0 = \Sigma l_{xmx}$. A net reproductive rate of 1.0 indicates that a population is neither increasing nor decreasing but replacing its numbers exactly. This rate indicates population stability. Any number below 1.0 indicates a decrease in population, while any number above indicates an increase. For example, the net reproductive rate for the Galapagos cactus finch (*Geospiza scandens*) is 2.101, which means that the population can more than double its size each generation (Table 7.1).

Table 7.1. Life Table for One Darwin Finch, the *Galapagos cactus* finch (*Geospiza scandens*)

Age class (x)	Probability of surviving to age x (l_x)	Average number of fledgling daughters (m_x)	Product of survival and reproduction ($\Sigma l_x m_x$)
0	1.0	0.0	0.0
1	0.512	0.364	0.186
2	0.279	0.187	0.052
3	0.279	1.438	0.401
4	0.209	0.833	0.174
5	0.209	0.500	0.104
6	0.209	0.833	0.174
7	0.209	0.250	0.052
8	0.209	3.333	0.696
9	0.139	0.125	0.017
10	0.070	0.0	0.0
11	0.070	0.0	0.0
12	0.070	3.500	0.245
13	0	—	—
			$R_0 = 2.101$
Net reproductive rate = $R_0 = \Sigma l_x m_x = 2.101$			
Mean generation time = $T = (\Sigma x l_x m_x)/(R_0) = 6.08$ years			
Intrinsic rate of natural increase of the population = r = approximately $1nR_0/T = 2.101/6.08 = 0.346$			

Source: Adapted from Peter R. Grant & Rosemary Grant, B. (1992). "Demography and the Genetically Effective Sizes of Two Populations of Darwin's Finches," *Ecology, 73*(3), Copyright © 1992 The Ecological Society of America, used by permission.

The other value needed to calculate the rate at which the population can grow is the mean generation time (*T*). Generation time is the average interval between the birth of an individual and the birth of its offspring. To determine the mean generation time of a population, the age of the individuals (*x*) is multiplied by the proportion of females surviving to that age (l_x) and the average number of offspring left by females at that age (m_x). This calculation is performed for each age group, and the values are added together and divided by the net reproductive rate (R_0) to yield the result

$$T = \frac{\sum x l_x m_x}{R_0}$$

For example, the mean generation time of the Galapagos cactus finch is 6.08 years.

Population biologists use another value to calculate the rate of increase in populations that reproduce within discrete time intervals and possess generations that do not overlap. This value is known as the intrinsic rate of natural increase (r), or the Malthusian parameter. Simply put, this rate can be seen as the number of births minus the number of deaths per generation time—basically, the reproduction rate minus the death rate. To calculate this value using a life table, you divide the natural logarithm of the net reproductive rate by the mean generation time:

$$r = \frac{\ln R_0}{T}$$

Values above zero indicate that the population is increasing, the higher the value, the faster the growth rate. The intrinsic rate of natural increase can be used to compare the growth rates of populations of a species that have different generation times. Some human populations have higher intrinsic rates of natural increase partially because individuals in those groups begin reproducing earlier than those in other groups. Mice have higher intrinsic rates of natural increase than elephants because they reproduce at a much earlier age and have a much shorter mean generation time.

If a population has an intrinsic rate of natural increase of zero, then it is said to have a stable age distribution and neither grows nor declines in numbers. A growing population has more individuals in the lower age classes than does a stable population, and a declining population has more individuals in the older age classes than does a stable population. Many human populations are currently undergoing population increase, far exceeding a stable age distribution. Although the global human population has increased almost continuously throughout history, it has skyrocketed since the Industrial Revolution, primarily because of a drop in death rates. No other species has shown such sustained growth (Table 7.2).

Table 7.2. The Intrinsic Rate Of Increase (r) Calculated for Populations of Species that Differ Greatly in Their Potential for the Rate of Population Growth

Species	**Intrinsic rate of increase (r)**
Elephant seal	0.091
Ring-necked pheasant	1.02
Field vole	3.18
Flour beetle	23
Water flea	69

Source: Adapted from Robert E. Ricklefs, The Economy of Nature, 3rd edition, copyright © 1993 by W.H. Freeman & Company, used with permission.

Remember

Populations grow through births and immigration and decline through deaths and emigration. When resources are unlimited, the growth is usually exponential but when resources become progressively limiting, the growth pattern turns logistic. In either case, growth is ultimately limited by the carrying capacity of the environment. The intrinsic rate of natural increase (r) is a measure of the inherent potential of a population to grow.

7.2.3. Exponential and Geometric Population Growth

In an ideal environment, one that has no limiting factors, populations grow at a geometric rate or an exponential rate. Human populations, in which individuals live and reproduce for many years and in which reproduction is distributed throughout the year, grow exponentially. Exponential population growth can be determined by dividing the change in population size (ΔN) by the time interval (Δt) for a certain population size (N):

$$\frac{\Delta N}{\Delta t} = rN$$

The growth curve of these populations is smooth and becomes increasingly steeper over time. The steepness of the curve depends on the intrinsic rate of natural increase for the population. Human population growth has been exponential since the beginning of the 20th century. Much concern exists about the impact this growth will have, not only on the environment but on humans as well. The World Bank projection for human population growth predicts that the human population will grow from 6.8 billion in 2010 to nearly 10 billion in 2050. That estimate could be offset by four population-control measures: (1) lower the rate of unwanted births, (2) lower the desired family size, (3) raise the average age at which women begin to bear children, and (4) reduce the number of births below the level that would replace current human populations (e.g., one child per woman).

Insects and plants that live for a single year and reproduce once before dying are examples of organisms whose growth is geometric. In these species, a population grows as a series of increasingly steep steps rather than as a smooth curve.

Keyword

Predation is a biological interaction where one organism, the predator, kills and eats another organism, its prey (Figure 7.2).

7.2.4. Logistic Population Growth

The geometric or exponential growth of all populations is eventually curtailed by food availability, competition for other resources, *predation*, disease, or some other ecological factor. If growth is limited by resources such as food, the exponential growth of the population begins to slow as competition for those resources increases. The growth of the population eventually slows nearly to zero as the population reaches the carrying capacity (K) for the environment. The result is an S-shaped curve of population growth known as the logistic curve. It is determined by the equation

$$\frac{\Delta\pm}{\Delta t} = rN\left(\frac{-}{K}\right)$$

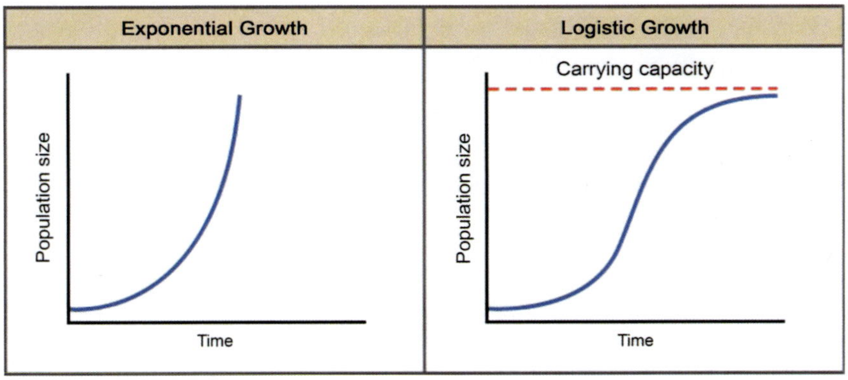

Figure 7.2. Carrying capacity and exponential versus logistic population growth.

*Source:*https://courses.lumenlearning.com/wm-biology2/chapter/environmental-limits-to-population-growth/.

Did you get it?

1. Based on the life history strategy, a population is divided into how many types?
2. How many types of survival curves are there?

Remember

Finally, when the population density reaches a certain point, the population stops growing and starts to decrease in size. Because the rate of growth of the population depends on the density of the population, logistic growth is also described as "density-dependent growth".

7.3. POPULATION INTERACTION

Learning Objectives

- Write the names of population interactions

Can you think of any natural habitat on earth that is inhabited just by a single species? There is no such habitat and such a situation is inconceivable. For any species, the minimal requirement is one more species on which it can feed. Even a plant species, that makes its own food, cannot survive alone; it needs soil *microbes* to break down the organic matter in soil and return the inorganic nutrients for absorption. And then, how will the plant manage pollination without an animal agent? It is obvious that in nature, animals, plants and microbes do not and cannot live in isolation but interact in various ways to form a biological community. Even in minimal communities, many interactive linkages exist, although all may not be readily apparent.

Table 7.3. Population Interactions

Interactions	Species A	Species B	**Examples**
Mutualism/ Symbiotic	+	+	Leguminous plants and nitrogen-fixing bacteria. Process of pollination in plants.
Commensalism	+	0	Remoras eat leftover food from sharks without depleting sharks' resources.
Amensalism	–	0	Shading out of one plant by a taller and wider one. Allelopathy – inhibition of one plant by the secretions of another.
Parasitism	+	–	Ticks, and the protozoan that causes malaria.
Competition	–	–	Lions and tigers are in the same niche.
Predation	+	–	Lion and zebra, bear and fish, fox and rabbit.

Source: https://compass.rauias.com/.

Interspecific interactions arise from the interaction of populations of two different species. They could be beneficial, detrimental, or neutral (neither harm nor benefit) to one of the species or both. Assigning a '+' sign for beneficial interaction, '-' sign for detrimental, and 0 for neutral interaction, let us look at all the possible outcomes of interspecific interactions (Table 7.3).

Both species benefit in mutualism, and both lose in competition in their interactions with each other. In both parasitism and predation, only one species benefits (parasite and predator, respectively), and the interaction is detrimental to the other species (host and prey, respectively). The interaction where one species is benefited and the other is neither benefited nor harmed is called commensalism. In amensalism, on the other hand, one species is harmed, whereas the other is unaffected. Predation, parasitism, and commensalism share a common characteristic – the interacting species live closely together.

- *Predation:* What would happen to all the energy fixed by autotrophic organisms if the community had no animals to eat the plants? You can think of predation as nature's way of transferring to higher trophic levels the energy fixed by plants. When we think of predator and prey, most probably it is the tiger and the deer that readily come to our mind, but a sparrow eating any seed is no less a predator. Although animals eating plants are categorized separately as herbivores, they are, in a broad ecological context, not very different from predators. Besides acting as 'conduits' for energy transfer across trophic levels, predators play other important roles. They keep prey populations under control. If not for predators, prey species could achieve very high population densities and cause ecosystem instability. When certain exotic species are introduced into a geographical area, they become invasive and start spreading fast because the invaded land does not have its natural predators. The prickly pear cactus introduced into Australia in the early 1920s caused havoc by spreading rapidly into millions of hectares of rangeland. Finally, the invasive cactus was brought under control only after a cactus-feeding predator (a moth) from its natural habitat was introduced into the country. Biological control methods adopted in agricultural pest control are based on the ability of the predator to regulate the prey population. Predators also help in maintaining species diversity in a community, by reducing the intensity of competition among competing prey species. In the rocky intertidal communities of the American Pacific Coast, the starfish Pisaster is an important predator. In a field experiment, when all the starfish were removed from an enclosed intertidal area, more than 10 species of invertebrates became extinct within a year, because of interspecific competition.

Keyword

Microbes are microscopic organisms that can exist as single cells or as colonies of cells.

If a predator is too efficient and overexploits its prey, then the prey might become extinct and following it, the predator will also become extinct due to lack of food. This is the reason why predators in nature are 'prudent'. Prey species have evolved various defenses to lessen the impact of predation. Some species of insects and frogs exhibit cryptic colors (camouflaged) to avoid being easily detected by the predator. Some are poisonous and therefore avoided by their predators. The Monarch butterfly is highly distasteful to its predator (bird) because of a special chemical present in its body. Interestingly, the butterfly acquires this chemical during its caterpillar stage by feeding on a poisonous weed.

For plants, herbivores are the predators. Nearly 25% of all insects are known to be phytophagous (feeding on plant sap and other parts of plants). The problem is particularly severe for plants because, unlike animals, they cannot run away from their predators. Plants have evolved an astonishing variety of morphological and chemical defenses against herbivores. Thorns (Acacia, Cactus) are the most common morphological means of defense. Many plants produce and store chemicals that make the herbivores sick when they are eaten, inhibit feeding or digestion, disrupt reproduction, or even kill them. You must have seen the weed Calotropis growing in abandoned fields. The plant produces highly poisonous cardiac glycosides, which is why you never see any cattle or goats browsing on this plant. A wide variety of chemical substances that we extract from plants on a commercial scale (nicotine, caffeine, quinine, strychnine, opium, etc.) are actually produced by them as defenses against grazers and browsers.

- *Competition*: When Darwin spoke of the "struggle for existence" and survival of the fittest in nature, he was convinced that interspecific competition is a potent force in organic evolution. It is generally believed that competition occurs when closely related species compete for the same resources that are limiting, but this is not entirely true. Firstly, totally unrelated species could also compete for the same resource. For instance, in some shallow South American lakes, visiting flamingoes and resident fishes compete for their common food, the zooplankton in the lake. Secondly, resources need not be limiting for competition to occur; in interference competition, the feeding efficiency of one species might be reduced due to the interfering and inhibitory presence of the other species, even if resources (food and space) are abundant. Therefore, competition is best defined as a process in which the fitness of one species (measured in terms of its 'r' the intrinsic rate of increase) is significantly lower in the presence of another species. It is relatively easy to demonstrate in laboratory experiments, as Gause and other experimental ecologists did, that when resources are limited the competitively superior species will eventually eliminate the other species, but evidence for such competitive exclusion occurring in nature is not always conclusive. Strong and persuasive circumstantial evidence does exist; however, in some cases. The Abingdon tortoise in the Galapagos Islands became extinct within a decade after goats were introduced on the island, apparently due to the greater browsing efficiency of the goats. Another piece of evidence for the occurrence of competition in nature comes from what is called 'competitive release'. A species whose distribution is restricted to a small geographical area because of the presence of a competitively superior species is found to expand its distributional range dramatically when the competing species is experimentally removed. Connell's elegant field experiments showed that on the rocky sea coasts of Scotland, the larger and competitively superior barnacle *Balanus* dominates the intertidal area and excludes the smaller barnacle *Chathamalus* from that zone. In general, herbivores and plants appear to be more adversely affected by competition than carnivores.

Gause's 'Competitive Exclusion Principle' states that two closely related species competing for the same resources cannot co-exist indefinitely, and the competitively inferior one will be eliminated eventually. This may be true if resources are limiting, but not otherwise. More recent studies do not support such gross generalizations about competition. While they do not rule out the occurrence of interspecific competition in nature, they point out that species facing competition might evolve mechanisms that promote co-existence rather than exclusion. One such mechanism is 'resource partitioning'. If two species compete for the same resource, they could avoid competition by choosing, for instance, different times for feeding or different foraging patterns. MacArthur showed that five closely related species of warblers living on the same tree were able to avoid competition and co-exist due to behavioral differences in their foraging activities.

Keyword

Brood parasitism is a subclass of parasitism, a phenomenon, and a behavioral pattern of certain animals, brood parasites that rely on others to raise their young.

• *Parasitism*: Considering that the parasitic mode of life ensures free lodging and meals, it is not surprising that parasitism has evolved in so many taxonomic groups from plants to higher vertebrates. Many parasites have evolved to be host-specific (they can parasitize only a single species of the host) in such a way that both host and the parasite tend to co-evolve; that is, if the host evolves special mechanisms for rejecting or resisting the parasite, the parasite has to evolve mechanisms to counteract and neutralize them, in order to be successful with the same host species. In accordance with their life styles, parasites evolved special adaptations such as the loss of unnecessary sense organs, presence of adhesive organs or suckers to cling on to the host, loss of digestive system and high reproductive capacity. The life cycles of parasites are often complex, involving one or two intermediate hosts or vectors to facilitate parasitization of its primary host. The human liver fluke (a trematode parasite) depends on two intermediate hosts (a snail and a fish) to complete its life cycle.

The malarial parasite needs a vector (mosquito) to spread to other hosts. The majority of the parasites harm the host; they may reduce the survival, growth and reproduction of the host and reduce its population density. They might render the host more vulnerable to predation by making it physically weak.

In contrast, endoparasites are those that live inside the host's body at different sites (liver, kidney, lungs, red blood cells, etc.). The life cycles of endoparasites are more complex due to their extreme specialization. Their morphological and anatomical features are greatly simplified while emphasizing their reproductive potential.

Brood parasitism in birds is a fascinating example of parasitism in which a parasitic bird lays its eggs in the nest of its host and lets the host incubate them. During the course of evolution, the eggs of the parasitic bird have evolved to resemble the host's egg in size and color to reduce the chances of the host bird detecting the foreign eggs and ejecting them from the nest. Try to observe the movements of the cuckoo (koel) and the crow in your neighborhood park during the breeding season (spring to summer) and witness *brood parasitism* in action.

- *Commensalism*: This is the interaction where one species benefits while the other is neither harmed nor benefited. For instance, an orchid growing as an epiphyte on a mango branch or barnacles growing on the back of a whale benefit, while neither the mango tree nor the whale seems to gain any advantage. Another example is the cattle egret and grazing cattle in close association, a common sight in farmed rural areas. The egrets benefit from the insects stirred up by the cattle as they graze. Similarly, the interaction between a sea anemone with stinging tentacles and the clown fish that lives among them showcases commensalism, where the fish gains protection from predators while the anemone doesn't seem to benefit from hosting the clown fish (Figure 7.3).
- *Mutualism*: This interaction benefits both species involved. Lichens show a close mutualistic bond between a fungus and photosynthetic algae or cyanobacteria. Similarly, mycorrhizae are partnerships between fungi and higher plant roots. The fungi assist the plant in nutrient absorption from the soil, while the plant provides the fungi with energy-yielding carbohydrates. Plant-animal relationships showcase remarkable examples of mutualism. Plants rely on animals for pollination and seed dispersal, offering rewards like pollen, nectar, and nutritious fruits. However, this mutually beneficial system must guard against 'cheaters', such as animals that consume nectar without aiding in pollination. Plant-animal interactions often involve the co-evolution of mutualists, where the evolution of the flower and its pollinator species are closely intertwined. In certain fig tree species, there is a specific one-to-one relationship with pollinator wasps. Each fig species can only be pollinated by its 'partner' wasp species. The female wasp uses the fruit for egg-laying and nourishing its larvae. In exchange for pollination, the fig provides the wasp with some developing seeds as food for the larvae.

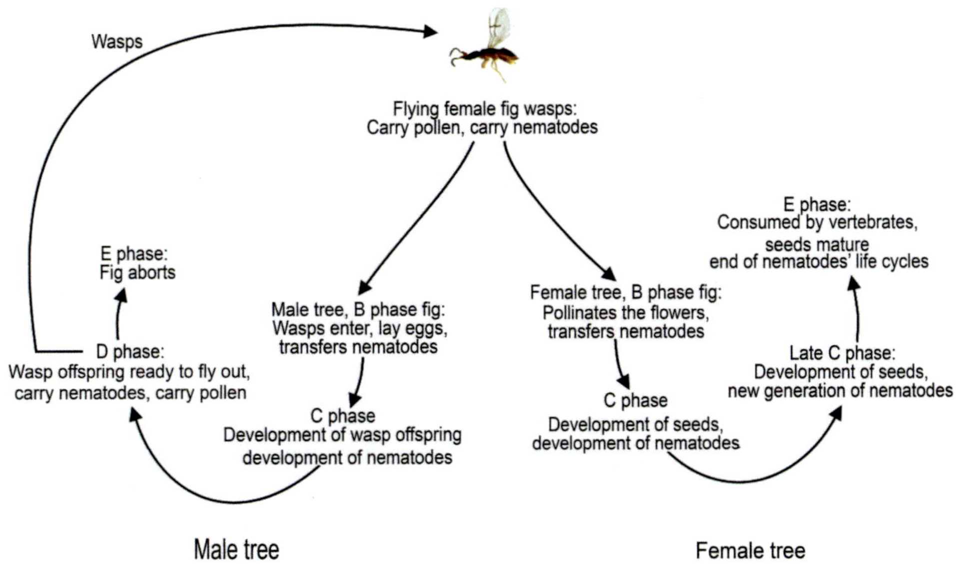

Figure 7.3. The mutual relationship between a fig tree and a wasp.

Source: Jauharlina, Oktarina, H., Sriwati, R., Sayuthi, M., Kanzaki, N., Quinnell, R. J., Compton, S. G. (2022). Association of Fig Pollinating Wasps and Fig Nematodes inside Male and Female Figs of a Dioecious Fig Tree in Sumatra, Indonesia. *Insects*. *13*(4):320. https://doi.org/10.3390/insects13040320.

Figure 7.4. Showing bee-a pollinator on orchid flower.

Source: By Bernard DUPONT from FRANCE – Bee Orchid (*Ophrys apifera*), CC BY-SA 2.0, https://commons.wikimedia.org/w/index.php?curid=48715235.

Orchids show a bewildering diversity of floral patterns, many of which have evolved to attract the right pollinator insect (bees and bumblebees) and ensure guaranteed pollination by it. Not all orchids offer rewards. The Mediterranean orchid Ophrys employs 'sexual deceit' to get pollination done by a species of bee. One petal of its flower bears an uncanny resemblance to the female of the bee in size, color, and markings. The male bee is attracted to what it perceives as a female, 'pseudocopulates' with the flower, and during that process is dusted with pollen from the flower. When this same bee 'pseudocopulates' with another flower, it transfers pollen to it and thus pollinates the flower. Here you can see how co-evolution operates. If the female bee's color patterns change even slightly for any reason during evolution, pollination success will be reduced unless the orchid flower co-evolves to maintain the resemblance of its petal to the female bee (Figure 7.4).

Did you get it?

1. **What are the names of population interactions?**

7.4. INTRA-SPECIFIC COMPETITION

Learning Objectives

- Describe the types of intra-specific competition
- Discuss the consequences of intra-specific competition
- Explain the significance of intra-specific competition

Individuals in a population grow, reproduce, and die. Individuals within a population have very similar requirements for survival, growth, and reproduction. This often leads to intra-specific competition.

This competition can take two forms:

- *Exploitation competition* – the interaction between individuals is indirect. For example, each individual is affected by the amount of resources that are left after being exploited by others. For instance, animals gathering at a watering hole during a drought. Here, individuals are affected by the indirect effects of others.
- *Interference competition* – Here, individuals interact directly with one another. One individual interacts directly in competition for limiting resource(s). For mobile animals, this could take the form of defending a territory that restricts access to resources. For sessile organisms where space may be the limiting factor, competition may involve direct interference by preventing access to space.

Competition will reduce the fitness of all individuals and often lead to differences in fitness among individuals. Competition will negatively affect all individuals since in the absence of competition, resources are not limiting. Therefore, it is relative fitness that is important to consider.

Because competition is often more intense as population size increases (and/or resources diminish), the effect of competition is often density-dependent; that is, at higher population density, competition increases. It adversely affects survivorship and births, i.e., population size.

Examples of intra-specific competition include:

- Larger, dominant grizzly bears occupying the best fishing spots on a river during the salmon spawning season.

- Songbirds like Eastern Towhees defending territories from which they exclude their neighbors in an effort to secure resources.
- Barnacles competing for space on rocks, from which they filter water to obtain their food.
- Plants using chemical compounds to discourage competitors, even those from the same species, and prevent them from growing too close.
    - *Types of Intra-specific Competition:* Scramble competition occurs when individuals obtain a declining fraction of the available resources as the number of competitors increases. Every individual suffers from limited food, water, or space, with consequences for survival and reproduction. This type of competition is indirect: for example, deer feed on woody browse all winter long, putting individuals in indirect competition with each other for a resource they cannot defend from others and keep for themselves.

Contest (or interference) competition is a direct form of interaction when resources are actively defended from other competitors. Examples include a song sparrow defending a territory, or an oak spreading its crown to gather as much light as possible, elbowing a spot within a forest canopy.

- *Consequences of Intra-specific Competition:* Intra-specific competition can suppress growth. For example, tadpoles take longer to mature when they are crowded, and foresters know that thinned-out tree plantations will lead to bigger trees than those left alone to grow at high density (density is the number of individuals per unit of area). Similarly, it is quite common for animals to experience a decrease in the number of young they can produce at a high population density.

To avoid high-density situations, many juvenile animals will have a dispersal phase when they move away from the areas where they were born. By striking out on their own, they increase their chances of finding more abundant resources with less competition. It comes at a cost, though, as there is no guarantee their new digs will have sufficient resources to raise a family of their own. Dispersing young animals are also at increased risk of predation as they travel through unfamiliar territory.

Some individual animals can exert social dominance over others to ensure better access to resources. That dominance can be applied directly by having better fighting abilities. It can also be demonstrated through signals, like coloration or structures, or behaviors like vocalizations and displays. Subordinate individuals will still be able to access resources but will be relegated to less abundant food sources, for example, or to areas with inferior shelter.

Dominance can also be expressed as a spacing mechanism, including by establishing a pecking order. Instead of competing directly over resources with other individuals of the same species, some animals protect a space from others, claiming property over all the resources within. Fighting can be used to establish territory boundaries, but given the risks of injuries, many animals use ritualistic, safer alternatives like displays, vocalizations, mock fighting, or scent marking.

Territoriality has evolved in several animal groups. In songbirds, territories are defended to secure food resources, nesting sites, and young-rearing sites. Most of the springtime birds singing we hear is evidence of male birds advertising their territory. Their vocal displays serve to attract females and to announce the location of their territorial boundaries.

In contrast, male bluegills will only defend a nesting site, where they will encourage a female to lay eggs which he then fertilizes.

- Significance of Intra-specific Competition

For many species, intra-specific competition has strong effects on how population size varies over time. At high density, growth is reduced, fecundity is suppressed, and survival is affected. As a result, the size of the population increases more slowly, stabilizes, and then eventually starts declining. Once the population size reaches lower numbers again, fecundity picks back up and survival is increased, putting the population back in a growth pattern. These fluctuations keep the population from getting too high or too low, and this regulating effect is a well-demonstrated consequence of intra-specific competition.

Did you get it?

1. **What are the two forms of intra-specific competition?**
2. **What is the significance of intra-specific competition?**

7.5. INTER-SPECIFIC COMPETITION

Learning Objectives

- Describe the types of inter-specific competition
- Learn about apparent competition
- Discuss the consequences of inter-specific competition

Inter-specific competition, in ecology, is a form of competition in which individuals of different species compete for the same resources in an ecosystem (e.g., food or living space). This can be contrasted with mutualism, a type of symbiosis. Competition between members of the same species is called intra-specific competition.

If a tree species in a dense forest grows taller than surrounding tree species, it is able to absorb more of the incoming sunlight. However, less sunlight is then available for the trees that are shaded by the taller tree, thus inter-specific competition. Leopards and lions can also be in interspecific competition, since both species feed on the same prey, and can be negatively impacted by the presence of the other because they will have less food.

Competition is only one of many interacting biotic and abiotic factors that affect community structure. Moreover, competition is not always a straightforward, direct, interaction. Interspecific competition may occur when individuals of two separate species share a limiting resource in the same area. If the resource cannot support both populations, then lowered fecundity, growth, or survival may result in at least one species. Interspecific competition has the potential to alter populations, communities and the evolution of interacting species. On an individual organism level, competition can occur as interference or exploitative competition.

7.5.1. Types

There are two types of inter-specific competition in ecology. These are those interspecies competitions that are based on the mechanism of competition and those based on the outcome of the competition. The mechanism is the way that the competition occurs, and the outcome is the result of the competition happening, the aftermath.

7.5.1.1. Based on Mechanism

Exploitative competition occurs as species indirectly compete for the same resources in an ecosystem. This would be like plants that all grow in the same area. However, some plants can absorb water and nutrients at a quicker pace than others. Another form of competition based on the mechanism is the interference mechanism. This is direct and can even be termed aggressive behavior. This is because these competing species will physically fight each other for whatever they are competing for, whether it is a resource or a potential mating partner.

For instance, aphids often need to defend their feeding sites from other competing animals. They do this by sending out their younger aphids from the better sites. This can also happen within the plant communities.

In fact, within the forests, the top or emergent layer of trees controls the nutrients and how they are distributed. This is because they are massive and often need more of all the resources than shrubs and other grass-like plants (forest floor). Even their branches and leaves determine how much sunlight the smaller plants will get by blocking access to the sun. All these competitive relationships are based on the way competition happens and how it occurs (Figure 7.5).

Remember

Exploitative competition, also referred to as resource competition, is a form of competition in which one species consumes and either reduces or more efficiently uses a shared limiting resource and therefore depletes the availability of the resource for the other species. Thus, it is an indirect interaction because the competing species interact via a shared resource.

Figure 7.5. Trees in the layers of the rainforest canopy.

Source: By © Vyacheslav Argenberg / http://www.vascoplanet.com/, CC BY 4.0, https://commons.wikimedia.org/w/index.php?curid=102119666

7.5.2. Based on Outcomes

The two types of competition based on outcomes in an ecosystem are scramble and contest competition. Scramble competition occurs between species when they compete for the same resource. However, the resources may be evenly shared among the species individually. As the population increases, there are not enough resources for numerous individuals to thrive optimally. In contrast, contest competition happens when some individuals control or take over the resources, ensuring their survival regardless of population growth or decline. In scramble competition, the competition for a finite resource that becomes inadequate over time may lower the survival chances of all competing species or lead to mass extinction in extreme cases. This is because the resource is not contested but rather partitioned equally among the competing species.

In intra-specific competition, scramble competition is exhibited by a flock of sheep grazing in a limited pasture area. As the flock multiplies over time, there are more mouths to feed but the same area of pasture, leading to a battle for the limited amount of grass. The white-faced monkey provides a clear example of contest competition, where alpha males take control of the group's resources. These dominant males, being larger and stronger, easily secure a mate and ensure their survival by dominating the pack. Regardless of internal group dynamics or external factors, the alpha males typically survive due to this advantage.

When it comes to interspecific competition, a scramble competition could occur for a limited resource. Imagine two different species of aerobic bacteria thriving in a limited source of oxygen. Both bacterial species could eventually perish as the oxygen depletes over time. For example, consider sheep grazing on pasture. If cows were introduced to the field, there would be interspecific scramble competition. This is because the sheep and cows would now have to compete for the grass not only among their own species but also with the other species. As for contest competition, a hypothetical scenario is when two different species compete for the same limited resource, and only one of them emerges as the "winner". This type of competition is commonly observed in the animal kingdom. For instance, two predators may "compete" for prey, with the stronger or more adapted one prevailing.

7.5.3. Apparent Competition

Apparent competition can be defined as a type of indirect competition that occurs between two organisms. These organisms are both of completely different species but are both preyed upon by the same predator. These prey indirectly compete against each other in terms of survival. Often this can occur when one of the prey has an abundant or sudden growth in their population. This causes the predator to have more food since there is more prey around. The predator population will then grow as well because of the accessibility to more food. However, this is often detrimental to the other prey whose population does not grow but now has more predators present to hunt it. This is the indirect competition between the two prey as now they must both fight to not be eaten by the predator.

We see apparent competition within ecosystems through both plants and animals. In plants, it is more common – but not limited to – when an invasive species comes into a niche in abundance, causing an herbivore to have more options for food. In animals, it could be a random population boom or decline via more or fewer resources or also an invasive species.

Apparent competition can often be hard to distinguish in nature as it can be very complex and overlap with other kinds of competition. For instance, both chipmunks and squirrels are prey for the hawk. Initially, an increase in the chipmunk population will benefit the squirrels as there will be more chipmunks for the hawks to feed on, resulting in fewer attacks on the squirrels. However, as the hawks obtain more nutrients from the increased chipmunk population, they will thrive and their population will rise. Consequently, increased predation will lead to a decrease in the populations of both chipmunks and squirrels, but since squirrels have a lower population to begin with, they will suffer more. Conversely, if the chipmunk population decreases, squirrels will become the primary source of nutrients for the hawks, leading to a decrease in the squirrel population.

7.5.4. Consequences

All interspecific interactions have consequences. Interspecies competition can have negative impacts on one or all of the species involved. These impacts may lead to one species being compelled to leave the ecosystem, organisms having to seek new ways to utilize resources, or even the complete extinction of a species' population. These and various other consequences influence the community and its future development.

7.5.5. Competitive Exclusion

The competitive exclusion principle arises when two species that rely on the same limited resource cannot coexist in an environment. This principle was established through extensive mathematical analysis and study models, concluding that eventually, one of the organisms must vacate the area for both to survive. If they persist together, one species will perish. This is also referred to as Gause's Law.

The dominant species often exploit resource utilization to drive away the other species. The inferior species will suffer a significant decline in population and eventually move away before facing endangerment. In a scenario where a forest houses numerous carnivorous animals but has a limited prey population, there will eventually be insufficient prey to meet the nutritional needs of every species. The superior species will persist while the others will be compelled to relocate to ensure their survival, seeking adequate food in a different location.

7.5.6. Niche Differentiation

Niche differentiation is referred to as the process by which species that compete for the same resources begin to use their environments in different ways so that they are now able to coexist. If the species do not undergo niche differentiation, the competitive exclusion will force one of the species out of the environment. If the organism then does not want to move, nor does it want to die, it must begin to use other resources in the ecosystem or use them in other ways so that it can continue to thrive.

In the Caribbean, Central, and South America, there are thousands of species of lizards among which many may end up living in the same ecosystems. They will all share the same niche and resources. For them to be able to survive, the lizards have adapted to their surroundings each in their unique way physically, so that they all have their specific microhabitats. This allows the reptiles to become "masters" of their own spaces and live in harmony with the rest of the lizards while still being able to obtain all their necessary resources.

7.5.7. Local Extinction

Once competition exclusion and niche differentiation do not work, it is inevitable that a species will become extinct if it remains in an intense form of interspecific competition. Too much pressure on one species will cause a rapid population decline and eventually, the population of that organism will die out. This has been seen before with zooplankton who lived in the same rock pool. Since it is a rockpool, often there is nowhere else to go (so competition exclusion cannot occur) and a limited number of ways resources can be used (no niche differentiation either). Hence, eventually, species of zooplankton will die out and go extinct due to the lack of availability of resources.

7.5.8. Impacts on Communities

All the effects mentioned above and many others can contribute to the numerous impacts that interspecific interactive competition can have on communities within an ecosystem. These competitions have serious and long-term effects on how communities continue or change as they happen. Consider organisms that rely on seeds to feed on. These organisms come every year and eat many seeds from plants, which results in plants not being able to continue to reproduce. To combat this, many plants have adapted to the process of mast seeding, where the plants overproduce seeds. This allows the animals to be able to eat all the seeds they need while the plants are still able to get seeds to reproduce.

Did you get it?

1. **What are the consequences of inter-specific competition?**
2. **What is niche differentiation?**

7.6. POPULATION DYNAMICS

Learning Objectives

- Understand the density-dependent regulation
- Describe the density-independent regulation and interaction with density-dependent factors

A population describes a group of individuals of the same species occupying a specific area at a specific time. Some characteristics of populations that are of interest to biologists include the population density, the birth rate, and the death rate. If there is immigration into the population or emigration out of it, then the immigration rate and emigration rate are also of interest. Together, these population parameters, or characteristics, describe how the population density changes over time. The ways in which population densities fluctuate—increasing, decreasing, or both over time—are the subject of population dynamics.

Population density measures the number of individuals per unit area, for example, the number of deer per square kilometer. Although this is straightforward in theory, determining population densities for many species can be challenging in practice.

7.6.1. Density-Dependent Regulation

Most density-dependent factors are biological in nature (biotic) and include predation, inter- and intra-specific competition, accumulation of waste, and diseases such as those caused by parasites. Usually, the denser a population is, the greater its mortality rate. For example, during intra- and inter-specific competition, the reproductive rates of the individuals will usually be lower, reducing their population's rate of growth. In addition, low prey density increases the mortality of its predator because it has more difficulty locating its food source.

An example of density-dependent regulation is shown in Figure 7.6 with results from a study focusing on the giant intestinal roundworm (*Ascaris lumbricoides*), a parasite of humans and other mammals. Denser populations of the parasite exhibited lower fecundity: they contained fewer eggs. One possible explanation for this is that females would be smaller in more dense populations (due to limited resources) and that smaller females would have fewer eggs. This hypothesis was tested and disproved in a 2009 study which showed that female weight had no influence. The actual cause of the density dependence of fecundity in this organism is still unclear and awaiting further investigation.

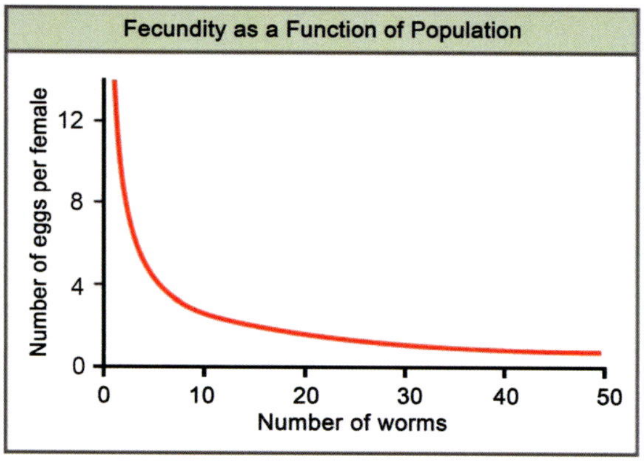

Figure 7.6. In this population of roundworms, fecundity (number of eggs) decreases with population density.

Source: lumenlearning.com (https://courses.lumenlearning.com/wm-biology2/chapter/population-dynamics-and-regulation/).

7.6.1.1. Density-Independent Regulation and Interaction with Density-Dependent Factors

Many factors, typically physical or chemical in nature (abiotic), influence the mortality of a population regardless of its density, including weather, natural disasters, and pollution. An individual deer may be killed in a forest fire regardless of how many deer happen to be in that area. Its chances of survival are the same whether the population density is high or low. The same holds true for cold winter weather. In real-life situations, population regulation is very complicated and density-dependent and independent factors can interact. A dense population that is reduced in a density-independent manner by some environmental factor(s) will be able to recover differently than a sparse population. For example, a population of deer affected by a harsh winter will recover faster if more deer are remaining to reproduce.

Density-dependent limiting factors tend to be biotic—living organism-related—as opposed to physical features of the environment. Some common examples of density-dependent limiting factors include:

- *Competition within the population*. When a population reaches a high density, more individuals are trying to use the same quantity of resources. This can lead to competition for food, water, shelter, mates, light, and other resources needed for survival and reproduction.
- *Predation*. Higher-density populations may attract predators that wouldn't bother with a sparser population. When these predators eat individuals from the population, they decrease their numbers but may increase their own.
- *Disease and parasites*. Disease is more likely to break out and result in deaths when more individuals are living together in the same place. Parasites are also more likely to spread under these conditions.
- *Waste accumulation*. High population densities can lead to the accumulation of harmful waste products that kill individuals or impair reproduction, reducing the population's growth.

Did you get it?

1. **What is density-dependent regulation?**
2. **Write one example of density-dependent limiting factors.**

CLOSER LOOK

POPULATION FLUCTUATIONS

Density dependence tends to push populations toward carrying capacity, K. Because the environment is variable, K is also variable and hence populations often don't rest at K too, which means density dependence doesn't always lead to a static equilibrium. Populations show ups and downs and always try to reach K; these ups and downs are referred to as population fluctuations. Population fluctuations can be erratic (irruptive) or they can be periodic (cyclic). Erratic fluctuations are mostly due to variations in density-independent environmental factors that have a large, immediate impact on population size (e.g., fires, catastrophes). While cyclic fluctuations, also known as oscillations, are the result of time lags in the responses of populations to their own density. Populations acquire "momentum" when high birth rates at low densities cause the populations to overshoot K, which causes very low survival and birth rates; consequently, the population falls below K, and recovery occurs when birth rates again increase due to lowered density conditions. Population cycles result from time delays in the responses of birth and death rates to current environmental conditions which try to either undercompensate or overcompensate for population size. The nature of the cycle depends on the population's resilience. Resilience decides how fast the population will regain equilibrium. The reproductive rate has a very strong influence on the resilient capacity of the population. Population fluctuations occur over many different time scales, ranging from millions of years on a geological time scale to years, seasons, or weeks on a short time scale. Fluctuations at short time scales are important to be studied in population dynamics.

Types of Fluctuations

i. *Stable:* the population size fluctuates around carrying capacity slightly above and below and is characteristic of many species living under fairly constant environments, like conditions in Tropical rainforests.

ii. **Irruptive**: The population is normally fairly stable; it occasionally explodes (irrupts) to peak, then crashes to a level below carrying capacity. This occurs due to a factor (e.g., resource availability) that temporarily increases carrying capacity. This is characteristic of short-lived, rapidly reproducing species.

iii. **Irregular** : No apparent recurring pattern is observed, and irregular, chaotic behavior is seen in population size. The cause for this behavior is poorly understood; some scientists attribute irregular behavior to chaos in the system.

iv. **Cyclic** Fluctuations occur over a regular time period, generally a multiple-year cycle.

Population Fluctuations and Species Conservation

Climate change is imposing a huge pressure on the biosphere. Species are either becoming extinct or are under threat of extinction. Habitat destruction is occurring at a very fast rate on a global scale. The populations left on fragmented islands are isolated populations that fluctuate markedly and may ultimately disappear; their replacement by recolonization is unlikely. Frequent fluctuations in such conditions increase the risk of extinction, making it urgent to support the processes of dispersal and interbreeding. Preserving the largest possible population size becomes a prime requirement as it will not succumb to extinction even if normal fluctuations occur.

ROLE MODEL

THOMAS MALTHUS

Malthus was interested in everything about populations. He accumulated figures on births, deaths, age of marriage and childbearing, and economic factors contributing to longevity. His main contribution was to highlight the relationship between food supply and population. Humans do not overpopulate to the point of starvation, he contended, only because people change their behavior in the face of economic incentives. Noting that while food production tends to increase arithmetically, the population tends to increase naturally at a (faster) geometric rate, Malthus argued that it is no surprise that people thus choose to reduce (or "check") population growth. People can increase food production, Malthus thought, only by slow, difficult methods such as reclaiming unused land or intensive farming; but they can check population growth more effectively by marrying late, using contraceptives, emigrating, or, in more extreme circumstances, resorting to reduced health care, tolerating vicious social diseases or impoverished living conditions, warfare, or even infanticide. Malthus was fascinated not with the inevitability of human demise, but with why humans do not die off in the face of such overwhelming odds. As an economist, he studied responses to incentives. Malthus is arguably the most misunderstood and misrepresented economist of all time. The adjective "Malthusian" is used today to describe a pessimistic prediction of the lock-step demise of humanity doomed to starvation via overpopulation. When his hypothesis was first stated in his best-selling An Essay on the Principle of Population (1798), the uproar it caused among non-economists overshadowed the instant respect it inspired among his fellow economists. So irrefutable and simple was his illustrative side-by-side comparison of an arithmetic and a geometric series—food increases more slowly than population—that it was often taken out of context and highlighted as his main observation. The observation is, indeed, so stark that it is still easy to lose sight of Malthus's actual conclusion: that because humans have not all starved, economic choices must be at work, and it is the job of an economist to study those choices. Malthus addressed many other issues. His Principles of Political Economy (1820) was the first text to describe a demand schedule as separate from the quantity demanded at a given price. His exposition of demand curves clarified the debate on Say's law and gluts (to which he objected in the long run on the grounds that markets self-adjust). His work centered on contrasting the long run, as exemplified by population growth, with the short run, reflected by cyclical events such as those affecting agriculture. Writing before the Industrial Revolution, Malthus did not fully appreciate the impact of technology (i.e., pesticides, refrigeration, mechanized farm equipment, and increased crop yields) on food production. Malthus died in 1834, before seeing economics characterized as the "dismal science." That phrase, coined by Thomas Carlyle in 1849 to demean John Stuart Mill, is often erroneously thought to refer to Malthus's contributions to the economics of population growth.

SUMMARY

- A population is a group of individuals of the same species living together in a specific area.
- Population ecology focuses on studying populations, especially their abundance and how they evolve over time. It delves into the factors influencing the distribution and number of animal and plant populations. A population consists of a portion of individuals from one species inhabiting a defined geographic region and, in sexually reproducing species, engaging in interbreeding.
- The three primary population properties are size (number of individuals in an area), density (individuals per unit or volume), and dispersion (arrangement of individuals).
- A population's size is dynamic, influenced by factors like food availability, predation, and weather.
- Natality is the number of births in the population during a specific period that are added to the initial density.
- Mortality refers to the number of deaths in the population during a given period.
- Immigration is the number of individuals of the same species that have come into the habitat from elsewhere during the time period under consideration.
- Emigration refers to individuals leaving their habitat for another location.
- Interspecific competition in ecology involves different species competing for shared resources in an ecosystem, such as food or living space, contrasting with mutualism, a type of symbiosis. Intra-specific competition occurs when members of the same species compete.
- A population refers to a group of individuals of the same species in a particular area at a specific time. Biologists find characteristics like population density, birthrate, and death rate intriguing.

MULTIPLE CHOICE QUESTIONS

1. **A population is a group of**
   a. individuals in a species
   b. species in a community
   c. communities in an ecosystem
   d. individuals in a family
2. **What is the most important factor for the success of the animal population?**
   a. natality
   b. adaptability
   c. unlimited food
   d. inter species activity

3. **The formula for exponential population growth is**
   a. dN/dt = rN
   b. dt/dN = rN
   c. dN/rN = dt
   d. rN/dN = dt
4. **The human population growth curve is a:**
   a. S-shaped curve
   b. parabola curve
   c. J shaped curve
   d. zig zag curve
5. **Exponential growth occurs when there is**
   a. a great environmental resistance
   b. no environmental resistance
   c. no biotic potential
   d. a fixed carrying capacity
6. **In a population, unrestricted reproductive capacity is called as**
   a. carrying capacity
   b. biotic potential
   c. birth rate
   d. fertility rate
7. **The concept that 'population tends to increase geometrically while food supply increases arithmetically' was put forward by**
   a. Adam Smith
   b. Charles Darwin
   c. Thomas Malthus
   d. Stuart Mill
8. **Two opposite forces operate in the growth and development of every population. One of them is related to the ability to reproduce at a given rate. The force opposite to it is called**
   a. fecundity
   b. mortality
   c. environmental resistances
   d. biotic control
9. **The carrying capacity of a population is determined by its**
   a. population growth rate
   b. natality

c. mortality

d. limiting resources

10. A population that exhibits birth rates that are identical to the death rate implies

a. Initial growth

b. Plateau phase

c. Acceleration Phase

d. None of the above

REVIEW QUESTION

1. Describe the properties of population ecology.
2. Write about the genetic variation within local populations
3. Define the effects of the modes of reproduction: sexual and asexual.
4. Discuss the life histories and the structure of populations.
5. Evaluate exponential and geometric population growth.
6. What are the types of intra-specific competition?
7. Compare and contrast density-dependent growth regulation and density-independent growth regulation, giving examples.

Answers

1. (a)	2. (b)	3. (a)	4. (c)	5. (b)
6. (c)	7. (c)	8. (b)	9. (d)	10. (b)

REFERENCES

1. Abrams, P. A., & Ginzburg, L. R., (2000). The nature of predation: Prey dependent, ratio dependent or neither? *Trends in Ecology & Evolution, 15*(8), 337–341.

2. Begon, M., Townsend, C. R., & Harper, J. L., (2006). *Ecology: From Individuals to Ecosystems* (4th edn.). Oxford, UK: Blackwell Publishing.

3. Benton, M. J., (2001). "Biodiversity on land and in the sea. *Geological Journal, 36*(3, 4), 211–230.

4. Brockhurst, M. A., Fenton, A., Roulston, B., & Rainey, P. B., (2006). The impact of phages on interspecific competition in experimental populations of bacteria. *BMC Ecology, 6*, 19.

5. Brotz, L., Cheung, W. W. L., Kleisner, K., Pakhomov, E., & Pauly, D., (2012). Increasing jellyfish populations: Trends in large marine ecosystems. *Hydrobiologia, 690*(1), 3–20.

6. Chakraborty, K., Das, K., & Kar, T. K., (2013). An ecological perspective on marine reserves in prey–predator dynamics. *Journal of Biological Physics, 39*(4), 749–776.

7. Gotelli, N. J., (2008). *A Primer of Ecology* (4th edn.). Sinauer Associates, Sunderland, MA, USA.
8. Pianka, E. R., (2000). *Evolutionary Ecology*. New York: Addison Wesley Longman.
9. Ricklefs, R. E., & Gary, L. M., (2000). *Ecology* (4th edn.). New York: W. H. Freeman.
10. Turchin, P., (2001). Does population ecology have general laws? *Oikos, 94*(1), 17–26.
11. Vandermeer, J. H., & Goldberg, D. E., (2003). *Population Ecology: First Principles.* Woodstock, Oxfordshire: Princeton University Press.

CHAPTER

8 Community Ecology

LEARNING OBJECTIVES

After studying this chapter, you will be able to:

- Understand community structure
- Explain the concept of community changes
- Describe species interactions in communities
- Know the term ecological succession
- Define climax communities

INTRODUCTORY EXAMPLE

Community ecology is the study and theory of how populations of organisms interact with each other and react to their nonliving surroundings. As a subset of the general study of ecology, this field of specialization explores the organization and functioning of biological communities.

Community ecologists protect the environment and save species from extinction by assessing and monitoring environmental conditions such as global warming.

Community ecology encompasses many types of ecological interactions that continue to change over time. A forest community includes the plant community, all trees, birds, squirrels, deer, foxes, fungi, fish in a forest stream, insects, and all other species living there or migrating seasonally. Similarly, a coral reef community includes a vast number of different species of corals, fish, and algae. Abundance and distribution are strong forces that shape the biological community.

Community ecology focuses on how interactions between different species affect the health, growth, dispersion, and abundance of the ecological system. At the community level, species are often interdependent. Several short food chains are common in most biological communities. Food chains often overlap and form food webs of producers and consumers.

INTRODUCTION

In ecology, a community—also referred to as a biocenosis, biotic community, biological community, ecological community, or life assemblage—is a collection or association of populations of two or more distinct species that are coexisting in the same geographic area. There are several ways to use the term "community." In its most basic sense, it describes populations of living things in a particular location or period, as in "the fish community of Lake Ontario before industrialization."

The study of communities, which are collections of interacting populations of a species that live in a specific region or habitat, is known as community ecology.

Biological communities are created when species interact with one another in populations. These communities' abundance of interacting species and the intricacy of their interactions serve as prime examples of what is meant to be understood by the term "biodiversity." As species interact, guilds, food webs, and other interactive webs are formed, and structures emerge within communities. As a result of species' mutual adaptation to one another during the process of coevolution, these relationships alter over the course of evolution.

The study of interactions between species in communities on a variety of spatial and temporal scales, including distribution, structure, abundance, demography, and interactions between coexisting populations, is known as community ecology, or synecology. Community ecology primarily studies the interactions between populations based on particular genotypic and phenotypic traits.

In addition, abiotic variables that affect species interactions or distributions are considered in community ecology, such as soil pH or the annual temperature. For instance, due to variations in annual precipitation, the plant communities found in tropical rainforests and deserts are very different from one another. Through habitat disturbance, such as the introduction of invasive species, humans can also have an impact on community structure.

The significance and utility of the community concept in ecology are debatable at a deeper level. Traditionally, communities have been conceptualized at a fine scale in terms of local processes creating (or destroying) an assemblage of species; this includes the way that the composition of grass communities is expected to change due to climate change. This emphasis on the local community has drawn criticism recently. Drawing on evolutionary taxonomy and biogeography, Robert Ricklefs has argued that it is more useful to consider communities on a regional scale, where some species or clades evolve and others go extinct.

8.1. COMMUNITY STRUCTURE

Learning Objectives

- Explain how we measure community structure.
- Illustrate the patterns of community structure

The kinds and quantities of species found in various ecological communities can vary greatly. Certain tropical rainforest communities contain a vast number of species per cubic meter, whereas certain Arctic communities only contain a small number of species.

The various structures in communities could be one way to characterize this variation. All of a community's physical characteristics, such as the number of species present and their relative abundances, make up its structure. It can also have a broader interpretation that considers all of how these various species interact.

How do we measure community structure?

The terms species richness and species diversity are used by ecologists to characterize the makeup of a community.

8.1.1. Species Richness

A community's species richness refers to the variety of species present in it. The species richness of the second community would be significantly higher than that of the first if we discovered 300 species in one community and 30 in the first.

The equatorial regions, which have high levels of solar radiation (which sustain high primary productivity), warm temperatures, copious amounts of rainfall, and minimal seasonal variation, are typically home to the species richest communities. Because the poles receive less solar radiation and are drier, colder, and less conducive to life, these communities have the lowest species richness. The pattern for mammalian species richness—which is the species richness calculated exclusively for mammal species and not for all species—is shown below. A community's species richness can be influenced by a wide range of other variables besides latitude.

8.1.2. Species Diversity

One indicator of the complexity of a community is species diversity. It depends on the total number of species (species richness) and the relative abundances of those species

(species evenness) in the community. Higher species diversity is a result of both greater species diversity and more uniform species abundance. For example:

- If the species of trees were even more abundant in both scenarios, a forest community with 20 different types of trees would have more species diversity than one with only 555 types of trees.
- Greater species diversity would be found in a forest community with 20 different species of trees growing in even abundances as opposed to one with the same number of species growing in extremely uneven abundances (for example, with 90% of the trees belonging to a single species).

Ecologists generally believe that ecological communities with greater diversity are more stable—that is, better equipped to bounce back after a disturbance. In the ecological networks video, you can learn more about the reasons behind this. The relationship between diversity and stability isn't always true, though, and there are situations in which variables other than species diversity have a greater bearing on the stability of communities and ecosystems.

Keyword

Community is a group of interacting species that inhabit a particular location at a particular time.

8.1.2.1. What Factors Shape Community Structure?

A community's structure is the product of numerous interacting abiotic (nonliving) and biotic (related to living organisms) factors.

Here are some

- The climate patterns of the community's location.
- The geography of the community's location.
- The heterogeneity (patchiness) of the environment
- The frequency of disturbances, or disruptive events.
- Interactions between organisms.

Incidents that happen by chance over time in a community can also impact its structure. For instance, imagine a seed lands in a particular spot. If it manages to grow, the species could thrive and eventually become dominant (aside from closely related species). If the seed doesn't sprout, another similar species might take its place and thrive instead.

8.1.3. Foundation and Keystone Species

Certain species have remarkably powerful effects on the structure of communities, maintaining the community's equilibrium or even enabling it to exist. Keystone and foundation species are among these "special" species.

Keyword

Coral reef is an underwater ecosystem characterized by reef-building corals.

8.1.3.1. Foundation Species

A foundation species is special and vital to the formation and identity of a community. Foundation species frequently work by altering the surroundings to make it more hospitable for the other creatures that make up the community.

In the kelp forests off the coast of California, kelp, or brown algae, is a foundation species. The kelp forest community is made up of various organisms that can survive thanks to the environments that kelp creates.

Another foundation species in a *coral reef* is the coral itself. The majority of the reef structure, which shields other species from waves and ocean currents, is made up of the exoskeletons of both dead and living coral. Beavers are another example of a foundation species because they alter their surroundings by creating dams.

8.1.3.2. Keystone Species

A keystone species is one that, in relation to its biomass or abundance, has an outsized impact on community structure. Two main characteristics set keystone species apart from foundation species: first, they are more likely to be top predators at higher trophic levels, and second, they exhibit a wider range of behaviors than foundation species, which typically alter their surroundings.

Perhaps the most well-known illustration of a keystone species is the intertidal sea star *Pisaster ochraceus,* which is found in the northwest of the United States. The intertidal zone in which the sea stars resided was purposefully removed from them in a traditional community ecology experiment. As a result, the number of their prey (mussels) increased, changing the community's species composition and drastically lowering species diversity. About 25 different species of barnacles and algae were found in the lower intertidal zone when sea stars were present; however, when they were absent, the population of mussels expanded downward and nearly completely replaced these other species.

The removal of a keystone species frequently results in this kind of abrupt collapse of community structure or sharp reduction in diversity. In this instance, species that would typically be able to survive because sea stars kept the mussels in check were displaced by the mussels, resulting in a loss of diversity.

8.1.4. Patterns of Community Structure

8.1.4.1. Stratification and Gradation

As species adapt to their habitat, the community structure can become stratified both horizontally and vertically during the succession process. This fractionation is caused by changes in environmental parameters such as light, temperature, or water. The different levels of light that the various strata in a forest receive cause vertical stratification; the taller the plant and the more foliage it produces, the more light it can intercept. A forest typically contains three or more vertical layers of plants: a layer of herbs, a layer of shrubs, a layer of small trees, and a layer of canopy trees. This plant life stratification has an impact on animals. Even though they are quite adept at switching between layers, they frequently stay put in one area for feeding, mating, or other activities. Additionally, gradients in the physical environment can give rise to horizontal patterns among species. The distribution of plant and animal species can be impacted by variations in the region's nutrient or water availability. Different plant and animal species can be found on a mountain at different elevations as well as on the east, west, north, and south slopes. While gradual changes can result in a more integrated flow of species, drastic differences in some factors over very short distances can produce sharp boundaries between communities. The regional biodiversity is preserved in part by these gradients.

8.1.4.2. Ecotones

Ecosystems are typically composed of a patchwork of communities existing at various stages of succession. Different ecosystems experience disturbances at different sizes, frequencies, and intensities, which result in variations in what are known as the patch dynamics of communities. Ecotones are regions found at the borders of each patch. Both species from each of the overlapping communities and some species that have evolved especially for living in these zones can be found in these junction zones. The edge effect refers to the fact that there are frequently more species and higher population densities inside the ecotone than there are outside it.

In ecotones between mature forests and earlier successional patches in North America, brown-headed cowbirds (*Molothrus ater*) are particularly prone to parasitize bird nests. Cowbirds are mostly active in early successional patches and lay their eggs in other birds' nests. Compared to birds living in ecotones, forest birds with nests deep within mature forests are less vulnerable to attacks. Mature forest removal has expanded the area of ecotones, which has led to an increase in cowbird parasitism rates throughout North America.

Remember

Currently held community ecology theory is based on the supposition that ecological communities are the result of different types of assembly processes. Assembly processes involve adaptation, speciation in evolutionary biology, competition, colonization, altitude, climate, habitat disturbances, and ecological drift.

8.1.4.3. Ecological Niches

An ecological niche includes a species' lifestyle. In essence, it describes how a species interacts with or blends into its surroundings. Natural selection helps a species develop specialized traits that enable it to uniquely utilize the resources in its environment as it adapts to the physical conditions and biota of the community. The niche is shaped by the region's physical characteristics, such as temperature, topography, or nutrient availability, and by biological limitations, such as competition, predation, or resource scarcity, which restrict how a species uses its surroundings. For instance, different plant species have different needs in terms of light, nutrients, and microbes, as well as different capacities to repel rivals and herbivores. Only a portion of the plants found in a community can be consumed by herbivore species, and only a portion of the numerous possible prey species can be caught by predators. As a result, the species "carves out" a place in the community.

One such niche is occupied by the critically endangered Kirtland's warbler *(Dendroica kirtlandii)*, which is found in North America. It only builds its nests in uniform stands of young, 2–4 meter (6–13 foot) tall jack pines (*Pinus banksiana*). Periodic fires are allowed to occur on these trees, which is essential for the jack pine seeds to germinate. Additionally, because these fires consistently create large new areas with young trees, the warblers are able to move their nesting locations over time to stay inside the preferred height of jack pine stands.

Keyword

Natural disasters are major events that occur when the Earth's natural processes cause loss of life, property damage, and economic damage.

The niche of a species changes in response to biological and physical changes in its environment, assuming these changes occur slowly enough for the species to adjust. No two species in a community can occupy the same niche, which is the primary restriction on this evolution. Thus, coexisting species benefit selectively from specialized modes of existence, which mitigates direct competition for resources.

8.1.4.4. Biodiversity and the Stability of Communities

Species create niches and guilds as a result of their mutual adaptation and community adaptation. It is possible for more species to coexist when more complex structures are developed. The community becomes more stable as a result of the rise in species richness and complexity acting as a buffer against environmental stresses and *natural disasters*.

8.1.4.5. Community equilibrium and species diversity

Certain environments experience a climax in succession, resulting in a stable community dominated by a few key species. It is believed that this equilibrium, known as the climax community, arises when the network of biotic interactions gets so complex that no additional species can fit in. In other settings, a variety of species coexist in communities created by persistent small-scale disturbances, and any species has the potential to become dominant. This nonequilibrium dynamic demonstrates the impact that unforeseen disruptions can have on the composition and structure of a community. There are hundreds of different tree species in a square kilometer of some species-rich tropical forests. The area left by a dead tree that falls to the ground is open for grabs. In a similar vein, hundreds of fish species can be found in some coral reefs; the species that succeeds in colonizing a new disturbance patch will win. Every minor disruption triggers a fresh attempt at dominance.

Communities that are diverse are generally healthier. According to extensive ecological research, communities with a higher species richness can bounce back from disturbances more quickly than communities with a lower species richness. In contrast to species-poor grasslands, species-rich grasslands in the Midwest of the United States continue to have higher primary productivity. The community's ability to withstand drought is progressively impacted by each new species that disappears from these grasslands. Similar to this, more diverse plant communities in Yellowstone National Park exhibit higher species composition stability than less diverse communities during periods of extreme drought. Furthermore, the more diverse communities in Africa's Serengeti grassland exhibit greater biomass stability throughout the seasons and greater capacity for post-grazing recovery.

The necessity to preserve the highest level of richness within biological communities is highlighted by the connection between species diversity and community stability. A rich interactive web of native species that have had time to adapt to one another is very different from a field of weeds that contains species that have only recently been introduced to the community. Communities rich in species that are not disturbed possess the resilience needed to maintain a healthy ecosystem, which is essential for life. Additionally, these communities are better equipped to withstand the effects of introduced foreign species, which, while sometimes unintentionally, can cause significant ecological and economic disruptions in less stable communities. Natural biological communities are made up of intricate networks of interactions that support both community stability and biodiversity.

8.1.4.6. Biogeographic Aspects of Diversity

The study of species distribution in a region is known as biogeography (see Biogeographic Region: General Features). Islands have been used to observe the factors that affect species diversity because they offer a controlled study area. The size of an island, its distance from other islands or the mainland, and the variety of species that live there are the three factors that affect how quickly an island is colonized. Based on this data, the theory of island biogeography can be used to estimate how many species will exist on a particular island. The species diversity of "islands" on land, such as mountaintops, lakes, and remnants of forest left over after logging, can also be explained by this theory. According to the theory of island biogeography, the number of species is proportionate to the size of the island and inversely proportional to its distance from the mainland where immigration and extinction rates are equal.

Six months had passed since the 1963 volcanic eruption on the Icelandic island of Surtsey before a few bacteria, molds, insects, and birds had taken over the island. After a volcano on the tropical Pacific island of Krakatoa erupted in 1883, a few grass species, insects, and vertebrates became established within a year or so. A few decades were all it took for hundreds of species to arrive on both Surtsey and Krakatoa. The island communities gradually stabilized into a dynamic equilibrium, though not every species is able to take root and become firmly established.

Did you get it?

1. **What is meant by species diversity?**
2. **What factors influence the community structure?**

8.2. COMMUNITY CHANGES

Learning Objectives

- Explain the history of the study of succession
- Describe the types of succession
- Discuss the models of succession

In ecology, the makeup of a community changes over time. This transition, influenced by factors like the environment, interactions among organisms, and dispersal, is the focus of succession research. You can observe changes in species residing in various locations over the years, such as your backyard, nearby forest, grassland, pond, abandoned farm (or urban lot), or park. It's common to see certain plants disappear and new ones appear, along with changes in the animal kingdom. Communities evolve in terms of composition, structure, diversity, and productivity, with changes ranging from subtle to noticeable. Throughout history, human populations have manipulated habitats (e.g., through controlled fires) to enhance specific plant and animal species. Ecologists have studied these community changes for over a century, identifying reasons, trends, underlying mechanisms, and developing theories to explain them. Succession refers to the gradual community transformation. Some definitions include adjectives like "orderly," "directional," and "stabilizing," which have sparked ongoing debates.

8.2.1. The History of the Study of Succession

Previous European studies describing succession on coastal dunes served as the inspiration for Henry Cowles's dissertation research on sand dunes around Lake Michigan. Even though Cowles did not invent the concept of succession, his work defined the laws that underlie patterns of community change, and as a result, it had an impact on generations of ecologists. Henry Cowles established the still-held belief that plant communities are dynamic in terms of both the species composition and the underlying physical environment. The basic idea was that the ages of the plant communities in sand dunes revealed the ways in which the communities had changed over time. The chronosequence approach has been utilized by Cowles and subsequent scholars to gain a deeper comprehension of the succession process. To infer the successional sequence, this method substitutes spatial differences (locations of different ages) for time periods, as it would be challenging to follow a successional sequence that lasted decades or centuries. A sere is a particular stage in a successional sequence. In the field of ecology, particularly plant ecology, the question of how this process operated has always been prominent. Since they dominate the landscape, plants have received the majority of attention in the last century.

8.2.2. Types of Succession

Two types of successional processes have been identified, based on the characteristics of the habitat in which they take place. The colonization of lifeless habitats in a newly created habitat that has never had any vegetation before is known as primary succession. Volcanism, the creation of dunes and barrier islands, and glaciation are the causes of these new land surfaces, and they necessitate the influx of outside species. Because soil development is a laborious process necessitated by both abiotic factors and the activities of organisms, the rate of primary succession is typically slow. For example, lichens frequently colonize rock locations. These lichens work in tandem with physical processes (e.g., wind, and rain), and alter the substrate in a way that makes it possible for additional colonists to survive (e.g., plants that are vascular and bryophytes). When an ecosystem is disturbed and leaves behind seeds, spores, or the underground parts of plants, secondary succession takes place. Any environmental event, usually abiotic, that causes a community to deviate from its equilibrium is referred to as a disturbance. This is an erratic event that may involve mortality or the removal of biomass.

8.2.3. Models of Succession

Numerous models that explain and forecast the succession process have been created. The work of Frederic Clements from the early 1900s was one of the most important. Succession was understood as a serial stage sequence that culminated in the climax community, following a deterministic and orderly order. According to this theory, every stage was made up of a collection of species that were thought of as a single super-organism that collaborated to develop on a location before being replaced by a new stage. The habitat's climax community, which was considered to be the "expression of the climate," was eventually reached by succession. Henry Gleason, on the other hand, offered an alternative interpretation of succession that placed an emphasis on the individualistic characteristics (abiotic tolerances and dispersal) of species and how they interact with their surroundings. Not because they are a component of a cohesive "superorganism," but rather because of their abiotic tolerances or niche requirements, species coexist in a given area. According to this viewpoint, succession is more dependent on random dispersal events and is less deterministic. The possibility of various successional trajectories leading to a distinct climax community is one of the effects. For many years, Clements' incorrect beliefs dominated ecology, but in the 1950s, research that upheld Gleason's individualistic viewpoint and early criticisms by eminent ecologist Arthur Tansley started to challenge this. This work included the continuum and gradient approaches of John Curtis and Robert Whitaker, which shed light on the idea of individualistic species segregation along environmental gradients that did not match the consistent associations that Clements had predicted. During this period, Clements' influence could also be observed in the advancement of ecological perspectives on succession. Within the context of the Clementsian successional paradigm, Eugene Odum offered an ecosystem perspective on succession. Expectations regarding energy, life history, biogeochemical processing, and selection pressure (i.e., r- and K-selection), numbers, and the composition of communities evolve with time. Connell and Slatyer finally reworked a number of the succession models and concepts. They offered three models that explained processes and offered theories regarding the course of succession. All three believed that early pioneer species that produced large numbers of seeds and germinated early were responsible for determining the community's original makeup.

The life-history traits that cause later successional species to grow more slowly and eventually supplant earlier successional species are what propel the second pathway, tolerance. Finally, the inhibition model predicts that long-lived late successional species will eventually supplant early colonists, who will remain dominant until a disturbance or mortality eliminates them. David Tilman offered a different mechanistic explanation of succession using the resource-ratio theory. He provided precise forecasts regarding adjustments to the resource ratios because species became specialized on different ratios of those resources, community composition can change over time due to factors like light or nitrogen availability. The current perspective recognizes the wide range of intricate variables and events that affect succession.

The replicas (such as Markov Models) and techniques have advanced through the incorporation of numerous contributing elements, such as (but not restricted to) spatial processes (e.g., dispersal), life cycles and characteristics of species, intricate relationships between species (e.g., gradients in the environment, disturbance characteristics, and direct and indirect).

8.2.4. Factors Affecting Change

Disturbances are one of the many reasons why a community's composition can shift over time. The characteristics of disturbances, such as their frequency, extent, and severity, can affect their effects. A disturbance in the succession context can move the community backward or forwards in time. For instance, fire can change the availability and utilization of nutrients in a community and eliminate the biomass of plants that are above ground. Fire-adapted species may have seeds that sprout again from subterranean portions or that germinate in reaction to fire. On the other hand, species that cannot withstand fire might have to relocate from areas that have burned. The rate at which the habitat is altered is another element that may have an impact on succession. For instance, some species' ability to coexist in their habitat may be restricted by the rate at which soil forms. The rate of succession can also be determined by the rates at which species disperse into a site. In addition, this rate may interact with arrival time and order to create patterns of shifting community composition. Priority effects, for instance, can happen, in which the emergence of a species depends on early arrival — "first come, first served.".

In fact, there are more similarities than differences between the concepts and procedures covered in the field of community "assembly" and the framework of "succession" when it comes to forecasting how communities will evolve over time in response to biotic interactions, dispersal, and abiotic factors. Though attempts to reconcile these two approaches are relatively new, many of their differences are artificial constructs of the ecology field. Remarkably, both strategies have been used to direct land management and habitat restoration. For instance, both paradigms discuss how species interactions and dispersal affect how communities change over time. The synthesis of these two theoretical bodies ought to continue in future work.

Did you get it?

1. **Who discovered succession?**
2. **What are the types of succession?**

8.3. SPECIES INTERACTIONS IN COMMUNITIES

Learning Objectives

- Describe the term 'biological interaction'
- Define competition

There are numerous interactions between different species. Some interactions occur directly, such as when lions kill and devour gazelles or when clownfish hide inside sea anemone tentacles. However, many interactions are less obvious because they are indirect. For example, lions depend directly on the plant-eating animals they hunt, which in turn rely on the plants they need for sustenance. Therefore, despite not consuming plants, lions are indirectly reliant on the type and quantity of plants. Changes in one species, such as population size, health, or geographic range, will likely affect other species and elements of the Earth system due to both direct and indirect interactions between species.

Species influence each other through interactions, which can be direct or indirect and have positive or negative outcomes. The most visible interactions are direct ones, like when a giraffe consumes leaves from a tree. The giraffe benefits from the interaction by obtaining nourishment from the tree, while the tree suffers from losing leaves essential for photosynthetic energy production. However, giraffes also benefit trees directly by depositing droppings on the ground, enriching the soil with nutrients and enhancing its quality, which promotes tree and plant growth. Giraffes are among the herbivores that depend on this plant growth for sustenance, indirectly supporting the survival of other herbivorous species. The growth of plants introduces more species to the network of indirect species interactions, providing habitats for smaller animals like insects and birds to live and nest. The biomass and productivity of an ecosystem are influenced by species interactions, which in turn impact the size of species populations. Evolutionary changes that impact an organism's capacity for long-term survival and reproduction are also influenced by species interactions. Learn more about interactions between species by looking through the examples below.

a. Earth System Models of Species Interactions

Remember

Competitive interactions are in many cases indirect, such as when two species both consume the same resource but do not directly interact with each other. Instead, they affect each other by reducing the availability of the resource.

Modeling species interactions within the biosphere can become very complex very quickly because species interact with other species both directly and indirectly. For instance, photosynthesis produces oxygen, which is used by almost all organisms, including those that are photosynthetic, to convert food into energy through a process known as respiration. Consequently, practically every species has indirect interactions with photosynthetic species, making photosynthetic species important in all Earth system models pertaining to the biosphere.

It should be noted that we only include interactions that have the greatest direct impact on the relevant species, ignoring many other direct and indirect interactions that have a significant impact on the species in the model.

The Earth system focuses on how interactions between species can impact columbine flower evolution. Knowing that columbines, like the majority of flowering plants, have evolved flower traits (colors and shapes) to attract pollinators—insects and birds that move pollen from plant to plant—will help you understand how this works. Pollinators visit columbine flowers in search of food, which the plant provides in the form of nectar. Each pollinator unintentionally brushes up against the stamens as it feeds, which transfers *pollen* to the pollinator's body. Then, some of this pollen rubs off onto other flowers as the pollinator travels from flower to flower in order to feed. Consequently, the pollinator inadvertently aids in flower pollination as it moves from flower to flower while consuming nectar. To find out more information about each topic, click on the icons or hover over them for a brief explanation.

Keyword

Pollen is a powdery substance produced by most types of flowers of seed plants for the purpose of sexual reproduction.

b. How Human Activities Influence Species Interactions?

Interactions between species are impacted by human activity both directly and indirectly. Large-scale farming and agricultural practices are one of our direct effects; they boost the populations of food species and reduce those of species that occupy the habitats that these practices have altered. The numbers of the exploited species' populations are also decreased by hunting and fishing. Smaller-scale hunting for food can also have an impact on humans, albeit in a negative way. For instance, eating wild-caught animals can spread dangerous animal diseases like HIV and Ebola to human populations. Urbanization and large-scale farming practices alter habitat, which have indirect effects on interactions between species. The impact of climate change through processes like ocean acidification and extreme weather, such as drought and fire, was also considered an indirect effect.

8.3.1. Biological Interaction

A biological interaction in ecology refers to the impact that coexisting organisms have on one another. Interspecific interactions refer to interactions between different species, while intraspecific interactions involve the same species. Both of these effects—which can be long-term or short-term, like pollination and predation—strongly impact the species in question's evolutionary history. A symbiosis is a long-term relationship. Mutualism, which benefits both parties, and competition, which hurts both parties, are two different types of symbioses. Indirect interactions can occur when individuals interact through intermediaries like common enemies or resources. This kind of relationship can be demonstrated by the net effect, which is based on the individual effects that the relationship has on both organisms.

According to a number of recent studies, mutualisms and habitat modification among non-trophic species may play a significant role in determining the structures of food webs. It is still unknown, though, if these results hold true for all ecosystems and whether non-trophic interactions have a random effect on food webs or have an impact on particular trophic levels or functional groups..

8.3.1.1. Short-Term Interactions

In ecology and evolution, short-term interactions such as pollination and predation are crucial. In terms of the length of a single interaction, these are brief—a predator kills and consumes its prey, and a pollinator spreads pollen from one flower to another—but they are incredibly persistent in terms of their impact on the evolutionary paths of both partners. The partners consequently coevolve.

8.3.1.2. Predation

Predation is the process by which an organism, known as the predator, kills and consumes its prey. Predators have keen senses like vision, hearing, and smell, and they are frequently highly specialized for hunting. Sharp jaws or claws are used by many predatory animals, both vertebrate and invertebrate, to seize, dismember, and kill their prey. Additional adaptations that increase hunting efficiency include aggressive mimicry and stealth. Predation exerts strong selective pressure on prey, leading to the development of antipredator adaptations like chemical and/or color changes, warning calls and other signals, camouflage, and defensive spines. Since at least the Cambrian era, predators have played a significant role in the evolution of species.

The ability of several intriguing microbes to prey on others has allowed microbiologists to survive over the last few decades. Members of the genera Bdellovibrio, Vampirococcus, and Daptobacter are among the best examples.

Bdellovibrios are highly mobile, active hunters that swim around in search of vulnerable Gram-negative bacterial prey. A Bdellovibrio cell detects such a cell and swims faster until it runs into the prey cell. It then penetrates its prey's outer membrane to reach the periplasmic space. It develops into a long filament as it grows, which eventually splits into septae and releases baby bacteria. New Bdellovibrio cells are released during the prey cell's lysis. It has never been observed that Gram-negative prey bacteria develop resistance to Bdellovibrio, and the bacteria will not assault mammalian cells. This has sparked interest in treating infected wounds with these bacteria as a "probiotic." Though it hasn't been tested yet, it makes sense that, given the rise in pathogens resistant to antibiotics, these kinds of treatments might be seen as workable substitutes.

8.3.1.3. Pollination

To facilitate fertilization, pollinators—such as insects (entomophily), certain birds (ornithophily), and certain bats (chiropterophily)—transfer pollen from a male flower part to a female flower part in exchange for pollen or nectar. Over geological time, the partners have coevolved; for insects and flowering plants, this coevolution has lasted for more than 100 million years. Flowers that are pollinated by insects have developed shaped structures, vivid colors, patterns, scents, nectar, and sticky pollen to draw in insects, assist them in gathering and depositing pollen, and compensate them for their labor. Pollinator insects, such as bees, are specialized to recognize color, pattern, and scent in flowers; to gather and carry pollen (e.g., by using bristles on their hind legs to form pollen baskets); and to gather and process nectar (e.g., honey, which bees make and store). Natural selection has shaped the adaptations on both sides of the interaction based on how well they pollinate. These adaptations match the adaptations on the other side.

8.3.1.4. Seed Dispersal

Seed dispersal is the process by which seeds are moved, dispersed, or transported from their parent plant. Since they can only move very slowly, plants rely on a range of abiotic (such as the wind) and biotic (such as birds) dispersal vectors to spread their propagules. Both individually and collectively, as well as in terms of time and space, seeds can be dispersed from the parent plant. In addition to having significant effects on migration patterns, species interactions, and the demographic and genetic makeup of plant populations, the dispersal mechanism plays a major role in determining seed dispersal patterns. There are five primary ways that seeds can spread: through gravity, wind, ballistics, water, and animals. Certain plants only scatter their seeds in response to external stimuli because they are serotinous. A diaspore that has been dispersed must release itself from its parent plant.

8.3.1.5. Symbiosis: Long-term Interactions

Mutualism, commensalism, parasitism, neutralism, amensalism, and competition are the six types of symbiosis that can occur. The degree of advantage or disadvantage these have for each partner sets them apart.

8.3.1.6. Mutualism

A relationship in which two or more species cooperate to achieve a common goal—such as a higher carrying capacity—is known as mutualism. Cooperation is the term used to describe similar interactions within a species. Mutualism can be categorized according to how closely an association is formed; symbiosis, which is frequently mistaken for mutualism, is the closest. There's a chance that one or both of the interacting species are obligate, meaning they couldn't endure either temporarily or permanently without the other species. Even though mutualism is a topic that is crucial to ecology, it has historically received less attention than other interactions like predation. Examples comprise gut flora, Müllerian mimicry, cleaning symbiosis, and nitrogen fixation by bacteria in legume root nodules.

8.3.1.7. Commensalism

Commensalism is advantageous to one organism while having no negative effects on the other. It happens when an organism interacts with another and gains something from it without affecting the host organism. A remora coexisting with a manatee is a prime example. Remoras consume the excrement of manatees. This interaction has no effect on the manatee because the remora does not deplete its resources.

8.3.1.8. Parasitism

A relationship between species known as parasitism occurs when one organism, called a parasite, lives on or inside another organism, called a host, harming it occasionally and developing a structural adaptation to this way of life. When it comes to intestinal parasites, the parasite either feeds on the host or partly consumes its food.

8.3.1.9. Neutralism

The relationship between two species that interact but do not affect one another is referred to as neutralism, a term that was first used by Eugene Odum. It is nearly impossible to find examples of true neutralism; in reality, the term is used to characterize circumstances in which interactions are minimal or inconsequential.

Keyword

Amensalism is a type of behavior where one organism harms another without any apparent gains or benefits for the organism itself.

8.3.1.10. Amensalism

An interaction known as amensalism occurs when one organism harms another without incurring any expenses or reaping any rewards for itself. The negative impact one organism has on another is referred to as amensalism. This is a one-way process that occurs when an organism releases a particular compound that harms another. The microbial production of antibiotics, which can inhibit or kill other, susceptible microorganisms, is a classic example of *amensalism.* When sheep or cattle trample the grass, it is an obvious example of amensalism. Although the presence of the grass has very little negative impact on the hoof of the animal, the grass becomes crushed. Amensalism is frequently used to characterize highly asymmetrical competitive interactions, like those that have been noted between the Spanish ibex and Timarcha weevils, which both consume the same kind of shrub. Weevils hardly affect the availability of food, but their presence greatly reduces the number of weevils because ibex eat a lot of plant matter and inadvertently ingest the weevils that are present in it.

Amensalism associations are often very intricate. An actinomycete and a parasitic fungus in the genus Escovopsis interact in a way that Attine ants (a family of ants from the New World) can benefit. The ant and members of the fungal genus Leucocoprinus can maintain a mutualistic relationship because of their amensalistic relationship. It is astounding to learn that these ants grow a garden of Leucocoprinus fungi for their own food. The ants also encourage the growth of an actinomycete of the genus Pseudonocardia, which generates an antimicrobial compound that inhibits the growth of the parasitic fungus Escovopsis, to keep it from completely devastating their fungal garden.

8.3.1.11. Non-Trophic Interactions

Interactions that are not trophic include competition for space, mutualism, and habitat modification. Recent research has indicated that non-trophic interactions may influence the species within the network and the strength of trophic links, which in turn may have an indirect impact on trophic dynamics and food web topology. The necessity of integrating trophic and non-trophic interactions in ecological network analysis has been highlighted by a number of recent theoretical studies. The limited number of empirical studies that tackle this issue indicate that species interactions occurring beyond the trophic network can have a significant impact on food web structures, or network topologies. It's unclear how broadly these findings can be applied because these studies only cover a small sample of coastal systems. It is unclear whether non-trophic interactions generally impact particular species, trophic levels, or functional groups within the food web, or if they mediate species and their trophic interactions across the network in an indiscriminate manner. While some studies indicate that non-trophic facilitation is beneficial to all species, others suggest that it is more beneficial to sessile species with generally lower trophic levels.

A 2018 study by Borst tested the general hypothesis that foundation species, or spatially dominant habitat-structuring organisms, modify food webs by increasing their size, as indicated by the number of species, and their complexity, as indicated by the number of links, through species facilitation, regardless of the type of ecosystem. They also tested the hypothesis that any changes in food web properties brought about by foundation species happen either randomly through species facilitation throughout the entire food web, or specifically through species facilitation of particular species that fall into particular trophic levels or functional groups. The findings indicate that the trophic level and average chain length of foundation species are higher than those predicted by random facilitation. Specifically, carnivores are more strongly facilitated in these food webs than are species at the base of the food chain. This shows that by non-trophic facilitation of species throughout the entire trophic network, foundation species significantly increase the complexity of the food web.

While foundation species, like all other species, are a part of the food web (e.g., both as prey and predator), a number of studies have demonstrated that they significantly support the related community by generating new habitats and reducing physical strain. There is a broad range of ecosystems and environmental conditions in which foundation species can provide this type of non-trophic facilitation. Seagrasses, mangroves, corals, kelps, mussels, oysters, salt marsh plants, and seagrasses all help organisms in harsh coastal zones by reducing waves and currents, supplying aboveground structure for attachment and shelter, concentrating nutrients, and/or lessening desiccation stress at low tide. Foundation species have also been shown to play a significant role in habitat structuring in less harmful systems. Examples of these species include trees in forests, savannah grasses and shrubs, and macrophytes in freshwater systems. All foundation species eventually improve the availability and complexity of their habitat, dividing and expanding the niche space that is accessible to other species.

8.3.2. Competition

Competition can be broadly defined as the direct or indirect interaction of organisms that leads to a change in fitness when they share a resource. Usually, it involves individuals competing for a limited resource, with weaker competitors being negatively affected. There are three main types of competition. Real competition includes exploitation competition and interference competition. Apparent competition, the third type, does not. While apparent competition and exploitation competition occur indirectly between individuals, interference competition happens directly between them (Figure 8.1).

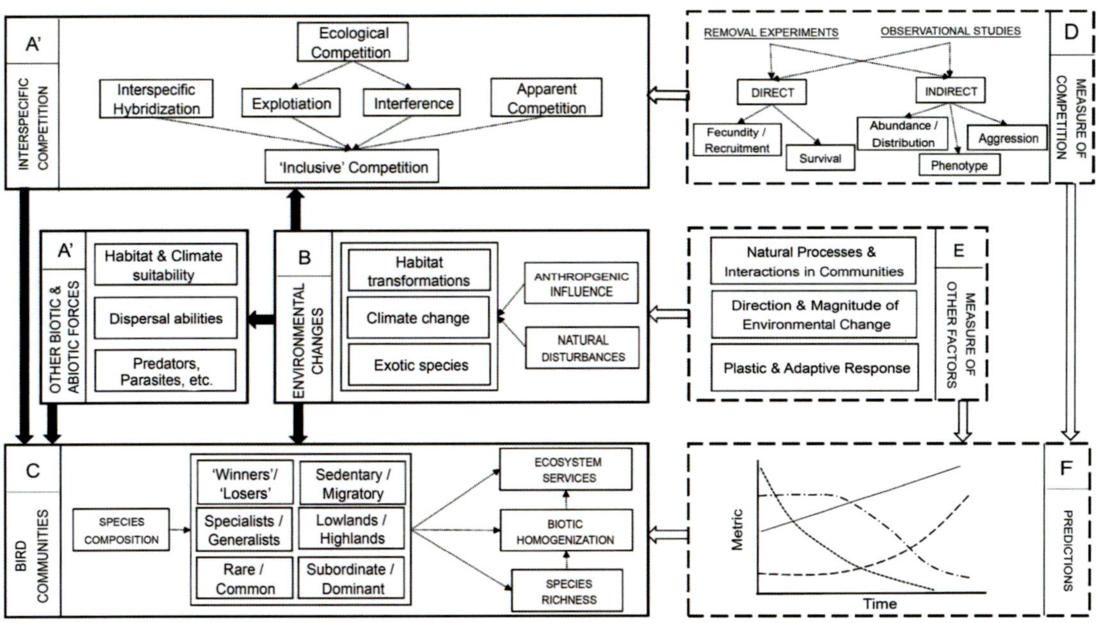

Figure 8.1. The three major types of competitive interactions.

Source: Guillaumet, A., Russell, I. J. (2022). Bird Communities in a Changing World: The Role of Interspecific Competition. *Diversity*. *14*(10):857. https://doi.org/10.3390/d14100857.

Interference competition occurs when one individual intentionally modifies another individual's resource-attaining behavior. For instance, a dominant male gorilla directly modifies the mating behavior of other males when he uses physical aggression or other aggressive displays to prevent other males from accessing a mate. Another illustration of an intra-specific interaction is this. The phenomenon of exploitation competition arises when species engage in indirect interactions while vying for shared resources such as territory, prey, or food. In other words, the resource will become less available for use by others when it is used by one individual. A superior competitor can eventually drive an inferior one out of the market through manipulation or outright exploitation, leading to competitive exclusion. Equations can be used to predict the results of competition between two species; the Lotka-Volterra model is one of the most well-known examples. This model incorporates both the overall impact of two species on each other as well as the relationship between population density and carrying capacity. The four outcomes of this model are:

1) Species A competitively excludes species B.
2) Species B competitively excludes species A.
3) Either species wins based on population densities.
4) Coexistence occurs.

Species can survive together if intra-specific competition is stronger than inter-specific competition. This means that each species will inhibit its own population growth before it inhibits that of the competitor, leading to coexistence. Adopting alternative life history and dispersal strategies—which are typically reinforced by natural selection—is another way to avoid competitive exclusion. Through this mechanism, chances for new colonization and nutrient acquisition are increased while competitive interactions are decreased. Events (like tidal, flood, or fire disturbances) that produce chances for dispersal and nutrient acquisition are frequently necessary for this to succeed. Let us consider the following: Plant B is a better disperser, but Plant A is more effective at absorbing nutrients. In this case, nutrients are the resource that is under competition, but obtaining nutrients is dependent on their availability. Plant B is anticipated to move in first and stay in the community until Plant A shows up and starts to compete with Plant B if a disturbance creates a new space for colonization. Plant A will eventually surpass Plant B in terms of growth rate, possibly due to Plant A's superior ability to absorb nutrients. The Plant B population will decrease as the Plant A population grows, and with enough time, it will be able to be eliminated from that area. Plant B's exclusion can be prevented if a local disturbance occurs. Since this frequently occurs in nature, disturbance can prevent competitive exclusion and balance competitive interactions by forming patches that are more easily colonized by species with more effective dispersal mechanisms. However, the frequency and spatial proximity (or closeness) of disturbance events in relation to the dispersal rates of individuals of the competing species determine the outcome of the dispersal versus nutrient acquisition trade-off. When disturbances happen frequently enough or far enough apart to keep the weaker but frequently more effective dispersing competitor within a habitat, coexistence can be attained. The inferior competitor (better disperser) prevails if the disturbance occurs too frequently; however, if the disturbance occurs infrequently, the superior competitor gradually outcompetes the inferior competitor, leading to competitive exclusion. The intermediate disturbance hypothesis is the name given to this.

When two individuals who do not directly compete with one another for resources nevertheless have an indirect impact on one another through sharing a predator, this is known as apparent competition. Think about a hawk that eats mice and squirrels. In this relationship, an increase in the squirrel population may have a positive effect on the mouse population because more squirrels will be available for the hawks to prey on. Nevertheless, as the squirrel population decreases, a greater number of hawks may eventually need more prey, which would negatively impact the mice through increased predation pressure. The predator's food supply may decline, which would have the opposite effect. Since mice will be the hawks' more plentiful food source, a decline in the squirrel population may also indirectly result in a decline in the mouse population. In nature, apparent competition can be hard to spot, frequently due to the intricacy of indirect interactions involving several species and shifting environmental factors.

Did you get it?

1. **How do human activities influence species interactions?**
2. **What are the types of competition?**

8.4. ECOLOGICAL SUCCESSION

Learning Objectives

- Explain the history of ecological succession
- Discuss the factors of ecological succession
- Describe the types of ecological succession

The change in species structure over time in an ecological community is called ecological succession. It can last from a few decades (after a wildfire, for example) to several decades.

The community starts out as a relatively small group of pioneering plants and animals and grows more complex over time until it reaches a climax community, which is stable and self-sustaining. The effects of established organisms on their own environments are the "engine" of succession, the reason behind changes in ecosystems. Living has the effect of changing one's surroundings, sometimes subtly and other times overtly.

It is the process through which an ecological community changes, after a disturbance or after the first colonization of a new habitat, in a relatively orderly and predictable manner. A community may be disturbed by a fire, a strong windstorm, logging, or the creation of new, unoccupied habitat as a result of a landslide or lava flow. These events can also start a succession. Primary succession refers to a succession that emerges in new habitats independently of pre-existing communities, while secondary succession is a succession that occurs after a pre-existing community is disrupted.

One of the earliest ecological theories to be proposed was succession. The Indiana Dunes in Northwest Indiana are where ecological succession was first observed, and it is still a crucial ecological subject of research.

8.4.1. History

The concept of ecological succession has its roots in early 19th-century thought. Poplars naturally come before oaks and beeches in the evolution of a forest, as the French naturalist Buffon observed as early as 1742. Later on, Buffon was compelled to renounce many of his theories by the University of Paris theological committee because they ran counter to the biblical account of Creation. The term "succession" was first used to describe the development of vegetation following forest clear-cutting by the French naturalist Adolphe Dureau de la Malle and the Swiss geologist Jean-André Deluc. In a speech titled "The Succession of Forest Trees," written in 1859, Henry David Thoreau discussed the succession of oak-pine forests. Although observers have long known that squirrels burrow nuts in the ground, no one has used this information to explain why forests regularly change. In 1863, the Austrian botanist Anton Kerner released a study on the plant succession in the Danube River basin.

According to Ragnar Hult's 1885 study on the stages of forest development in Blekinge, heath emerges from grassland before forests do. In the early phases of forest development, birch predominated, followed by pine (on dry soil) and spruce (on wet soil). It eventually becomes beechwood if oak replaces the birch. Swamp vegetation progresses from moss to sedges to moor vegetation, then birch to spruce.

8.4.1.1. H. C. Cowles

Henry Chandler Cowles, at the University of Chicago, created a more formal understanding of succession between 1899 and 1910. Cowles investigated the growth of vegetation on sand dunes along the shores of Lake Michigan, specifically the Indiana Dunes, after being inspired by Eugen Warming's studies of Danish dunes. He realized that the vegetation on different-aged dunes could be seen as distinct phases of a larger trend in the development of vegetation on dunes (a method of studying vegetation change that was later dubbed space-for-time substitution, or chronosequence studies). The original version of this work, titled "The ecological relations of the vegetation of the sand dunes of Lake Michigan," was published in the Botanical Gazette in 1899. He developed the concepts of primary succession and a sere—a recurring series of community changes unique to certain environmental conditions—in this seminal work and other papers that followed.

8.4.1.2. Gleason and Clements

The theories of Frederic Clements, a contemporary of Cowles, however, dominated knowledge of succession from roughly 1900 to 1960. Clements maintained that seres were highly predictable and deterministic and that they converged on a stable climax community determined by climate, regardless of initial conditions. Clements' model, which is sometimes referred to as the pseudo-organismic theory of community ecology, explicitly compared the ontogenetic development of individual organisms with the successional development of ecological communities. A sophisticated taxonomy of communities and successional pathways was created by Clements and his adherents.

In the 1920s, Henry Gleason provided an alternative framework. The Clementsian model was much less deterministic and the Gleasonian model was more complex. Its primary differences from the Clementsian perspective are that it denies the existence of coherent, sharply bounded community types and suggests a much greater role for chance factors. Communities were best understood as artifacts of the juxtaposition of species distributions, according to Gleason, who also claimed that species distributions responded individualistically to environmental factors. After being published in 1926, Gleason's theories were mainly disregarded until the late 1950s.

8.4.1.3. Modern Era

The work of Robert Whittaker and John Curtis in the 1950s and 1960s generally marked the beginning of a more rigorous, data-driven testing of community theory and successional models. Since then, succession theory has evolved to become more nuanced and less rigid. J. Connell and R. Slatyer tried to use mechanisms to codify successional processes. The idea of a stable climax vegetation has largely been abandoned by ecologists in North America and Britain, and successional processes are now viewed as considerably less deterministic, with significant roles for historical contingency and alternative pathways in the actual development of communities. There is still debate regarding the relative significance of equilibrium vs. random processes and the general predictability of successional dynamics. Fakhri A., a former Harvard professor. Bazzaz brought up the topic of scale, arguing that while processes are random and patchy at the local or small area scale, there are undeniable trends when larger regional areas are considered.

Remember

Primary succession starts with the introduction of organisms and species, usually on newly exposed rock. Pioneer species like lichens on rock come first.

8.4.2. Factors

Site characteristics, the nature of the events that start succession (perturbations), interactions between the species present, and more random elements like the availability of colonists or seeds or the meteorological conditions at the time of disturbance can all have an impact on the successional change trajectory. While some of these factors add more probabilistic elements, others help to make succession dynamics more predictable. Climate change and human activity are two major perturbation factors in the modern world.

Early succession communities are typically dominated by quickly expanding, widely distributed species (opportunist, fugitive, or r-selected life histories). These species will typically give way to more competitive (k-selected) species as succession moves forward.

Although some trends in ecosystem and community properties have been proposed, most of them don't seem to be widespread. For instance, species diversity almost always rises in the early succession as new species appear, but it may fall in the later succession when opportunistic species are eliminated by competition and locally superior competitors take the lead. Depending on the specific system and site, net primary productivity, biomass, and trophic properties all exhibit variable patterns over succession.

Ecological succession was previously thought to have a stable final stage known as the climax, which was largely influenced by the local climate and occasionally referred to as the "potential vegetation" of a site. Modern ecologists have largely abandoned this theory in favor of theories of ecosystem dynamics that do not require equilibrium. A "climax" community is unachievable in the majority of natural ecosystems due to the rate of disturbance. A climax state is frequently avoided by climate change because it happens frequently enough. Communities can also be continuously reshaped by adding new species through introductions and range expansions.

Community characteristics impact the development of some ecosystem attributes, like nutrient cycles and soil qualities, which in turn impact subsequent successional development. This process of feedback may only happen over millennia or centuries. In addition to the random character of disruption occurrences, other extended (e.g., climatic) shifts, it is questionable if the "climax" idea ever applies or is especially helpful when considering real vegetation.

8.4.3. Types

8.4.3.1. Primary Succession

Successional dynamics that start when an ecological community settles in a previously uninhabited area, like recently exposed sand or rock surfaces, lava flows, recently exposed glacial tills, etc., are known as primary succession. Pioneer microorganisms, plants (lichens and mosses), the grassy stage, smaller shrubs, and trees are among the stages of primary succession. When food becomes available, animals start to return. The community reaches the pinnacle stage when it becomes a fully functional ecosystem.

8.4.3.2. Secondary succession

Secondary succession refers to successional dynamics that occur after a pre-existing community is severely disrupted or eliminated. Pre-disturbance conditions, such as soil development, seed banks, remaining organic matter, and residual living organisms, have a significant impact on secondary succession dynamics. Early in secondary succession, community change can happen relatively quickly due to pre-existing organisms and residual fertility.

Studying and observing secondary succession is more common than primary succession. Secondary succession is particularly common in response to human-caused disturbances like logging and agriculture, as well as natural disturbances like fire, flood, and strong winds. The purpose of secondary succession is to allow the new community to rebuild itself while preserving the ecosystems and soils.

For example, woody plants "colonized more rapidly (per unit area) on large and nearby patches" in a fragmented old field habitat created in eastern Kansas.

A landscape can undergo significant changes through secondary succession. A wildfire in Acadia National Park in the 1900s destroyed much of the surrounding area. Initially, the terrain was covered in evergreen trees. It took at least a year for shrubs to sprout after the fire. Over time, deciduous trees began to replace the evergreens.

8.4.3.3. Seasonal and Cyclic Dynamics

In contrast to secondary succession, these kinds of vegetation changes are periodic in nature, resulting from shifting species interactions or recurrent events rather than being dependent on disturbance. The climax idea is changed in these models to reflect dynamic states.

8.4.3.4. Causes of Plant Succession

Changes in the soil brought about by the organisms living there can result in autogenic succession. These alterations can be caused by plants growing in the soil, shifting soil nutrients, or building up organic matter in the humic layer or litter. The community can also be changed by the plants' own structural makeup. Larger species, like trees, for instance, cast shade on the growing forest floor as they mature, which tends to keep out species that need light. Shade-loving plants will spread throughout the region.

The cause of allogenic succession is not the vegetation itself, but rather factors from the surrounding environment. For instance, alterations to the soil brought about by leaching, erosion, or the deposition of silt and clays can change the ecosystems' water and nutrient relationships. Animals are herbivores, seed dispersers, and pollinators, which means they contribute significantly to allogenic changes. Additionally, they have the power to alter the soil's composition (as termites, ants, and moles do) and add more nutrients to specific regions of the ecosystem. Regeneration sites that favor particular species may result from this.

Perhaps more so than any other factor, but on a far longer timescale, are climate-related factors. Variations in rainfall and temperature will encourage changes within communities. At the conclusion of each ice age, significant successional changes occurred as the climate warmed. The tundra vegetation and bare glacial till deposits underwent succession to mixed deciduous forest. In the upcoming century, significant Allogenic changes are probably going to occur due to the greenhouse effect's increased temperature. Allogenic changes are also brought about by meteors, earthquakes, avalanches, floods, fires, and strong winds, among other geological and climatic disasters.

8.4.3.5. Mechanisms

A descriptive theory of succession was published by Frederic Clements in 1916, and it was then promoted as a broad ecological idea. His succession theory had a significant impact on ecological philosophy. The idea of Clements is commonly referred to as classical ecological theory.

According to Clements, succession is a process involving several phases:

- Nudation: Succession begins with the development of a bare site, called Nudation (disturbance).
- Migration: refers to the arrival of propagules.
- Ecesis: involves the establishment and initial growth of vegetation.

- Competition: as the vegetation becomes well established, grows, and spreads, various species begin to compete for space, light, and nutrients.
- Reaction: during this phase autogenic changes such as the buildup of humus affect the habitat, and one plant community replaces another.
- Stabilization: a supposedly stable climax community forms.

Did you get it?

1. **Who first used the term succession?**
2. **Name the two types of succession.**

8.5. CLIMAX COMMUNITIES

Learning Objectives

- Discuss Frederic Clements' use of "climax"
- Learn about continuing usage of "climax"

A community of plants, animals, and fungi that have reached a steady state through the process of ecological succession in the development of vegetation in an area over time is referred to as a climax community, also known as a climatic climax community in scientific ecology. It was previously believed that this equilibrium would arise because the species in the climax community are the most suited to the local average conditions. The phrase is occasionally used in reference to soil development. However, it has been discovered that, especially over extended periods of time, a "steady state" is more apparent than real. Nevertheless, it's still a good idea.

a. Frederic Clements' Use of "Climax"

According to Clements, the ontogenetic development of individual organisms and the successional development of ecological communities are similar. Clements only proposed analogies with extremely basic organisms. The notion that an ecological community is a "superorganism" was furthered by later ecologists, who also attempted to identify a single climax-type for every region and occasionally asserted that communities could be homologous to complex organisms. The concept of "polyclimax," or multiple steady-state endpoints in a given climatic zone that are determined by edaphic factors, was developed by the English botanist Arthur Tansley. These endpoints were referred to by different names by Clements, who believed that they were unstable because climax vegetation is inherently the most suited to a particular region's climate. Research in the 1950s and 1960s significantly supported Henry Gleason's early criticisms of Clements' organism simile and other methods of characterizing vegetation, which were largely ignored for a number of years (see below). In the meantime, vegetation management and theoretical ecology both heavily incorporated climax theory. Terms by Clements include plagioclimax, pre-climax, and post-climax.

Although the opinions are sometimes credited to him, Clements never maintained that plant communities have distinct boundaries in space or time, that climax communities must always exist, or that the various species within an ecological community are tightly integrated physiologically. Instead, he used the concept of a climax community—a type of vegetation best suited to an idealized set of environmental circumstances—as a conceptual springboard for characterizing the vegetation in a particular region. There is good reason to think that when certain conditions arise, the species best adapted to them might appear there. However, Clements spent a lot of his time characterizing what happens in the absence of those perfect circumstances. Under those conditions, vegetation other than the perfect climax will frequently arise. These various vegetation types can still be characterized as departures from the ideal climax, though. In order to describe the numerous non-climax states that vegetation adopts as a result of its development, Clements created a very extensive lexicon of theoretical terms. Defining an ideal form was his way of addressing ecological complexity.

Did you know?

The idea of a single climatic climax, which is defined in relation to regional climate, originated with Frederic Clements in the early 1900s. The first analysis of succession as leading to something like a climax was written by Henry Cowles in 1899, but it was Clements who used the term "climax" to describe the idealized endpoint of succession.

b. Continuing Usage of "Climax"

Even though climax theory was generally abandoned, some theoretical ecologists started using climax concepts again in the 1990s. A lot of writers and environmentalists still use a watered-down version of the term "climax" to describe what is otherwise known as mature or old-growth communities. For marine macroinvertebrate communities, a late successional stage is referred to as the "climax" in some instances.

Furthermore, some modern ecologists still refer to an ecosystem dominated by invasive species that compete with native species for reintroduction as "disclimax." This idea is derived from Clements' original definition of climax, which is related to an ecosystem that is resistant to alien species colonization. Although disclimax is an anthropogenic phenomenon that hinders the facilitation and succession to a true climax community, Clements used the term in context, and it is one of the few examples of climax that can be observed in nature.

Climax has the following important characteristics:

(a) Unitary:

A climax is a unit. To illustrate the climate, all species are considered as a cohesive group. Stated differently, a climax is a measurement that serves as a gauge for the local climate.

(b) Stability:

The climax community's shape is stable in relation to the weather. Competition cannot supplant the climax community.

(c) Origin and Organism Concept:

Climax communities are thought to be superorganisms. Like an organism, this community is born, grows, and develops. Climax communities experience alterations due to variations in time and climate.

There are three theories about the climax community:

(a) Mono-climax theory:

Federick Clements put forth this theory in 1916. In a climatic region of a geographical area, this theory states that only one type of climax community develops. Subordinate communities are those that exist in that area in addition to the climax community.

The subordinate communities may be:

- Pro-climax
- Post-climax,
- Pre-climax.

It would be incorrect to consider climate as a determining factor for climax formation, since different types of climax communities can develop in an area of uniform climate. Furthermore, the climax state is not the final modified form of the succession; rather, it is a form of minimum changes. Thus, the mono-climax theory is not without its detractors.

(b) Poly climax theory:

In 1935, Braun-Blanquet and Tansley put forth this theory in response to criticism of the monoclimax theory. This theory states that in addition to the local climate, a variety of other factors also influence the climax communities. So the climax stage may be named depending upon the nature of the factor in stabilization:

(i) *Topographic Climax:*

Hills, mountains, slopes, and other topographical features of the area stabilize the climax community. It is referred to as a topographical climax.

(ii) *Biotic Climax*:

The terms "anthropogenic climax," "grazing climax," and "zootic climax" refer to the points in a biotic community where living factors such as animals, herbivores, and humans stabilize the community.

(iii) *Edaphic Climax*:

The edaphic climax is the climax community that stabilizes as a result of the soil factor.

(iv) *Fire-Climax*:

A fire climax is the culmination of repeated fire.

(c) Climax-pattern hypothesis:

It was R. H. Whitaker who put forth this hypothesis in 1953. The hypothesis states that under specific climatic conditions, a variety of climax communities may form in a given area. In other words, communities adapt themselves to the varying environmental conditions of their habitat, changing the way they live.

Did you get it?

1. **What is one of the characteristics of a climax community?**
2. **What are the three theories about the climax community?**

CLOSER LOOK

NEIGHBORHOOD MODELS OF COMPETITION AMONG PLANTS

Plant ecologists were never very comfortable with Lotka-Volterra's models of competition, which were designed largely with mobile animals in mind. The sessile nature of plants means that an individual plant generally interacts only with its close neighbors, rather than with the entire population of competitors. Consequently, competition among plants might be best portrayed by models that describe how individual plants respond to variation in the abundance of their immediate neighbors. Such neighborhood models have been developed to describe intra- and interspecific competition among annual plants. The models have been set up in two ways: as computationally intensive computer simulations that keep track of the spatial locations of individual competitors within a given plot of ground, and as analytical models that use assumptions about spatial distributions to capture the essence of spatially constrained competition. The computer simulation models are quite complex because they keep track of the positions of all the individual plants in the modeled population. Analytical models that make use of the average spatial features of interacting plant populations are considerably simpler, yet they yield many of the same predictions as the more complex simulation models.

At their simplest, neighborhood models of competition among plants assume that a focal plant only responds to the number of competitors found within a certain radius, or neighborhood of the plant. Competition is manifested by the ways in which neighbors affect fecundity and survival. In simpler models, survival is independent of the neighborhood density of competitors, but fecundity is not. The number of seeds set per plant depends on the number of intraspecific neighbors within a 5 cm radius for the plant *Arabidopsis thaliana*. This kind of relationship is termed a fecundity predictor. A fecundity predictor that describes this sort of relation can be modeled as Me^{--cn}, where M is the number of seeds produced by a plant without neighbors, e is the base of the natural logarithms, n is the number of neighbors affecting a particular plant, and c describes the intensity of neighborhood competition.

For intraspecific competition, Pacala and Silander (1985) have developed a neighborhood model that predicts the density of annual plants present in one year from the density of plants present in the previous year. Unlike the models described previously, which used differential equations, this model uses a different mathematical framework using difference equations. Models using difference equations describe how processes operating on populations during one discrete time interval (say the current year, time t) will affect those populations in the next time interval (say next year, time t + 1). Consequently, population size in year t + 1, N_{t+1}, is a function of population size the previous year, N_t, or $N_{t+1} = f(N_t)$, where f is a function relating density at time t to density at time t + 1.

The neighborhood model assumes that the number of individuals per neighborhood area in the next generation, N_{t+1}, is based on the number of seeds in the previous generation, N_t, the probability that those seeds will germinate, g, the probability that a germinated seed survives to adulthood and reproduces, P, and the way that the number of neighbors affects the seed output, fecundity, of each plant. For simplicity, Pacala and Silander assume that the probabilities of germination and survival do not depend on the density of neighbors. Neighbors only affect reproduction, through their effects on the fecundity predictor described above. The single species analytical model, assuming a Poisson spatial distribution of seeds, and neighbors, is

$$N_{t+1} = N_t gP \sum_{n=0}^{\infty} (e^{-PN_t g} \cdot (PN_t g)^n / n!) \cdot Me^{-cn}$$

The term in the summation gives the probability that an individual plant will have a given number of neighbors, n, and multiplies that probability by the expected fecundity for a plant with n neighbors, Me^{--cn}, to arrive at an expected or average value of fecundity. This is based on the probability density function of the Poisson distribution. The summation reduces to $Me^{(--gPNt\gamma)}$, where g is the probability of germination, P is the probability that a germinated seed survives to adulthood, M is the number of seeds produced by a plant without neighbors, and $\gamma = (1 - \exp(-c))$. The model for intraspecific competition then reduces to

$$N_{t+1} = gPN_t \, Me^{(-gPNt\gamma)}.$$

The single-species model produces increasingly variable dynamics as the average fecundity of a plant without neighbors, M, increases. This is precisely what one would expect for a discrete-time logistic equation incorporating rates of increase that generate chaotic dynamics.

Pacala (1986a) developed a similar analytical model for neighborhood competition among two annual plant species without seed dormancy, which can be portrayed by the following pair of difference equations:

$$N_{1,t+1} = g_1 P_1 M_1 N_{1,t} \, e^{(-g_1 P_1 N_{1,t} \gamma_{11} - g_2 P_2 N_{2,t} \gamma_{12})}$$

$$N_{2,t+1} = g_2 P_2 M_2 N_{2,t} \, e^{(-g_2 P_2 N_{2,t} \gamma_{22} - g_1 P_1 N_{1,t} \gamma_{21})},$$

where $N_{1,t}$ and $N_{2,t}$ describe the density of seeds of species 1 and 2 at time t, g_i is the probability that a seed of species i will germinate, P_i is the density-independent survivorship of a seedling of species i, M_i is the fecundity of a plant of species 1 when it has no neighbors, and $\gamma_{ij} = 1 - e^{(-c_{ij})}$. Fecundity is related to the number of neighbors via a predictor, or submodel, given by $e^{(-c_{ij}n - c_{ij}m)}$, where n and m are numbers of species i and j within the neighborhood of a focal plant. The c_{ij}s are analogous to competition coefficients, in that they describe how the density of neighbors depresses fecundity. For randomly distributed plants that follow a Poisson distribution, the average value of the fecundity predictor is $M_i e^{(-g_i P_i N_{i,t} \gamma_{ii} - g_j P_j N_{j,t} \gamma_{ij})}$.

The model can also be expressed concisely in terms of adult plants. The number of adults at time t is $A_{i,t} = P_i g_i N_{i,t}$ in other words, the number of seeds multiplied by the probability that a seed will germinate multiplied by the probability that the seedling survives to adulthood. Then the model looks like

$$A_{1,t+1} = Q_1 A_{1,t}\ e^{(-A_{1,t}\gamma_{11} - A_{2,t}\gamma_{12})}$$

$$A_{2,t+1} = Q_2 A_{2,t}\ e^{(-A_{2,t}\gamma_{22} - A_{1,t}\gamma_{21})},$$

where $Q_i = M_i P_i g_i$, as in the model for seeds described above. Despite all of the complications imposed by the spatial components of the model, it reduces to a framework very similar to the Lotka – Volterra models! The general conclusions are the same. If the intensity of intraspecific competition is greater than the intensity of interspecific competition, then the species will coexist.

ROLE MODEL

ERNST HAECKEL

Ernst Heinrich Philipp August Haeckel (16 February 1834 – 9 August 1919) was a German zoologist, naturalist, eugenicist, philosopher, physician, professor, marine biologist, and artist. He discovered, described, and named thousands of new species, mapped a genealogical tree relating all life forms, and coined many terms in biology, including *ecology, phylum, phylogeny*, and *Protista.* Haeckel promoted and popularized Charles Darwin's work in Germany and developed the influential but no longer widely held recapitulation theory ("ontogeny recapitulates phylogeny") claiming that an individual organism's biological development, or ontogeny, parallels and summarizes its species' evolutionary development, or phylogeny.

The published artwork of Haeckel includes over 100 detailed, multi-color illustrations of animals and sea creatures, collected in his *Kunstformen der Natur* ("Art Forms of Nature"), a book that would go on to influence the Art Nouveau artistic movement. As a philosopher, Ernst Haeckel wrote *Die Welträthse*l (1895–1899; in English: *The Riddle of the Universe*, 1901), the genesis for the term "world riddle" (*Welträtsel*); and Freedom in Science and Teaching to support teaching evolution.

Haeckel was also a promoter of scientific racism and embraced the idea of Social Darwinism. He was the first person to characterize the Great War as the "first" World War, which he did as early as 1914.

Early Life and Education

Ernst Haeckel was born on 16 February 1834, in Potsdam (then part of the Kingdom of Prussia). In 1852 Haeckel completed studies at the *Domgymnasium*, the cathedral high school of Merseburg. He then studied medicine in Berlin and Würzburg, particularly with Albert von Kölliker, Franz Leydig, Rudolf Virchow (with whom he later worked briefly as an assistant), and with the anatomist-physiologist Johannes Peter Müller (1801–1858). Together with Hermann Steudner, he attended botany lectures in Würzburg. In 1857 Haeckel attained a doctorate in medicine, and afterward he received the license to practice medicine. The occupation of physician appeared less worthwhile to Haeckel after contact with suffering patients.

Career

Haeckel studied under Carl Gegenbaur at the University of Jena for three years. In 1861, he earned a habilitation in comparative anatomy and became a professor of zoology at the University of Jena, where he remained for 47 years, from 1862 to 1909. Between 1859 and 1866, Haeckel worked on many phyla, such as radiolarians, poriferans (sponges), and annelids (segmented worms). During a trip to the Mediterranean, Haeckel named nearly 150 new species of radiolarians. From 1866 to 1867, Haeckel made an extended journey to the Canary Islands with Hermann Fol. On 17 October 1866, he arrived in London. Over the next few days, he met Charles Lyell and visited Thomas Huxley and his family at their home. On 21 October, he visited Charles Darwin at Down House in Kent. In 1867, he married Agnes Huschke. Their son Walter was born in 1868, their daughters Elizabeth in 1871, and Emma in 1873. In 1869, he traveled as a researcher to Norway, in 1871 to Croatia (where he lived on the island of Hvar in a monastery), and in 1873 to Egypt, Turkey, and Greece. In 1907, he had a museum built in Jena to teach the public about evolution. Haeckel retired from teaching in 1909, and in 1910, he withdrew from the Evangelical Church of Prussia. On the occasion of his 80th birthday celebration, he was presented with a two-volume work entitled "*Was wir Ernst Haeckel verdanken*" (What We Owe to Ernst Haeckel), edited at the request of the German Monistenbund by Heinrich Schmidt of Jena.

Personal Life and Death

In 1864, his first wife, Anna Sethe, died. Haeckel dedicated some species of jellyfish that he found beautiful (such as *Desmonema annasethe*) to her.

Haeckel's second wife, Agnes, died in 1915, and he became substantially frailer, breaking his leg and arm. He sold his "Villa Medusa" in Jena in 1918 to the Carl Zeiss Foundation, which preserved his library. Haeckel died on August 9, 1919.

SUMMARY

- In ecology, a community—also referred to as a biocenosis, biotic community, biological community, ecological community, or life assemblage—is a collection or association of populations of two or more distinct species that coexist in the same geographic area. There are several ways to use the term "community."
- The species found in different ecological communities can vary greatly in kinds and quantities. Tropical rainforest communities often have a high species diversity per cubic meter, while Arctic communities tend to have fewer species.
- Species richness in a community refers to the variety of species present. If one community has 300 species and another has 30, the second community would have a significantly higher species richness.
- One indicator of the complexity of a community is species diversity, which is influenced by the total number of species (species richness) and the relative abundances of those species (species evenness) in the community. Higher species diversity is a result of both greater species diversity and more uniform species abundance.
- A community's structure is shaped by various interacting abiotic (nonliving) and biotic (related to living organisms) factors
- Certain species have quite a strong impact on how communities are structured, helping to keep the community in balance or even making it possible for it to thrive. Keystone and foundation species are some of these unique species.
- A foundation species plays a crucial role in shaping and defining a community. They often modify their environment to create a more welcoming habitat for other creatures within the community.
- A biological interaction in ecology is the impact coexisting organisms have on each other. Interspecific interactions involve different species, while intraspecific interactions involve the same species.
- Competition is when organisms interact directly or indirectly, affecting their fitness when sharing a limited resource. Weaker competitors are usually negatively affected.
- The process where an ecological community's species structure changes over time is called ecological succession. The duration can range from a few decades (like after a wildfire) to several decades.
- A climax community, also called a climatic climax community in scientific ecology, is a stable community of plants, animals, and fungi that develops through ecological succession in an area over time.

MULTIPLE CHOICE QUESTIONS

1. **The plants occurring together with a similar ecology in definite strata are called:**
   a. Periodicity
   b. Stratification
   c. Floristic
   d. Vitality
2. **The rhythmic phenomena related to seasonal changes is called periodicity.**
   a. Periodicity
   b. stratification
   c. Floristic
   d. Vitality
3. **The capacity of plants to complete their life cycle is called:**
   a. Stratification
   b. Periodicity
   c. Floristic
   d. Vitality
4. **The number of individuals of species in a unit area is called:**
   a. Abundance
   b. Density
   c. Frequency
   d. Cover
5. **The estimation of individuals of species is called:**
   a. Abundance
   b. Density
   c. Frequency
   d. Cover
6. **The degree of occurrence of individuals of a species within an area is called:**
   a. Frequency
   b. Cover
   c. Abundance
   d. Density
7. **The relative occurrence of species in an association or a group of related associations is:**
   a. Fidelity
   b. Dominance

c. Constancy

d. Presence

8. **A species that occurs in 90% or more of the stand is:**

a. Fidelity

b. Dominance

c. Constancy

d. Presence

9. **The species which have high numbers and large volumes are:**

a. Dominance

b. Fidelity

c. Constancy

d. Presence

10. **The phenomena of growth of propagules and their adjustments in the new surrounding area:**

a. Reaction

b. Aggregation

c. Competition

d. Ecesis

REVIEW QUESTIONS

1. What do you understand by community ecology? Explain.
2. How do we measure community structure?
3. Describe the patterns of community structure.
4. What do you understand by community changes?
5. Focus on species interactions in communities.
6. Give a detailed overview of ecological succession.
7. Discuss about climax communities.

Answers

1. (b)	2. (a)	3. (d)	4. (b)	5. (a)
6. (a)	7. (a)	8. (c)	9. (a)	10. (d)

REFERENCES

1. Audia, P. G., Sorenson, O., & Hage, J., (2001). Tradeoffs in the organization of production: Multiunit firms, geographic dispersion, and organizational learning. *Adv. Strateg. Manag., 18*, 75–105.
2. Cadotte, M. W., Cavender-Bares, J., Tilman, D., & Oakley, T. H., (2009). Using phylogenetic, functional, and trait diversity to understand patterns of plant community productivity. *PLoS (Public Library of Science) One [Online Serial], 4*(5), e5695.
3. Cam, E., Link, W. A., Cooch, E. G., Monnat, J. Y., & Danchin, E., (2002). Individual covariation in life-history traits: Seeing the trees despite the forest. *Am. Nat., 159*, 96–105.
4. Carroll, G. R., & Swaminathan, A., (2000). Why the microbrewery movement? Organizational dynamics of resource partitioning in the U.S. brewing industry. *Am. J. Sociol. 106*, 715–762.
5. Chase, J. M., & Leibold, M. A., (2003). *Ecological Niches: Linking Classical and Contemporary Approaches* (p. 212). University of Chicago Press, Chicago.
6. Chave, J., & Leigh, Jr. E. G., (2002). A spatially-explicit model of betadiversity in tropical forests. *Theor. Pop. Biol., 62*, 153–168.
7. Chave, J., Muller-Landau, H. C., & Levin, S., (2002). Comparing classical community models: Theoretical consequences for patterns of diversity. *Am. Nat., 159*, 1–23.
8. Chesson, P. L., (2000). Mechanisms of maintenance of species diversity. *Annu. Rev. Ecol. Syst., 31*, 343–366.
9. Cross, W. F., Baxter, C. V., Donner, K. C., Rosi-Marshall, E. J., Kennedy, T. A., Hall, Jr. R. O., Wellard, K. H. A., & Rogers, R. S., (2011). Ecosystem ecology meets adaptive management: Food web response to a controlled flood on the Colorado River, Glen Canyon. *Ecological Applications, 21*, 2016–2033.
10. Csillag, F., Boots, B., Fortin, M. J., Lowell, K., & Potvin, F., (2001). Multiscale characterization of boundaries and landscape ecological patterns. *Geomatica, 55*, 509–522.
11. Currie, D. J., (2007). Disentangling the roles of environment and space in ecology. *Journal of Biogeography, 34*, 2009–2011.
12. Dale, M. R. T., & Fortin, M. J., (2010). From graphs to spatial graphs. *Annual Review of Ecology, Evolution, and Systematics, 41*, 21–38.
13. Diniz-Filho, J. A. F., & Bini, L. M., (2005). Modeling geographical patterns in species richness using eigenvector based spatial filters. *Global Ecology and Biogeography, 14*, 177–185.
14. Dormann, C. F., et al., (2007). Methods to account for spatial autocorrelation in the analysis of species distributional data: A review. *Ecography, 30*, 609–628.
15. Dray, S., Legendre, P., & Peres-Neto, P. R., (2006). Spatial modeling: A comprehensive framework for principal coordinate analysis of neighbor matrices (PCNM). *Ecological Modeling, 196*, 483–493.

16. Richardson, J. E., Pennington, R. T., Pennington, T. D., & Hollingsworth, P. M., (2001). Rapid diversification of a species-rich genus of neotropical rain forest trees. *Science, 293*, 2242–2245.

17. Ricklefs, R. E., (2003). A comment on Hubbell's zero-sum ecological drift model. *Oikos, 100*, 185–192.

Index

FSC
www.fsc.org
MIX
Papier | Fördert gute Waldnutzung
FSC® C083411

Kolibri 360 GmbH
Ferdinand-Jühlke-Straße 7
99095 Erfurt, Deutschland
produktsicherheit@kolibri360.de